Tidal power: trends and developments

Proceedings of the 4th Conference on Tidal Power, organized by the Institution of Civil Engineers and held in London on 19–20 March 1992

 Thomas Telford, London

Conference organized by the Institution of Civil Engineers and co-sponsored by the Department of Energy, the Institute of Energy, the Institution of Mechanical Engineers and the Institution of Electrical Engineers.

Organizing Committee: R. Clare, Sir Robert McAlpine and Sons (Chairman); Dr R. Price, Energy Technology Support Unit; B. Madge, Bryan Madge Associates; C. J. A. Binnie, W. S. Atkins Consultants; E. A. Wilson, Mersey Barrage Company.

A CIP catalogue record for this publication is available from the British Library

ISBN 0 7277 1905 X

First published 1992

© The Authors and the Institution of Civil Engineers, 1992, unless otherwise stated

All rights, including translation, reserved. Except for fair copying, no part of this publication may be reproduced, stored in a retrieval system or transmitted in any form or by any means electronic, mechanical, photocopying, recording or otherwise, without the prior written permission of the Publications Manager, Publications Division, Thomas Telford Services Ltd, 1 Heron Quay, London E14 4JD.

Papers or other contributions and the statements made or opinions expressed therein are published on the understanding that the author of the contribution is solely responsible for the opinions expressed in it and that its publication does not necessarily imply that such statements and or opinions are or reflect the views or opinions of the organizers or publishers.

Published on behalf of the organizers by Thomas Telford Services Ltd, Thomas Telford House, 1 Heron Quay, London E14 4JD.

Printed in Great Britain by Redwood Press Ltd, Melksham, Wilts.

Contents

Opening address. Sir Peter MASEFIELD	1
1. Annapolis: the Straflo Turbine and other operating experiences. R. G. RICE and G. C. BAKER	3
2. Current status of tidal power in the Bay of Fundy. G. C. BAKER	15
Discussion on Papers 1 and 2	23
3. The Mersey Barrage — civil engineering aspects. B. I. JONES, C. D. I. MORGAN, D. PHILLIPS and M. W. PINKNEY	27
4. The Mersey Barrage — the impact on shipping. H. ALTINK and G. W. R. HAIGH	49
Discussion on Papers 3 and 4	63
5. The Mersey Barrage — mechanical and electrical development. R. D. CONROY, H. R. GIBSON and D. PHILLIPS	69
6. The Mersey Barrage — hydraulic and sedimentation studies. H. ALTINK, K. MANN and N. V. M. ODD	89
7. The Mersey Barrage — preparations for an environmental assessment. J. V. TOWNER and E. A. WILSON	105
Discussion on Papers 5–7	125
8. The Mersey Barrage — further assessment of the energy yield. R. POTTS and E. A. WILSON	129
9. The Mersey Barrage — indirect benefits: the economic and regional case. P. J. COCKLE and J. J. McCORMACK	145
10. The Mersey Barrage — finance and promotion: the way forward. C. J. ELLIOTT and J. J. McCORMACK	155
Discussion on Papers 8–10	167
11. Feasibility of a Conwy Barrage. M. E. MATTHEWS and R. M. YOUNG	169
12. Tidal energy from the Wyre. M. E. MATTHEWS and R. M. YOUNG	183

13. Environmental aspects of small tidal power schemes. R. M. YOUNG and M. E. MATTHEWS	**197**
Discussion on Papers 11–13	**211**
14. Prospective tidal power projects in the Kimbereley region of Western Australia. E. T. HAWS, N. REILLY and P. WOOD	**215**
15. Update on Severn Barrage studies. H. J. MOORHEAD, R. POTTS, T. L. SHAW and C. J. TEAL	**231**
16. The environmental effects of tidal energy. S. J. MUIRHEAD	**245**
Discussion on Papers 14–16	**257**
17. Composite construction of caissons. C. J. BILLINGTON and H. M. BOLT	**263**
18. Field, laboratory and mode feasibility of a tidal barrage in a muddy hypertidal estuary. K. W. OLESEN, A. J. PARFITT and H. ENGGROB	**279**
Discussion on Papers 17 and 18	**297**
19. The Rance Tidal Power Station: a quarter of a century in operation. M. RODIER	**301**
20. Tidal power in Russia. L. B. BERNSHTEIN	**311**
Discussion on Papers 19 and 20	**331**

Opening address

Sir PETER MASEFIELD

Two years ago, I addressed the Tidal Power conference just as the bids for the initial renewable tranche of the NFFO were to be submitted and I stressed the Government's view that renewable energy sources offered the potential to increase diversity of supply and would assist in reducing the threat and damage posed by the greenhouse effect.

We know that the first renewable tranche of the NFFO committed the Regional Electricity Companies to secure the supply of 152 MW of renewable power supplies, and last year Colin Moynihan announced that the second renewable tranche hoped to secure a further 457 MW.

We should welcome these developments and the Government's continuing Tidal Energy R&D Programme. Through this ongoing programme, many of the potential uncertainties and areas of concern in connection with tidal power costs, technical performance and regional and environmental effects have been considerably reduced.

The Mersey Barrage Company's Stage III report — released at the beginning of this conference — is an excellent example of a product of the R&D programme and shows how the public and private sectors can work together. (I must declare an interest as a Director of the Mersey Barrage Company). I believe that more needs to be done both by the public and the private sectors if we are able to announce a tidal scheme in the third or fourth renewable tranches of the NFFO.

The UK has the most favourable "physical" conditions in Europe for generating electricity from the tides, and tidal energy is reliable and large-scale. The Mersey Barrage at 251 MW declared net capacity would almost double the first tranche of the NFFO with a single project — the Severn Barrage would be six times again the maximum size of the second tranche of the NFFO — again with a single project rather than hundreds.

However, we need to ensure that the right economic conditions are created to take advantage of the physical conditions.

Tidal energy was considered "promising but uncertain". I hope that Colin Moynihan's Renewable Energy Advisory Group will have the courage, in the light of the excellent work undertaken on the Wyre, Conwy, Mersey and Severn Barrages and the supporting generic studies, to grasp the considerable opportunity available to them and confirm tidal energy as being "economically attractive" and that it should be developed.

Tidal power: trends and developments, Thomas Telford, London, 1992

TIDAL POWER

We must remove the 1998 cut-off date for renewables withtin the NFFO and we must direct the £1 billion which will become available in 1999 — following the termination of the nuclear NFFO projects — to large-scale renewable projects such as tidal power.

The institutional barriers which continue to hamper the development of tidal power — which is addressed in several of the Papers in this conference — must be overcome and a fair method of evaluation established to capture and develop "the most favourable conditions" in Europe for generating clean, reliable and predictable energy.

This conference comes at a crucial time for tidal barrage projects in the UK. I would like to thank the Institution of Civil engineers for organising this conference, the other sponsors and contributors and the DEn and ETSU for their continued support for the extensive R&D programme in conjunction with the Private sector.

1. Annapolis: the Straflo Turbine and other operating experiences

R. G. RICE, PEng. and G. C. BAKER, PEng

At the time the Annapolis project was undertaken (1980) there were many straight-flow turbines in operation, including several STRAFLO units, principally on European rivers, but they were all of relatively small diameter. There was considerable interest in the STRAFLO turbine for both river hydro and tidal sites in Canada, but for both purposes turbines with much larger diameter than any existing straight-flow turbine were required. There was, however, some concern that unforeseen defects or design handicaps might emerge in a scale-up from existing machines. It was therefore decided to demonstrate the commercial operation of a large STRAFLO turbine at Annapolis.

In this Paper, experience gained during design, construction and operation of the Annapolis project is summarized and a brief appraisal of the STRAFLO turbine as a prime mover in tidal power plants is presented.

Turbine characteristics

The machine installed at Annapolis has a diameter of 7.6 m, the same size as proposed for a large-scale tidal plant at site B9 in Minas Basin. Design data is summarized as follows

Throat diameter	7.6 m
Speed	50 rpm
Runaway speed	98 rpm
Fixed propeller blades	4
Wicket gates	18
Rated head	5.5 m
Maximum output at rated head	19.6 MW
Maximum flow at rated head	407.5 m^3/s
Maximum output at 5 m head	19.9 MW
Maximum flow at 6 m head	383.6 m^3/s
Rim seals	Hydrostatic
Sealing water requirement	0.33 litres/ms

(Efficiency curves are shown in Figs 1 and 2.)

The generator rotor is mounted on the turbine rim with the stator independently mounted outside the rotor. The mounting permits sliding the

TIDAL POWER

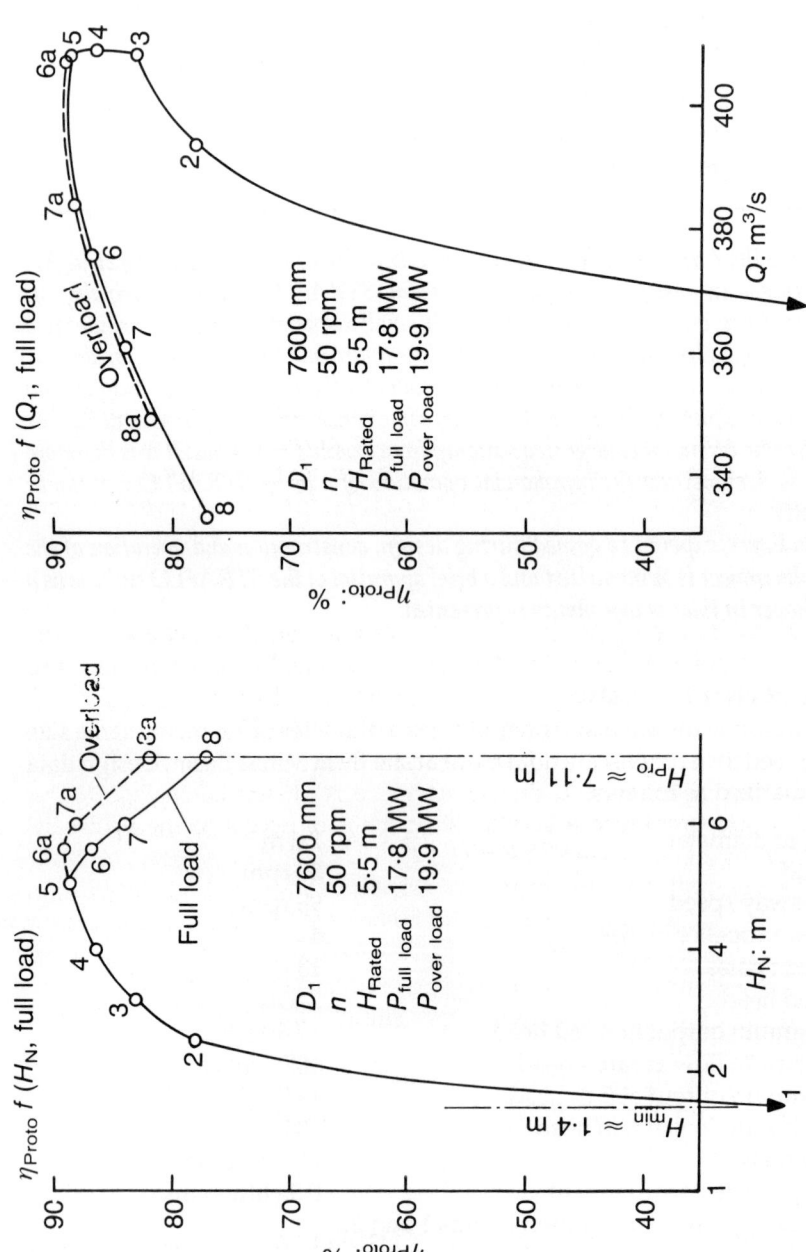

Fig. 1. Turbine efficiency curves

Fig. 2. Turbine efficiency curves

stator in an upstream direction for access to, or removal of, the rotor/runner assembly.

Generator data is as follows:

Rated output at 70°C over 30°C ambient temp.	19100 kW
Rated output at power factor	0.917190 kW
Maximum (150 minute) output at power factor 0.9	19,150 kW
Stator voltage	4160 V
No load loss	213 kW
Load loss	547 kW
Generator efficiency at 100% load	96.4%
Generator efficiency at 75% load	96.5%
Poles	144
Air gap (at pole centre)	11.7 mm

The generator is cooled by air circulating inside shrouds which cover the gap between stator and turbine casing. The air moves around the machine to heat exchangers located at 90° spacing. Salt water then carries heat from the exchangers to the sea. Salt water is also used for sealing water and for cooling of governor and bearing oil.

Setting

Due to the nature of the site, the unit was not placed in a caisson. A pit was dug in a strip of land separating the Annapolis River estuary and the sea. The powerhouse was constructed in the pit. After backfilling, intake and discharge canals were excavated to connect the powerhouse to the river estuary, which serves as a headpond, and the sea.

Positioning of the unit in the powerhouse is illustrated in Fig. 3. The original design employed a vertical steel shaft for access to the upstream bearing housing, which contains both thrust and guide bearings and through which field current is fed to the generator rotor. However, it was found that halving the intake and draft tube horizontal spans would considerably reduce the weight and size of stop logs. The vertical concrete divider at the intake was therefore extended downstream to the thrust bearing and provides the access to it. This arrangement slightly improved the hydraulic efficiency of the intake.

The setting was made as shallow as possible to limit the necessary depth of excavation and to expose any tendency of the runner to cavitate. No damage has occurred.

Erection

The erection procedure started with the installation of the inner and outer stay rings to which the distributor assembly was later attached. No difficulty was encountered with the inner stay ring, which was the first to be installed.

TIDAL POWER

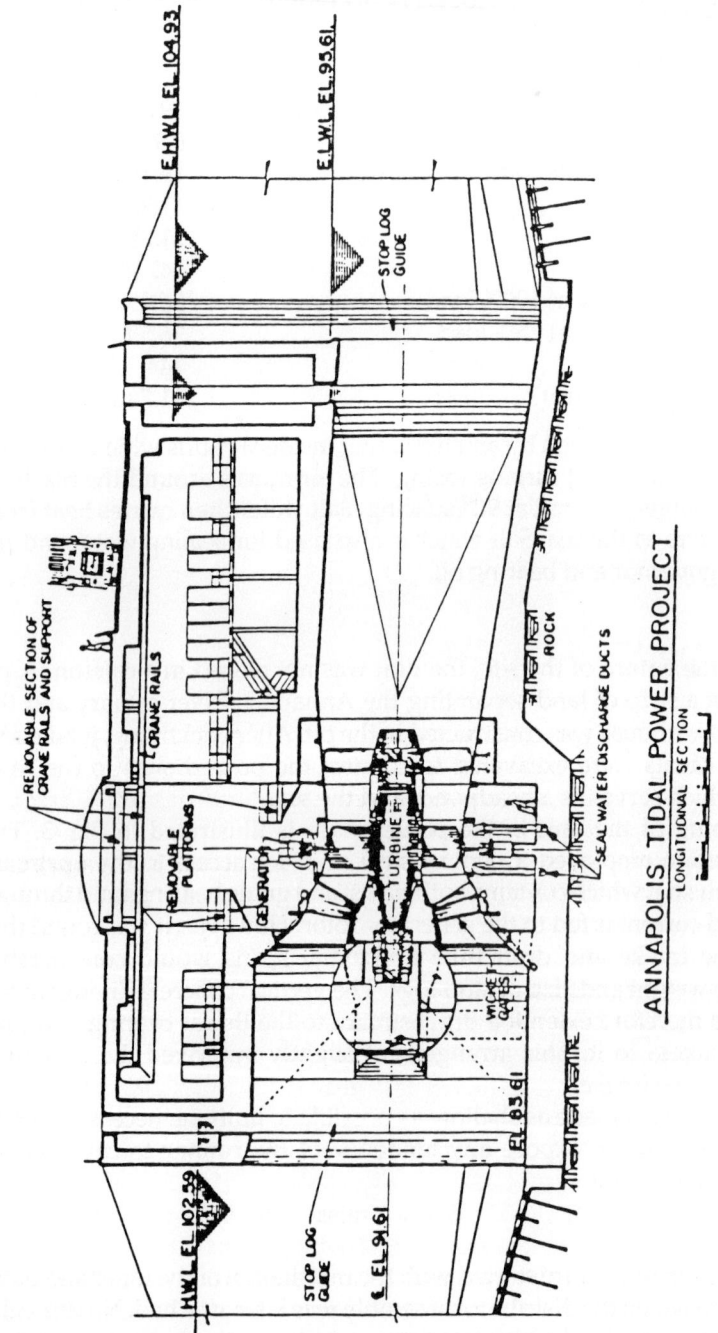

Fig. 3. *Annapolis Tidal Power Project longitudinal section*

However, the outer stay ring proved difficult to align. The outer stay ring is fastened to the powerhouse concrete by means of thirty-eight large pre-tensioned bolts equally spaced around the ring and machine axis (Fig. 4). The procedure required tensioning the bolts before grouting the ring in place.

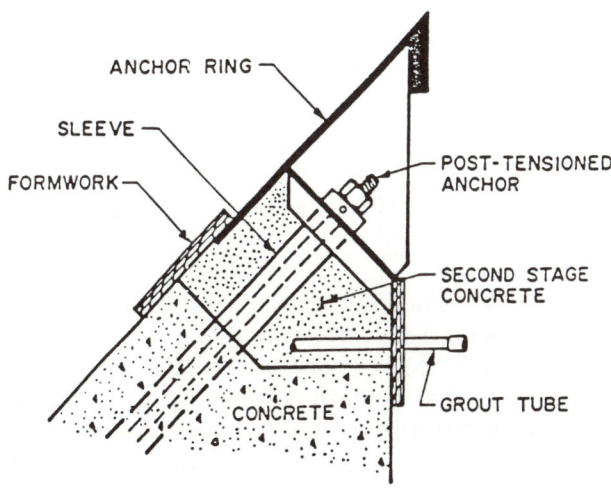

Fig. 4. Distributor anchor ring cross section

Application of tension tended to misalign the ring vertically, horizontally, axially or in skew. The task was finally accomplished, but it was tedious and time-consuming. This was one of the few aspects of design or erection procedure which need to be improved.

Subsequent steps in erection included the placement of upstream guide bearing and shaft, distributor, downstream bearing and shaft and finally the rotor. The generator stator was placed in its disengaged position before turbine erection. No difficulties were encountered.

All the major assemblies were lowered through a hatch in the powerhouse roof by means of a gantry crane designed for that purpose. After construction was completed, the crane was dismantled and stored on site in case it ever should become necessary to dismantle the machine. To date this has not been necessary.

Commissioning

Two problems arose during the commissioning state neither of which relate to the turbine. The generator field windings had been designed to achieve maximum heat dissipation and on this account no insulation was applied to the surface of the coils. Salt carried by the atmosphere soon resulted in ground faults. The exposed portions of the conductors were

TIDAL POWER

therefore coated with epoxy, which eliminated the trouble. However, the generator insulation on both stator and rotor windings is affected by humidity to a greater extent than is desirable or necessary. It has been found expedient to keep the generator warm when not in operation (by applying some field current) and so long as this is done, insulation resistance remains at satisfactory levels.

The other problem which arose during commissioning was plugging of the filter on the seawater intake. To prevent possible damage to the seals, which run with very little clearance, the filter apertures were less than 50 microns in diameter. The fouling was caused by a combination of organic and inorganic substances. An improved filter design with continuous backwash provided a permanent and satisfactory solution.

Operation

The turbine runs very smoothly and quietly, and has operated since it entered commercial service in 1984 with a high level of availability. It provides rated output at rated head and has required relatively little maintenance. In the project design, no trash racks were provided. The wisdom of this decision has been borne out by operating experience. The water passages are large enough so that most trash simply flows through the turbine blades. On one occasion, a number of large timbers, 30 cm by 30 cm, were carried into the machine. One of the timbers lodged in the wicket gates and bent a link-pin. No other damage has been caused by trash.

The powerhouse design does not include gates at either intake or draft tube. These were deemed to be unnecessary for unit protection due to the ability of the machine to withstand runaway speed for 2.5 hours and the fact that heads drop to zero at intervals of just over six hours.

When dewatering is necessary, stop logs are placed at both ends of the water passage. The unit has been dewatered three times to date At this frequency, the added convenience of gates could not be economically justified.

Active cathodic protection was provided to protect the wetted surfaces of turbine and draft tube from corrosion. Protection has proved to be completely adequate.

Corrosion has, however, occurred in one small area of the runner hub adjacent to the root of one of the propeller blades. The blades are stainless steel, while the hub is carbon steel. The junction between blade and hub was formed by welding in place several layers of metal with carefully graduated electrolytic potential. It is possible that the weldments were not accurately placed in this particular spot and that electrolytic action is causing the corrosion there.

Unfortunately, cathodic protection is necessarily limited to wetted surfaces. Inside the powerhouse, salt air and frequently humid conditions necessitate a considerable maintenance effort to prevent surface corrosion of pumps, valves, pipes and other equipment. This does not relate to the type of turbine, but rather may be a problem common to any tidal generating station located in a similar climate. It suggests the desirability of choosing non-corroding materials and finishes where possible. At Annapolis, fibreglass pipes have proven satisfactory for a number of applications.

Station personnel provide preventative maintenance and routine station checks during normal daytime working hours, and they respond to after hour station alarms. Operational control of the unit is by programmable logic controllers (PLCs) in the station and a system control and data acquisition (SCADA) terminal sends status information to Nova Scotia Power's Milton Area Control Centre about 100 km distant. The Milton system control operators continuously monitor the station and may take control at will.

Control of the unit by an on-site computer was discontinued in 1991 as the result of problems in developing a satisfactory model for tide prediction and headpond recuperation. Typically, the headpond filled completely on less than five tide cycles per month when control was by the computer.

The computer also could not be relied upon to ensure adequate drainage gradients are maintained for marshlands immediately adjacent to the headpond. Each of the individual marshes is relatively small with independent drainage systems. The runoff on these marshes is almost instantaneous, but changes in stream flow at the gauging station located upstream above the tidal influence are gradual and delayed by 24 hours or more. The only real-time data available to the computer was, therefore, not suitable for maintaining satisfactory drainage for the agricultural lands. It has been found that the only practical means of ensuring proper drainage was by operators making decisions based upon weather forecasts and initiating the lowering of the headpond not later than the commencement of the storm.

Under the current mode of control, PLCs initiate gate opening at the earliest opportunity in the tide cycle and initiate gate closing when the headpond has reached its designated full level. Logic does not exist to select the various combinations of gates and sluicing, to account for the river flow contribution, or to provide flood control. Decisions with respect to these matters are made by operating personnel. Normally operator decisions are not required from cycle to cycle and only required in the event of a marked change in runoff conditions.

The benefit of resorting to a substantially simplified control has been an increase in energy production in the order of ten to fifteen percent annually. Because operator intervention has always been necessary during significant weather events, the change resulted in little or no additional operator involvement.

Annual output has been less than the 50 GWh design estimate, due to a variety of factors. The headpond elevation has been kept lower than the

TIDAL POWER

design level, thus reducing the head on the machine. Hydraulic losses in intake and discharge have been higher than design estimates because canals were not excavated to design dimensions due to the fact that unexpectedly hard materials were encountered in some areas.

Also, tidal ranges have been less than expected. It was thought that the installation of a single unit would not appreciably alter the local tidal regime, but it may have had some effect.

The mean tidal range at Annapolis is about seven metres too small for economic exploitation of tidal power at energy prices existing in Nova Scotia. Therefore, even small head losses have an appreciable effect on output.

Longer term experience

One problem did not emerge until the plant had been in operation for almost four years. At that time, the sealing water discharge was found to be decreasing. The machine was stopped and the seals (Fig. 5) were removed. On inspection it was found that the internal passages carrying water through the seals to the operating face had become plugged or partly plugged by manganese deposits. The manganese, always present in sea water in trace amounts, had been precipitated by the small amounts of chlorine used to inhibit marine fouling of the plant water system, heat exchangers and sealing supply. Water passages in the seals were cleaned out and the seals, which showed absolutely no signs of wear, were returned to service. It is apparent that so long as sea water continues to be used, the seals will have to be removed and cleaned about once every three years. Fresh water is available in the Annapolis area. However, the design decision to use salt water was made because of the expectation that similar units might be used in a large-scale tidal plant where fresh water would not be so conveniently available. On the other hand, the standstill seals have not functioned well. The original design provided for four individual segments on each of the upstream and downstream sealing faces. Failure of all eight segments was initially believed to be the result of an accumulation of solid matter around the seals while the unit was unwatered, or that the segments adhered to the sealing faces and were torn when the unit rotated. Subsequently, the design was modified to a single continuous seal on each face. The single seals failed almost at once and have not been replaced. Sealing while the unit is at a standstill is satisfactorily accomplished with hydrostatic seals. The belief that the standstill seals would enable change out of the hydrostatic seal elements without dewatering was incorrect. The requirement for longitudinal jacking of the unit and personnel safety consideration makes dewatering of the unit mandatory when replacing seals.

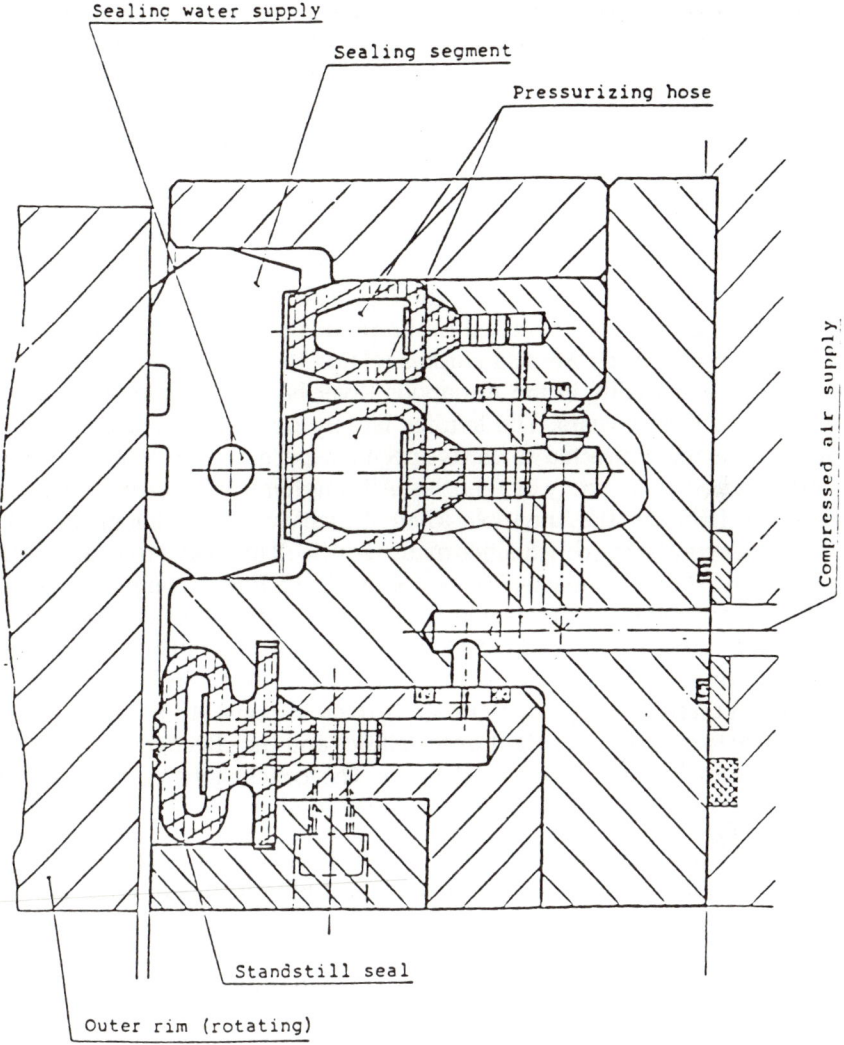

Fig. 5. Sectional drawing of hydrostatic seals

Environmental considerations

The three most significant environmental aspects to emerge from the development were the effects of operations on adjacent agricultural land, shoreline erosion, and fish mortality. Remedial work including improved drainage systems and tide gates, known locally as aboideaux, were provided for the tracts of land of least elevation. These works have mitigated the impairment of surface drainage which otherwise would have resulted from higher water levels in the river. Fears of salt intrusion and elevated groundwater levels were dispelled by the results of monitoring groundwater chemistry and movement prior to the development and for a period of five years following commencement of operations. There simply is no evidence of altered groundwater conditions since construction or operation of the station. An increase in the rate of shoreline erosion occurred immediately following the increase in water level in the headpond. Generally speaking, the river banks have now stabilized to the higher water levels and erosion rates have more or less returned to those prevalent before the development. In no case has erosion exceeded limits predicted by engineering studies. Nearly two dozen reports and studies related to fish habitat, populations, passage and mortality in the Annapolis River have been completed in the past ten to fifteen years. The studies largely dealt with single issues in isolation or were directed towards specific aspects of fish behaviour, and were initiated by a variety of agencies. Although much information existed about fish habitat in the Annapolis River, the varying methodologies and the sometimes conflicting conclusions made it difficult to make a reliable assessment of the stations impacts. In 1991, Nova Scotia Power commissioned an independent review of all relevant Annapolis River studies, as well as other pertinent studies completed in Nova Scotia and throughout North America. The objective was to identify the valid conclusions that may be made about the effects on fish passage by the station, the additional studies that may be necessary to remove important uncertainties and the effectiveness of deterrent devices being utilized at the site. The conclusions of the recently completed review may be summarized as follows:

(a) The estimate of turbine mortality for post-spawned shad (21.3%) of Hogan (1987) is reasonable and consistent with results of other studies using valid methodology. Various estimates of juvenile turbine mortality are considered highly unreliable.

(b) A reliable conclusion cannot be made with respect to which route, the turbine or the fishway, is favoured by adult alosids for passage through the barrage. Substantial juvenile recruitment in 1990 suggests that estimates of juvenile turbine mortality are too high and/or more juvenile fish are using fishways for passage than estimated.

(c) Large numbers of five year olds in the 1990 run indicates that substantial recruitment to the spawning population is continuing. However, the overall effects of the tidal station on the stock will not be determined without the knowledge of the proportion of the population using the turbine for passage.

(d) Acoustic devices in use at the station are effective in deterring adult fish from the intake, but little can be concluded about the effect on juveniles or whether fish become accustomed to the noise over time. A number of specific and expensive studies could be initiated to remove most of the remaining uncertainties, but the report concludes that population studies undertaken over three consecutive years will be the most effective means of assessing overall turbine mortality. Ultimately, turbine effects, if significant, will manifest themselves in the population.

An overview

The relatively few problems encountered in the seven years of operating experience at Annapolis have mainly involved auxiliary equipment rather than the turbine design as such. None of the problems have appreciably affected the operation of the plant. It therefore appears that the STRAFLO turbine is eligible for consideration as a prime mover in the large tidal power developments. However, this type of machine has one inherent handicap - adjustable blades are not feasible. The peak efficiency curve associated with a fixed blade runner is not a disadvantage where DC transmission is contemplated, as is the case for the proposed large plants in the Bay of Fundy. Under those conditions, a synchronous turbine operation carries no economic penalty and results in much the same characteristics as a double-regulated turbine. However, where pumping is possible and advantageous, the STRAFLO is unlikely to be a contender. The STRAFLO is a rugged, compact, yet easily accessible machine which in itself makes minimum demands on powerhouse space and cost. However, the installation at Annapolis advertises the fact that the STRAFLO does require more auxiliary equipment than other types of low-head machines. There is surely no one best turbine type for tidal applications. Choice is likely to depend on site characteristics, optimum plant operating strategy, and last but perhaps not least, on the prices offered by manufacturers. However, based on our experience at Annapolis and subject to all the foregoing qualifications, the STRAFLO turbine is, from the Canadian perspective, a strong contender.

2. Current status of tidal power in the Bay of Fundy

G. C. BAKER, PEng

Several feasibility studies of tidal power development in the Bay of Fundy have been undertaken over the past quarter-century, with the result that the potential of various sites is well known and near-optimum designs exist for plants at the most favourable locations. The environmental consequences of development have also been explored. While environmental studies have been more limited than technical studies in both scope and time span, they have nevertheless provided a fairly good understanding of the potential impacts. In this Paper the present technical and economic status of tidal power in Fundy is summarized; environmental impacts and benefits are outlined and prospects for development are assessed.

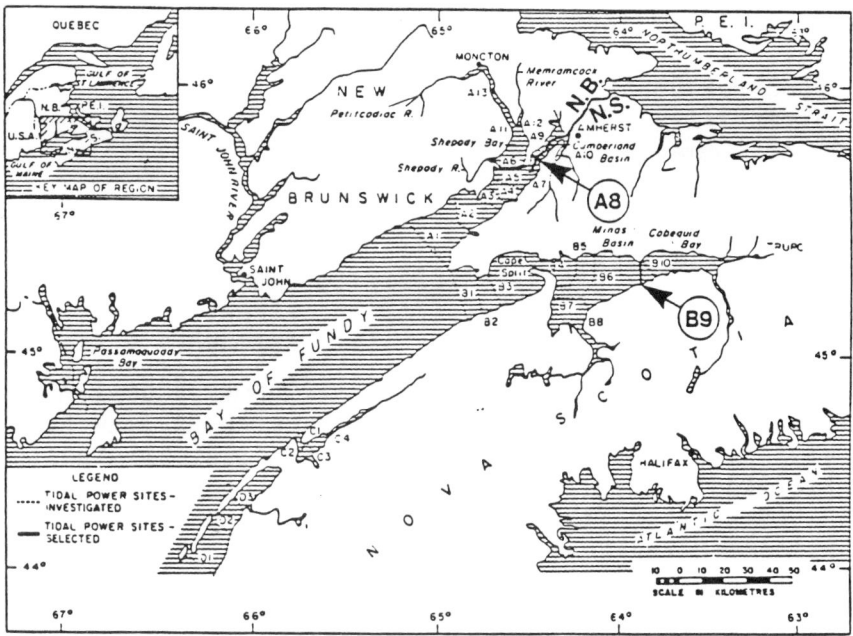

Fig. 1. Site location

TIDAL POWER

Technical

In general, feasibility studies of Fundy sites have found B9 in Minas Basin to be the most economical and A8 in Cumberland Basin to be the most developable.

Site A8 has an optimum installed capacity of about 1400 MW, a figure which has varied from study to study and obviously depends on the relative costs of electric energy, rotating machinery and civil works. 1400 MW would represent about a 5 to 1 scale-up from the largest existing tidal plant. Capital cost would be in the 3 to 4 billion dollar range, including transmission. In terms of both capacity and cost, A8 is thought to represent the prudent upper limit for the first full-scale plant in Fundy.

By comparison, B9 would involve about four times the capacity and capital cost of A8 and in addition is predicted to have much more significant effects on the tidal regime. For these reasons, it is not now viewed as a realistic candidate for initial development.

The latest design for A8 was produced in a 1984 study. It is based on a strategy of early closure, and makes extensive use of caissons. The operating scheme is single effect, ebb generation, with variable speed operation of fixed blade turbines and rectification of output for DC transmission to market.

Table 1. A8 design features

Number of turbines	42
Turbine rating	34 MW
Installed capacity	1428 MW
Sluice number	20
Sluice total throat area	3,680 m
No. of barrage caissons	52
Barrage length	2,560 m
Earth dyke	614 m
Annual energy, net of trans. loss	3307 GWh

A pertinent question is whether there is scope for significant cost reduction through improvements in design and construction techniques. In that context, past studies have tended to show that the choice between single effect and double effect is pretty much a toss-up for these sites. Probably this is in part due to the gently sloping shores typical of the headpond areas, which result in a significant reduction in basin area at lower water levels. In any case, the flood generation part of double effect output is quite small, and there is little or no increase in total energy output (compared to single effect) except at spring tides.

A second common strategy not used in the present design is pumping. It is of course a sound technical option, but where dykelands abut the headpond and where drainage and salt intrusion are perceived problems, raising the headpond level may cause more trouble than it is worth.

Overall, the present design is probably not optimum, but not so far from optimum that any major improvements in cost effectiveness can be expected. Further shortening of the construction period and refinement of caisson design to provide for complete-unit installation of prime movers are potentially useful targets for technical effort.

A marketing study conducted in 1984 indicated that the New England Power Pool would require firm capacity in any imports. This led to a preliminary study (ref. 1) of compressed air energy storage (CAES). Pumped storage as a method of retiming had been investigated in an earlier study (ref. 2). It is relatively unattractive because of high capital cost and a throughput loss of about 25%. The result of the CAES study showed that it would avoid the high losses of pumped storage, but would nevertheless be costly. This is partly because the natural gas Canada is selling to the US at bargain prices is not available in the Maritimes.

The best option for retiming tidal power is operation in synchronism with a hydro system having large storage capacity. Under these conditions, retiming would involve the capital cost of additional hydro plant capacity, but would incur small, or negative, losses. Hydro Quebec has suitable facilities but has been disinclined to make them available. Opposition to further developments in the James Bay area could alter this attitude.

Technical studies in Fundy have been limited for the most part to the consideration of commercially proven technology. However, some work was done (with National Research Council funding) on ultra low head tidal plants using vertical axis turbines of the Darrieus type. Such plants would be best sited in locations offering high currents rather than in locations offering high heads. They would provide two-way generation with annual capacity factors considerably higher than those achievable from present Fundy designs.

On the other hand, they would work on water of much lower energy density, requiring much larger turbine throat area per kilowatt of capacity and, given equal construction quality and durability, much higher unit capital cost.

Preliminary indications were that the tidal regime would be less affected by ultra low head plants than by those of conventional type. On the other hand, ultra low head plants would not be free of fish mortality, and fish diversion would be almost impossible. Because of the present complete lack of commercial experience and for other reasons, it would take many years to bring ultra low head designs to commercially proven status.

To summarize, one should not expect any great cost or performance breakthroughs from technical studies. Minor gains are possible in optimiza-

TIDAL POWER

tion through cycle improvements, turbine type selection, and powerhouse design. The main potential probably lies in cutting construction time.

Economic

Based on future energy costs as perceived in the 1970s and early 1980s, tidal plants in Fundy promised to be economic, with benefits exceeding costs by a margin depending on one's crystal ball.

The picture changed in the middle and late 1980s when oil prices plummeted, gas prices in North America were forced down by policies based on market competition, and interest rates were sustained at high levels by monetary policy. The perspective may be changing again, as we look at the growing concern over the emission of greenhouse gases.

The economic status of Fundy tidal power can be measured in two ways

(a) From the standpoint of internal costs of utilities, Fundy tidal power today is uneconomic, and

(b) From the standpoint of the benefits and costs to society as a whole, a framework which includes environmental and social, as well as purely economic, considerations.

Some comparisons based on internal costs are presented in Table 2. The figures indicate that A8 is not economic under present conditions as a source of displacement energy, but that it could be combined with other resources to supply the shoulder and peak portions of utility load curves.

Although tidal plus CAES would produce a premium product, it would be too costly. Tidal retimed by river hydro storage would be competitive under any circumstances, and tidal firmed by combustion turbines would be attractive at real interest rates up to about 5%. In terms of social cost, it is not easy and may be impossible to make a meaningful estimate of the dollar cost of environmental factors. Probably social policy in this regard will be applied through controls on the total amount of emissions, rather than through internalization of pollution costs relating to greenhouse gases. In that case, tidal power and nuclear could become the only major sources of additional electric energy in the Maritimes.

Environmental

Many environmental evils were conjectured to arise from tidal power plant developments when professional attention was first directed to this matter in 1977. Now, after studies which vastly increased knowledge of Fundy ecology, most of the conjectures have faded, while a few impacts have emerged as vitally important.

This is not to say that there would be few impacts; but only that most of the negative impacts would be minor, or could be mitigated, or could be compensated for.

However, two or three potential impacts are significant. One of these is the mortality inflicted by a tidal plant on fish and marine mammals. There is a great difference between river hydro plants and tidal plants in the Fundy headwaters. River plants are typically passed by spawning adults and juveniles making their way to the sea. This entails at the most one passage per year. Tidal plants in Fundy would have to cope with resident fish stocks likely to transit the turbines fifteen times or so per year. Obviously, a low single passage mortality rate, acceptable in a river plant, would be unacceptable in Fundy where multiple passages are expected.

The other aspect of the problem is that larger fish and marine mammals could be expected to have very high mortality rates if they passed through a turbine. The obvious answer is to make provision for keeping fish out of the turbines. Fish passages can be provided, but then the problem is to ensure that they are used. Tidal Power Corporation investigated behavioural devices for fish diversion, with promising results. However, there is not yet a good enough answer to this problem to make environmental approvals likely.

The other potentially grave impact relates to tidal ranges in the Gulf of Maine. According to the Greenberg Model, ranges would be increased slightly (about 3 cm on average) by a barrage at site A8 and considerably more (about 13 cm average) by a barrage at site B9.

These increases appear very modest, both in comparison to tide ranges in Fundy, and in comparison to the variability of the tides due to wind shear and storm surge. However, many installations along the US coast in the Gulf of Maine were built long ago. Since then, the coast has sunk and mean sea level has increased. As a result, safety factors have been eroded and the US Army Corps of Engineers now advises that the damage threshold is at spring high tide level.

Obviously, the situation is such that a tidal plant in Fundy might cause damage, or even if it did not, might well be subject to damage claims.

The Greenberg Model (ref. 3) was used to ascertain what could be done by way of changes in flow and phase, to reduce the effect on Gulf of Maine tides. It was found that simply opening all sluices and turbines, making the barrage as permeable as possible, reduced the tides almost to natural levels within one or two tidal cycles. Thus there is a defense against damage, given a sufficient advance warning of a storm surge in the Gulf. However, there is not a defense against damage claims.

Leaving the sluices open would result in almost complete loss of output, so this expedient could not be used as the regular mode of plant operation. Unfortunately, changes in the phase of plant discharge, which might be

TIDAL POWER

achieved with less loss of output, showed no appreciable effect on the tidal regime.

Table 2. Cost comparisons (This table gives estimated life-levelized unit costs stated in constant $C 1991)

	Real Interest Rate		
	4%	5%	6%
Displacement energy basis:			
A8 with transmission to Mass.[1]	5.42	6.58	7.82
A8 plant only	4.34	5.28	6.30
Displacement energy cost, no esc.[2]	23.29	3.29	3.29
Displacement energy cost, 1% esc.[2]	24.16	4.03	3.93
Firm capacity basis:			
A8 retimed by CAES[3]	9.58	10.75	12.02
A8 firmed by CTs, no esc.[4]	6.95	7.6	8.47
A8 retimed by river hydro[5]	5.74	6.75	7.85
Coal-fired thermal (80% CF)[6]	65.32	5.57	5.82
Coal-fired thermal (36% CF)[6]	67.81	8.37	8.92
Combustion turbine (15% CF)	11.18	11.46	11.76

Notes:

[1] Based on 75-year life. Cost of losses included.
[2] Based on estimated fuel & variable O&M for a 440 MW unit recently committed.
[3] Provides 1400 MW of firm capacity and energy for 12 hours per day, 5 days per week (4250 GWh per year).
[4] Provides 1400 MW of firm capacity as in 3 above but at lower load factor during peak period, plus displacement energy at other times (5218 GWh per year).
[5] Includes cost of extra 1400 MW of generating capacity plus 0.5cents/KWh service charge.
[6] Based on same unit as 2 above but includes capital and fixed O&M costs.

It should be noted that the Greenberg Model is the only source of predictions for tidal range and currents in the Fundy/Gulf of Maine system. While it is considered trustworthy by most Canadian modelling experts, there is no consensus on this point in the US Arguments can be, and have been, made that model results are subject to error because they are based on two-dimensional analysis of a system which is in fact three-dimensional. No effective way of truthing the Greenberg Model seems to be available, but there has been an urge in the US to develop a three-dimensional model, and this point will probably rise again if development appears imminent.

No means exist today for reflecting US interests in environmental regulation of a tidal project in Canadian waters. Assuming, as seems likely, that a considerable portion of the output would be marketed in the US at least during the early years of plant life, any US project would be at unacceptable risk from US action related to real or imagined effects on endangered species, and claims for real or imagined damages.

Sediment is also a matter of concern at site A8. To investigate this potential problem a three-dimensional numerical model of Cumberland Basin has recently been developed. Results are expected to be reported in the near future. However, accretion in the headpond would depend on whether the observed solids content seaward of the barrage originates in Cumberland Basin, or from adjacent areas of Chignecto Bay. The model cannot elucidate this point.

Summary

Construction of a tidal plant at A8 would be contingent upon attainment of economic viability and resolution of the remaining major environmental concerns. If these conditions were achieved, there is little doubt that the project would proceed.

Economic viability is likely to depend mainly on circumstance rather than on technical design. It would be realized if co-ordinated river hydro became available; if hydrocarbon-fuelled generation is further impacted by internalization of social costs, emission limits, or fuel price escalation; or if the real interest rate on long-term debt dropped another one percent or so. One or more of these conditions could be realized at any time and the probability of realization at sometime within the next decade seems high.

Environmental feasibility would require an adequate and demonstrated solution to fish mortality problems, resolution of remaining concerns about sedimentation and aqreement with the United States on trans-border impacts.

Much remains to be done on the environmental side. It should be done while awaiting the conditions of economic availability. Otherwise, no major tidal plant will be built in Fundy for a long time to come.

References

1. CANADIAN ATLANTIC POWER GROUP LTD. Assessment of retiming tidal power from Bay of Fundy using compressed air storage. *Energy, Mines & Resources*, Canada, 1988, p. 141.
2. ANON. *Feasibility of tidal power development in the Bay of Fundy*. 6 vols. Atlantic Tidal Power Programming Board, Ottawa, Canada, 1969.
3. GREENBERG, D. A. A numerical model investigation of tidal phenomena in the Bay of Fundy and Gulf of Maine. *Marine Geodesy*, 1979, Volume 2, No. 2; New York.

Discussion on Papers 1 and 2

M. JEFFERSON, World Energy Council

Mr Baker in his presentation did not list birds under his heading "Unresolved Problems", yet the Bay of Fundy is of great international ornithological importance, especially in the southward phase of the Eastern Migration Flyway. Shepody Bay particularly has been registered as being of importance by the Western Hemisphere Reserve Network. Is this an issue of importance in the Maritime Provinces?

G. C. BAKER

The principal feeding ground for migratory birds in the Fundy headwaters is in the Southern Bight of Minas Basin. Shepbody Bay is also of importance. Neither of these are located in the headpond areas of either A8 or B9, the only Fundy sites for which development has been proposed.

Observations by the Centre for Estuarine Research indicate that no increase in the density of feeding birds occurs, as bird populations increase 30-fold from July through September. The conclusion is forced that food and feeding space is ample even for years with peak migratory populations.

A tidal plant at site A8 or B9 would decrease the intertidal zone within the headpond, but would increase the far-field tidal range, increasing the area of intertidal zone over most of Fundy external to the headpond.

For these reasons, and subject to the findings of the environmental assessment which would necessarily precede any construction, impacts on migratory birds are expected to be close to neutral. The unresolved problems listed in the paper are those involving potentially major impacts. Impacts on migratory birds would certainly not be major.

H. MOORHEAD, Severn Tidal Power Group

In Table 2 of Paper 2 there is reference to unit costs based on firm capacity retimed by river hydro. It would be interesting to know how this would be achieved, as it would no doubt require an integrated approach to system planning and operation on a long-term basis.

G. C. BAKER

Retiming of tidal power by river hydro requires that the river hydro have large storage capacity compared to the tidal energy of a single tidal cycle, and preferably large compared to tidal energy over a lunar fortnight. For a large tidal plant, it would also be a requirement to add river hydro capacity equal to the maximum demand contracted to be supplied from the tidal plant.

In operation, the river hydro would supply power and energy in the interval between each successive pulse of tidal energy and the tidal plant would repay this energy by supplying an equal amount to the customers of the river hydro system from each operating cycle.

Obviously, if the tidal power were to be delivered at a high load factor, the contracted capacity could only be a correspondingly low fraction of installed capacity.

Operation would, as Mr Moorhouse assumes, require co-ordination of the two systems. However, planning of the hydro system would not be affected, apart from the capacity addition noted above, because except for minor fluctuations in storage level, the hydro would see no change in operating conditions or capability.

In most North American utility systems, generating units and circuit loadings are controlled by SCADA equipment, and centrally dispatched, often by optimization programs. Co-ordination would therefore not create undue difficulty.

R. PRICE, *Energy Technology Support Unit*

All recent designs for tidal energy schemes in Canada have simple ebb generation, while UK studies indicate that energy output can be increased by the use of flood pumping. Could Mr Baker indicate why Canadian experts have chosen not to pump?

G. C. BAKER

Considerable areas of dykeland abut the headponds at both site A8 and B9. Dykelands are the most productive soils available to Nova Scotia farmers and are therefore valuable resources. Omission of flood pumping results in lower average and lower maximum headpond levels and in consequence provides benefits in terms of flood abatement, reduced dyke maintenance costs, better dykeland drainage and avoidance of impact on aquifers.

These benefits, and the acceptance of tidal power as beneficial by most farming interests, are considered to outweigh the value of extra production through pumping.

P. HUNTER, *Sir Alexander Gibb*

In his Paper on the Annapolis project, Dr Baker refers to levels of groundwater and states that they were not changed by the construction of the barrage. Was this forecast before construction of the project, and if so did the measured levels confirm the forecast?

G. C. BAKER

The project environmental assessment predicted no deleterious changes in groundwater levels or salinity providing that certain conditions were observed. The prediction was based on hydrogeological considerations plus data from a few piezometers at sensitive locations.

Remedial works were constructed to realize the conditions on which the prediction was based. An initial array of 24 piezometers, later expanded to several times that number, was used to gather baseline data. Monitoring of the array for a few years after the plant became operational confirmed the prediction. Apart from normal fluctuations due to precipitation and season, no changes in groundwater elevation and no changes in salinity were observed.

M. KENDRICK, *Mersey Conservancy*

The Author reports that although 3D sediment model studies of Cumberland Basin have been made, prediction of disposition in the headpond depends on whether the sediment originates from the Cumberland Basin itself or from adjacent areas in Chignecto Bay. Have field programmes been devised to try to determine this and if so what are the findings so far?

G. C. BAKER

Sediment investigations in Chignecto Bay have been reported by a number of authors. The Author is not aware of the extent of the fieldwork involved. No fieldwork is scheduled or being conducted at this time.

Dr C. Amos (Atlantic Geoscience Centre) and Kim Tai Tee (Dalhousie University) found that the sediment concentration in the lower portion of Chignecto Bay is remarkably constant; that any deviation is met by erosion or deposition in more landward areas. The importance of Cumberland Basin, compared to other sources and sinks in the Chignecto area, is to the best of the Author's knowledge still an open question.

S. BENNETT, *Stopdate Ltd*

What was the time between starting construction and operation of the Annapolis Tidal Power Station, and now that you have operational experience of the Straflo turbines, what is the predicted life span of the equipment?

G. C. BAKER

Construction started in May 1980 and the unit entered commercial service in August 1984. Construction was delayed by a labour dispute between parties not involved in the project.

Hydro stations usually have an indefinitely long life. When components fail, they are simply replaced. Experience to date at Annapolis does not provide any data useful for estimating component life. There is no cavitation;

corrosion is insignificant and there is as yet no detectable wear on bearings or operating seals.

3. The Mersey Barrage – civil engineering aspects

B. I. JONES, B Eng, MICE, MIWEM, Director, Rendel Parkman,
C.D.I. MORGAN, BSc, Civil Engineering Manager, Mersey
Barrage Company, D. PHILLIPS, BSc(Hons), MICE, MIStructE, FIHT,
MHKIE, Study Manager, Mersey Barrage Company and
M. W. PINKNEY, MA, MSc, DIC, MIStructE, Technical Director,
Rendel Parkman

Over the period since the last Tidal Energy Conference in November 1989, considerable progress has been achieved on the Mersey Barrage Project, with significant design developments having taken place. This Paper summarises the results of recent design studies and provides a description of the current civil engineering and construction proposals for the barrage. A number of features which were considered to merit special attention are highlighted and probable costs and overall conclusions are given.

Introduction

It is now some 10 years since Merseyside County Council first annunciated the concept of a tidal energy project on the Mersey Estuary, to take advantage of a mean spring tide range of 8.4 metres (maximum 10.5 metres), with typical flood and ebb currents of some 5 knots, and an enclosed basin area of potentially 60 square kilometres. Having completed pre-feasibility studies, Merseyside County Council was dismantled in 1986, along with the other Metropolitan Counties, and responsibility for the project passed into the private sector through the formation of the Mersey Barrage Company. The Company has since carried out a series of comprehensive feasibility studies, with the assistance of the Department of Energy, and work is currently continuing.

At the time of the last ICE Tidal Energy Conference in November 1989, the Stage II Feasibility Studies were in progress and the information then available allowed the presentation of only broad indications of the manner in which the project was developing. The completion of the Stage II Studies, embracing engineering, environmental and economic considerations, allowed significant advances in project definition, with conclusions emerging on a number of key planning and design issues, as follows

TIDAL POWER

- *Preferred barrage location* - within the general Line 3 area, between New Ferry and Dingle, upstream of the main urban centres of Liverpool and Birkenhead, as illustrated in Fig. 1.
- *Barrage operating regime* - ebb generation with flood pumping
- *Turbine generators* - 8 m diameter, geared drive units of 25MW rating
- *Method of civil construction (turbines and sluice housings)* - prefabricated caisson structures, probably in reinforced concrete but steel remaining an option

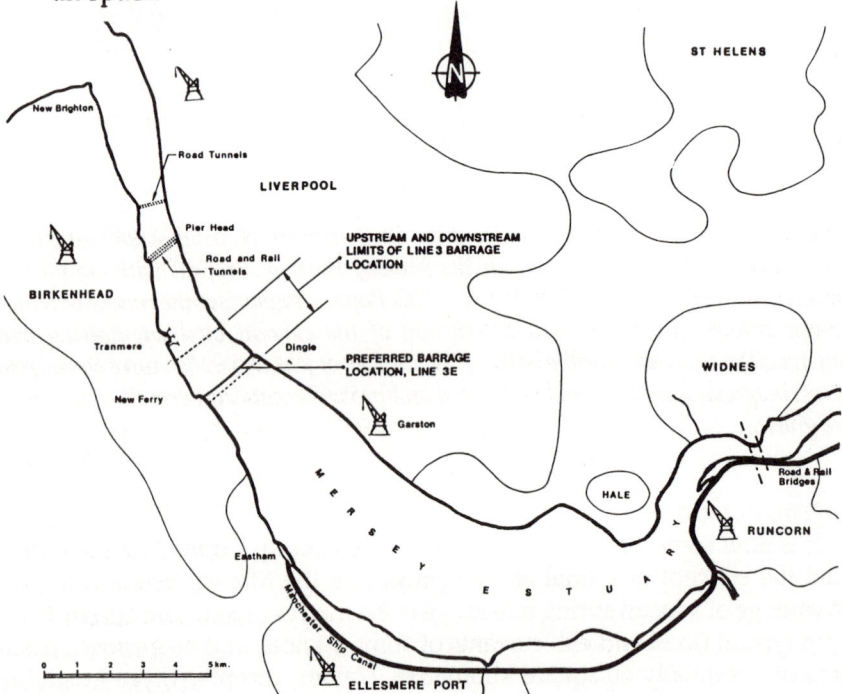

Fig. 1. Plan of the Mersey estuary

The Stage III Studies, completed during September 1991, included investigation of energy yield, hydraulics, sedimentation, shipping and ecological issues, together with project economics and financing. In the engineering context, further site investigations were undertaken and preliminary designs for civil, mechanical and electrical engineering components of the project were prepared, together with correspondingly detailed construction sequences and programmes, to support reliable estimates of construction, operating and maintenance costs. These procedures have led to a refinement of barrage location, to Line 3E, and the adoption of channel sluices in preference to the low level venturi sluices and rockfill closure embankments previously proposed.

The project management team for the environmental and engineering aspects of the study consisted of senior staff seconded from the Mersey Barrage Contractor's Group, namely: Tarmac, Cementation, Costain, HBM and NEI. The teams were located in MBCs Liverpool offices which allowed particularly close co-ordination of the M&E and civils works during the design, development and construction planning stages of the studies. Further benefits arose due to the close proximity to MBCs offices of the project office of Rendel Parkman, the civil engineering design and principal shipping consultants, which greatly assisted the integration of the design and construction requirements of the project.

Site conditions

In the general Line 3 zone, the Mersey Estuary is characterized by the presence of Devils Bank, an extensive drying sandbank which separates the Garston Channel, lying close to the eastern shore, from the central deep water areas. In this reach, the width of the estuary is some 1.9 kilometres and bed levels vary from -4 metres OD on Devils Bank and also on the western margins of the estuary, to approximately -15m OD in the deep water areas. Within these deep water areas, tidal variations result in minimum and maximum water depths of some 10 and 20 metres respectively.

Site investigations conducted between August and November 1990 confirmed sub-bed strata to comprise essentially a sequence of soft/loose alluvium (sands and silts), overlying glacial tills (medium sands and gravels, followed by stiff boulder clays), with Sherwood Sandstone of the Triassic Series forming the bedrock, as shown in Fig. 2. Over the western half of the estuary, the depth of the sandstone below estuary bed level does not generally exceed some 14 metres (i.e. approximately -19 metres OD). Within the eastern portion of the estuary, however, the depth to the bedrock increases significantly to a maximum of perhaps 65 metres (i.e. approximately -70 metres OD), due to the presence of a deep buried glacial channel which is infilled predominantly with boulder clay materials. It therefore follows from foundation considerations that, while the western half of the estuary is capable of accommodating a wide range of structures, as the sandstone bedrock is relatively accessible in these areas, the eastern sections are suitable only for relatively light structures, if complex foundation arrangements are to be avoided.

Barrage location and layout

Having established that the general Line 3 area, as shown on Fig. 1 provided the most favourable basis on which to develop the project, the selection of the preferred barrage location and layout within this zone

TIDAL POWER

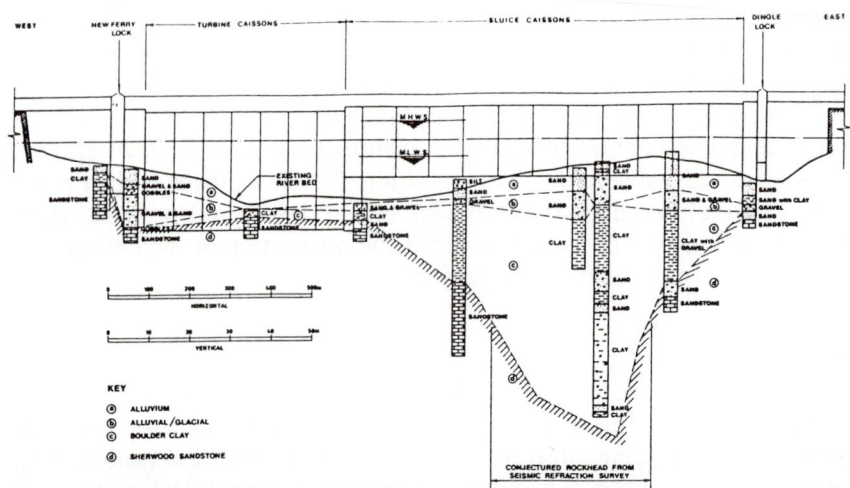

Fig. 2. Longitudinal section of the barrage showing sub-bed strata

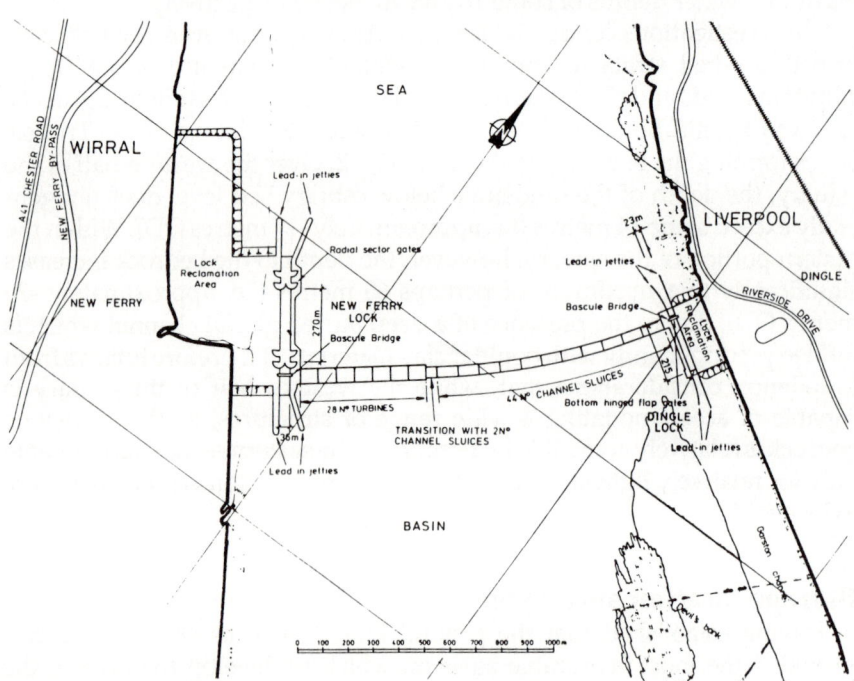

Fig. 3. Mersey Barrage Line 3E — general arrangement

depended on a variety of considerations, with major influences being related to geotechnics and foundations, the optimum hydraulic regime with particular reference to shipping, remoteness from the existing Tranmere Oil Jetties and landfalls, although energy yield, construction costs and project economics naturally proved equally dominant issues.

Following technical and economic appraisals of a number of project options, involving different locations within the Line 3 zone and various turbine-generator and sluice layouts, Line 3E was adopted as the preferred barrage location, with the corresponding general arrangement comprising

- Western reclamation area
- New Ferry Lock
- 28 turbine generators
- 46 channel sluices
- Dingle Lock
- Eastern reclamation area

The selected layout of the Line 3E barrage is shown on Fig. 3 and leading factors influencing this layout are discussed below.

The locks are located to coincide broadly with the existing shipping lanes, serving the upstream port facilities at Eastham and Garston.

Bearing pressure and settlement requirements dictated that the turbine-generator caissons should be founded directly on the Sherwood sandstone bedrock to avoid piling or other expensive foundation designs. As a result, the turbine generators are located towards the western side of the estuary, where rock is relatively accessible, with Line 3E affording the opportunity of accommodating the greatest number of turbine units. The westerly location of the turbine-generators also allows their position to coincide with the main deep water channel zone of the estuary, thereby minimising dredging requirements.

In connection with sluice arrangements, it became clear that submerged venturi sluices, as proposed for earlier barrage configurations, would not only involve substantial capital dredging, to provide satisfactory hydraulic conditions in the approach to and exit from the low level water passageways, but they would also require deep and probably complex foundations in the central area of the estuary. A fundamental re-assessment of sluice design was therefore undertaken and this revealed that channel sluices would provide a more economic solution than the alternative venturi sluices and rockfill closure embankments, while greatly improving the flow distribution acoss the estuary under flood tide conditions. The channel sluices take the form of relatively light caissons, founded on rockfill and extending as a continuous feature over the glacial deposits present in the eastern areas of the estuary. Energy modelling indicated that to provide the required hydraulic capacity, 46 No. channel sluices would be appropriate.

TIDAL POWER

With the lock locations described, significant quantities of dredged material can be deposited in the reclamation areas between the lock structures and the existing shorelines, thereby eliminating the necessity to dispose of dredged materials offshore. These areas are subsequently available to accommodate the various service and ancillary buildings required for the barrage.

Design and construction criteria

The overall project objectives, requiring that the barrage be installed as economically as possible, led to the adoption of the following construction principles.

(*a*) Maximise use of prefabricated elements, incorporating civil, mechanical and electrical components, to shorten the construction programme.

(*b*) Use of floating caissons wherever possible to minimise the estuary in-situ works.

(*c*) Maximise use of existing construction facilities around the estuary, whether of a permanent or temporary nature.

(*d*) Minimise disposal of dredged material outside the estuary, in order to reduce vessel movements within the estuary and to provide land areas for temporary construction yards and barrage operational support facilities.

(*e*)Use, wherever possible, of proven construction techniques, in order to limit any potential design and construction risk.

These principles formed the basis of the design development and construction planning phases of the recent civil engineering studies.

The barrage elements have been designed for strength and stability under a range of still water levels with appropriate wind and wave loads. Generally, hydraulic cut-offs have been included to control water flow and pressure under the structure bases but these were considered to be only partially effective when assessing stability. Water levels were taken from the two dimensional hydraulic model of the barrage operation, with allowances for modelling tolerance, hydraulic surge, atmospheric surge, and secular trends.

The caisson structures have been designed to withstand both global and local loads caused by

- uneven temporary and permanent support conditions
- operating and extreme hydrostatic heads
- differential water pressures on panels
- differential sand pressures
- floor loads

- plant and mechanical equipment loads (including, in the case of turbine caissons, short circuit loads on the generator and runner, together with dynamic loads on the distributor).

The caissons finally adopted have been designed in reinforced concrete, although steel alternatives were also considered. Crack widths were limited to 0.2 mm above -4 m OD under normal loads and 0.3 mm under extreme loading conditions elsewhere. The assessment criteria for caisson stability are shown in Table 1.

Table 1. Caisson stability criteria*

Loading Category	Factor of Safety			Maximum Bearing Stress Normalised to Normal Ops = 1.0
	Flotation	Sliding	Overturning	
1. Normal Operational	1.3	1.7	No negative heel pressure	1.0
2. Normal Maintenance	1.25	1.7	See Note	1.0
3. Extreme Operational	1.2	1.4	See Note	1.2
4. Accident- • Barrage maloperation	1.2	1.3	See Note	1.4
• Rapid shutdown	1.2	1.2	See Note	1.5
5. Temporary (Initial Ballasting)	1.1	1.3	See Note	1.0

* Negative heel pressure is permitted under maintenance, extreme, accident or temporary loadings provided that the result falls within the middle 50% of the base width. There must also be a minimum net bearing pressure of 10kN/m2 at the position of any hydraulic cut-off.

Design of main barrage elements
Turbine generator structures

The generator section of the barrage comprises five caissons, each containing four turbine units, with the remaining eight units being accommodated in an in-situ structure which is similar in size and layout to two caissons.

The layout and leading dimensions of the caissons are governed by the

TIDAL POWER

requirements of the 8 m diameter turbine units and the associated electrical equipment, coupled with the need to ensure overall stability and structural integrity under the anticipated construction and operational conditions. The resulting caisson layout is shown in Fig. 4, with each caisson measuring approximately 69 m x 62 m x 32 m high and weighing some 78000 tonnes at float-out.

On immersion of the caisson, downstands incorporated along the basin

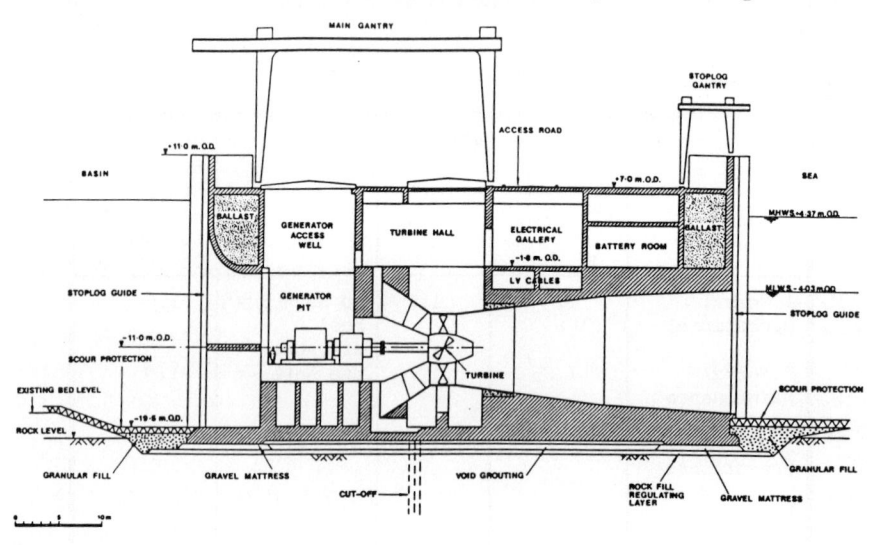

Fig. 4. Cross-section of turbine generator caisson

and sea edges of the base will land on 500 mm thick gravel mattresses overlying a regulating layer of rockfill placed on the dredged sandstone foundation. To allow for possible level tolerances on the placed foundation materials, the caisson has been designed to accommodate an out-of-plane deformation of 200 mm at any corner. In order to assess the effect of such distortion, the caisson was modelled as a series of membrane plates using the finite element method to obtain estimates of global stresses and deformations. In particular, the model was used to predict the deformation between a line projected normal to the turbine staying and the section where the steel liner of the turbine is to be concreted into the caisson draft tube, with the joint in the steel liner being designed to accommodate the predicted deformation of approximately 10 mm, prior to final securing.

To minimise the possibility of water flow below the foundation, which could cause seepage erosion, and to control hydrostatic uplift pressures, a three row grout curtain has been detailed. Grouting of this curtain together with the void beneath the caisson base, will take place from the turbine pits,

Sluice caissons

The sluices comprise a series of gated channels, of rectangular cross section and of width 17 m, separated by 3 m wide piers, a total of forty six channels being required. Discharge through the sluices will be controlled by hydraulically operated radial gates, capable of sustaining hydrostatic head in both the upstream and downstream directions. Access over the sluices will be provided by a multi-span simply supported precast concrete box girder bridge, the bridge void carrying power, control and instrumentation cabling and other services. Stop logs for installation and maintenance purposes can be installed at both the basin and sea ends of the water passageway but, in the event of an emergency, each channel can be closed by stop logs placed from the access bridge in slots between the gate and the bridge.

The sluices are to be formed as reinforced concrete cellular caissons, the standard caisson comprising four channels. A special caisson of only two channels is, however, required for installation over a transition structure designed to accommodate the change in foundation level at the interface of the turbine generator structures and the sluices. The transition structure essentially comprises a reinforced concrete, closed, cellular caisson, founded broadly at the same level and in the same manner as the adjacent turbine generator caissons and designed to provide support to the higher level sluice foundations.

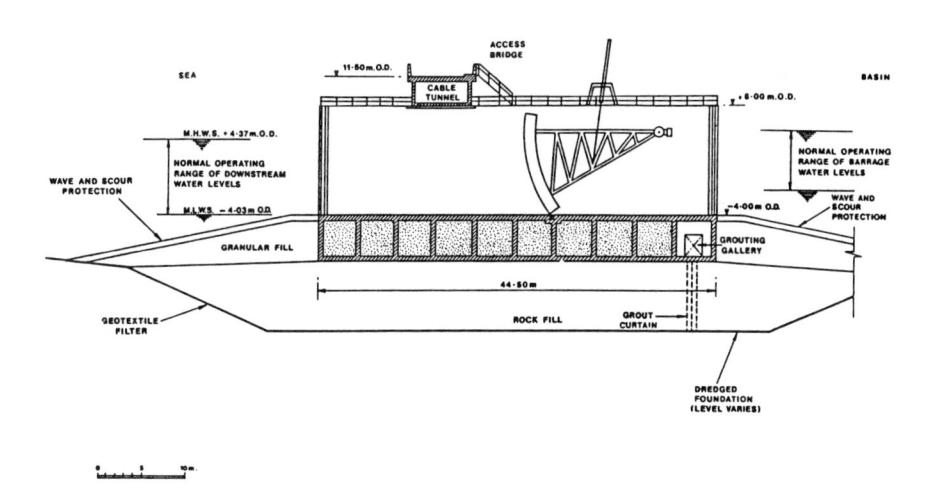

Fig. 5. Cross-section of sluice caisson

TIDAL POWER

The four channel sluice caissons each measure approximately 33 m x 44.5 m x 17 m high and weigh some 22000 tonnes at float out. They are to be installed on rockfill foundations which will be placed following the removal by dredging of soft alluvial deposits, as shown in Fig. 5. Water seepage through the rockfill under the caisson will be controlled by a three row grout curtain, while pore pressures in the underlying glacial till will be relieved by vertical drains installed at the sea side of the foundation, with piping being controlled by a geotextile filter below the rockfill.

Following installation, the cells of the caisson piers and base will be sand filled, except for the basin-side cell of the base, which provides a grouting gallery for both installation and maintenance of the grouted cut-off.

The design of the cellular base has been primarily governed by two load cases with conflicting requirements

- Strength and stiffness during installation
- Flexibility to allow the structure to conform to the foundation under various settlement conditions.

It was estimated that the total settlement above the centre of the buried glacial channel would be of the order of 200 mm. The caisson has therefore been designed to accommodate a combination of differential settlements and allowances for initial tolerances on the screeded rockfill foundation upstands which, together, it was estimated, could result in a maximum out-of-plane deformation of any corner of the base of 335 mm.

Locks

The lock arrangements at New Ferry and Dingle have been established to suit the requirements of shipping bound respectively for the Queen Elizabeth II Dock and Manchester Ship Canal Entrance at Eastham, on the west side of the estuary, and the Garston Docks on the east. Corresponding details are shown in Table 2.

	New Ferry	Dingle
Shipping Requirements	39,000 dwt oil tanker	2 No. 4,000 dwt coasters
Lock Dimensions: Length (metres)	270	215
Width (metres)	36	23
Floor Level (metres OD)	-12.5	-6.0
Maximum Water Depth (metres)	Approx 17.0	Approx 10.5

Table 2. Shipping requirements and dimensions of the barrage locks

PAPER 3: JONES, MORGAN, PHILLIPS AND PINKNEY

In the case of Dingle Lock, the design has been influenced by a requirement for early completion of the lock, to ensure satisfactory passage of Garston bound shipping during the barrage construction period. Lock construction utilising steel sheet piling has therefore been adopted, as shown in Fig. 6. The inner lock wall, retaining the Eastern Reclamation Area, will take the form of a single skin wall with ties extending to an anchor wall located within the placed fill, while the outer, free standing wall will comprise a double skin arrangement. The lock floor will be formed in articulated concrete block mattresses placed on granular fill.

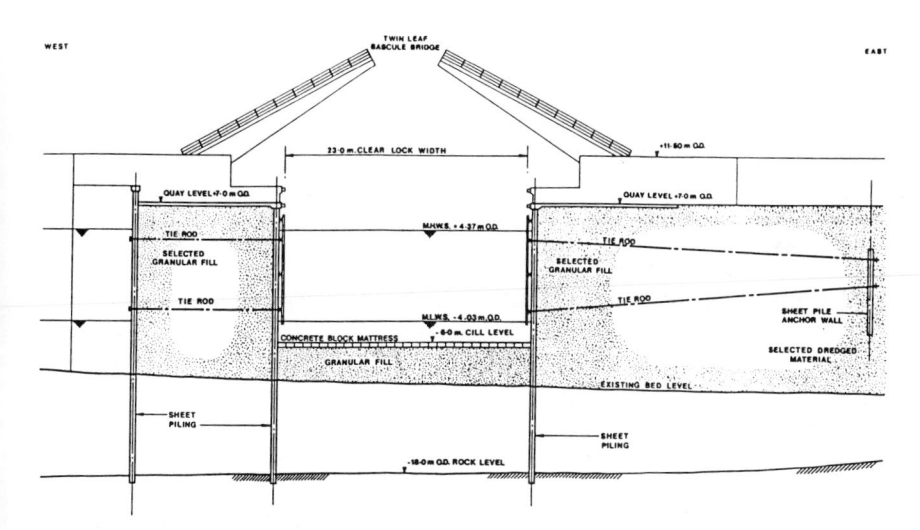

Fig. 6. Cross-section of Dingle Lock

For Dingle Lock, bottom hinged flap gates have been selected, to meet the requirement of reversing hydrostatic loads during the barrage operating cycle. These will be installed in reinforced concrete abutment/sill structures which will incorporate the associated levelling culvert systems.

The site of New Ferry Lock will be encompassed by the extensive casting basin required for the turbine and sluice caissons of the barrage, as shown on Fig. 8, and the lock can therefore be formed within the cofferdammed area using conventional construction techniques. From consideration of costs and the logistics of concentrating major construction activity in this single area, New Ferry Lock is to be formed in relatively simple, mass concrete gravity walls, founded directly on the Sherwood Sandstone bedrock, as illustrated in Fig. 7. The intervening lock floor will comprise a 600 mm thick reinforced concrete slab, incorporating substantial propping beams to carry the opposing horizontal thrusts which would be generated in the event that the lock

TIDAL POWER

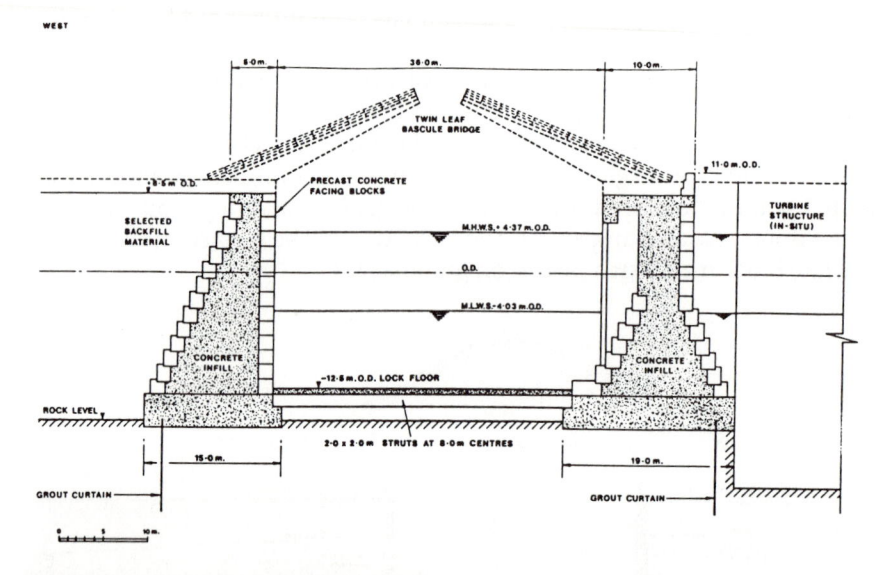

Fig. 7. Cross-section of New Ferry Lock

were to be de-watered for major maintenance activities, although this requirement will be re-appraised in future studies.

In the case of New Ferry Lock, radial sector gates have been selected, with two pairs of gates being installed at each end of the lock to ensure lock operation during periods of gate maintenance or repair. Both New Ferry and Dingle Locks will be provided with bascule bridges, to carry the barrage access road, together with lead-in jetties and the normal ship handling, fendering, navigation and safety facilities.

Ancillaries

The project involves a variety of ancillary works which include dredging, bed protection in the vicinity of sluices and turbines, peripheral rockfill bunds to the reclamation areas, access roads and various service and control buildings, together with the related infrastructure and security facilities.

In addition, a series of accommodations works will be required around the perimeter of the estuary, to maintain satisfactory operation of sewage and storm water outfalls and watercourses under the changed water level regime introduced by the barrage. The requirements for such works are, however, very dependent on the operating regime adopted for the barrage and, by refining current proposals, it is anticipated that not only may the extent of such works be reduced but flood protection in the upper reaches of the estuary may be significantly enhanced.

Construction planning
General approach

Having adopted the principle of caisson construction for the major elements of the barrage, further studies showed that it would be preferable to construct the necessary caissons locally within the Mersey Estuary and thereby avoid the costs, programme limitations and risks potentially associated with multiple sea tows of substantial units from a remote construction facility. As a result two separate casting basins will be established on the Wirral (west) foreshore, in which the turbine-generator and the sluice caissons will be constructed, prior to being floated out and warped into their final position. There will therefore be a concentration of construction activities on the western side of the estuary, the corresponding layout of the casting basins and related facilities being shown in Fig. 8.

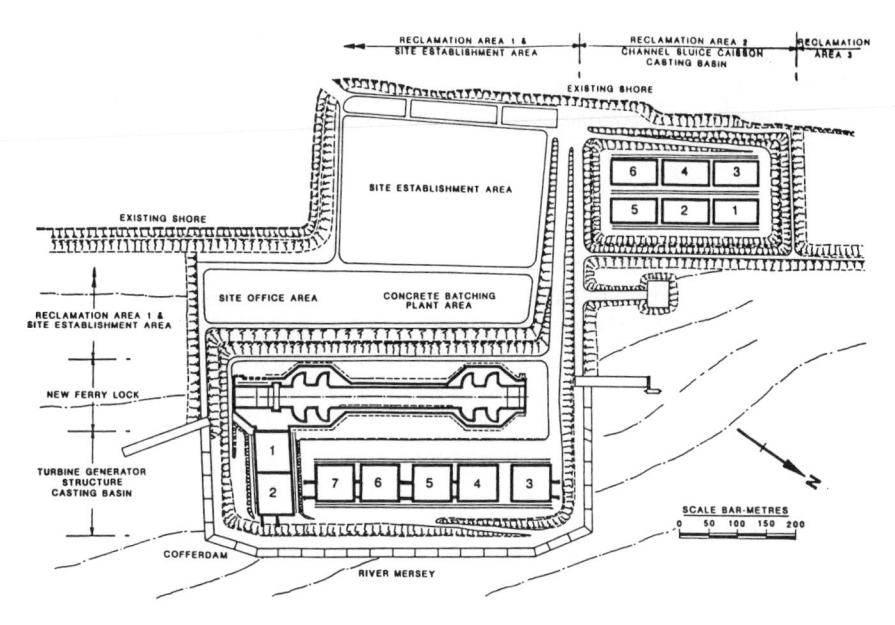

Fig. 8. *Layout of construction site, Wirral Shore*

Within the larger casting basin, which will be isolated from the estuary by a double skin sheet pile cofferdam, the construction of the five turbine-generator caissons, the in-situ turbine-generator structure and New Ferry Lock will take place. The smaller casting basin, formed with a rockfill/sand bund incorporating a sheetpile cut-off, will be used for the construction of the sluice caissons, together with the associated transition structure.

TIDAL POWER

Turbine-generator structures

To ensure that the overall project programme is as short as possible the following principles were adopted for the construction of the turbine-generator structures.

(a) For speed and ease of working, the 28 No. turbine housings have been designed to allow the maximum use of system formwork, prefabricated reinforcement and permanent precast concrete items.

(b) The turbine draught tube, which is conical in shape, has been designed to be cast in 10 No. circular segments which will be stressed together in a precast yard. The completed unit, weighing some 700 tonnes, will then be carried by a multi-wheeled transporter to the turbine generator structure and installed in a single operation of planned duration 3 days.

(c) The installation of the mechanical and electrical components in the form of large integrated units is envisaged. The units will be pre-assembled at a separate yard adjacent to the Mersey Estuary and transported by barge and multi-wheeled transporter into the structure, in a similar manner to that proposed for the pre-cast draught tube.

(d) In the case of the caisson structures, as much work as possible will be completed within the dry casting basin prior to float-out, including the installation and partial commissioning of all mechanical and electrical plant.

Following completion of caisson construction, plant installation and fitting out, the five caissons will be placed in turn, to progressively close the remaining construction gap. The placing of the caissons is critical and to ensure the required degree of control, this operation will be accomplished utilising winches positioned on the caisson, together with piled anchors in the river. The warping and immersion operation will be limited to neap tides, in order to limit the magnitude of current forces, which leads to float-out on a basic 14 day cycle. To alleviate the increase in flow velocity in the construction gap with the successive placing of caissons, the sluices and a number of the previously placed turbines, will be open to substantially maintain existing tidal flows in the estuary.

After immersion of the caissons, utilising a water ballast system, the sand and concrete permanent ballast will be pumped into the caissons and the grout curtain will be installed. The final operation prior to commissioning will be the dewatering of each water passage in turn and the completion of the joint between the steel runner/distributor tube and the concrete draught tube. This joint is to be maintained in a flexible state during caisson installation, to ensure that deflection of the caisson structure on landing and subsequent ballasting does not result in distortion of the turbine runner/distributor tube and the adjacent steel draught tube.

Sluice caissons

The Line 3E barrage involves the construction of eleven sluice caissons of 4 channels and one sluice caisson of 2 channels, together with the transition structure caisson which forms a low level sub-structure for the 2 channel sluice caisson. For economy and as a result of programme constraints, these caissons will be constructed in two batches, requiring a double use of the corresponding casting basin.

Due to the requirement that the caissons be transported and immersed during the summer, coupled with the need for the basin to be used twice, the period for the construction of each batch of six caissons will be limited to ten months. To achieve this rate of construction, the caissons will be laid out to allow the use of rail mounted tunnel forms to construct the base and climbing system formwork to construct the walls. In addition to the reinforced concrete structure, work carried out within the basin will include the installation of the radial gates and the associated electrical and mechanical equipment. On completion, the caissons will be warped from the casting basin to their final position in the barrage on neap tides, using equipment and techniques as described for the turbine-generator caissons.

The sluice foundations are to be formed in crushed rock, placed from side dump barges over a geotextile protected formation. The caisson structures will be ballasted onto prepared upstands, which will be formed and screeded on the upper surface of the general rockfill by means of a jackup barge and purpose made frame. After placing the caissons, the void formed beneath the caisson by the foundation upstands will be grouted in stages, the extent of the area grouted at any one time being controlled to prevent uplift of the caisson. On completion of the void grouting, sand ballast will be pumped from barges into the cellular structure. Finally the grout curtain will be installed from the gallery, to provide a vertical cut-off through the rockfill foundations.

The road bridge over the sluices will be precast in 20 metre sections and lifted into position with a large floating crane, each unit weighing approximately 400 tonnes. This will allow rapid installation of the bridges and thus provide access between the shore and the placed caissons at the earliest possible time. The final operations on the sluices will comprise the levelling of the gates, grouting of the gate bearings and the fixing of side sealer.

On completion of the sluices, the casting basin will be reclaimed using the arisings from the dredging of the turbine-generator foundations and the lock approaches.

Dingle Lock

Due to constraints arising from the overall barrage construction programme, Dingle Lock is required to be operational before Garston Channel is interrupted by the construction of the adjacent sluice caisson foundations (month 19). Accordingly, a method of construction is required which is rapid and which is largely independent of the main site mobilisation activities. As a result, Dingle Lock utilises heavy steel sheet piles driven in situ from marine plant.

The lock construction will commence with the local widening of the Garston Channel, to allow satisfactory passage for shipping while site operations are in progress, and the dredged material arising from these activities will be used to reclaim the landward side of the lock. The inner wall will then be constructed by driving a single row of sheetpiles which are to be tied to dead man anchor sheetpiles buried in the partially formed reclamation area. The outer wall of the lock, comprising a twin sheetpile wall with granular material infill, will then be constructed.

Following the installation of the inner and outer walls, the lock will be closed off at both ends with further twin sheetpile walls, to form a cofferdam which can be dewatered to allow the in-situ construction of the reinforced concrete flap gate structures and levelling culverts. Subsequently the floor of the lock will be constructed in the dry, using precast concrete blocks laid on a granular infill/bedding. Both the sea and basin approaches to the lock require lead-in jetties which will be constructed using marine plant concurrent with the construction of the gate structures.

New Ferry Lock

During the evaluation of construction methods for New Ferry Lock, programme considerations and the implications of the concurrent turbine and sluice caisson works, especially demands for the assembly and fixing of reinforcement, proved dominant in the selection of the mass concrete wall option as the most economic and practical approach. In addition, to obviate the need for formwork erection, the lock walls are to comprise precast block out facings, with an in-situ concrete hearting.

Construction of the lock will commence with the dredging of the lock area to the level of the underside of the floor construction with the arisings being used to reclaim the area for the site offices, yards and batching plants. These operations will be undertaken concurrently with the installation of the cofferdam to the main casting basin and, when this is complete and the casting basin is dewatered, the remaining excavation will be carried out using conventional plant.

Following the casting of the in-situ foundation to the lock walls, the sequential operation of placing the precast blocks and infilling with mass concrete will proceed on a cyclic basis, with the height and length of the concrete bays being restricted to 2 m and 15 m respectively. Heat generation will be limited by the maximised use of a partial cement replacement. All in-situ concrete will be delivered by pump, the anticipated maximum rate of placing being 27000 cubic metres per month. The precast facing blocks will

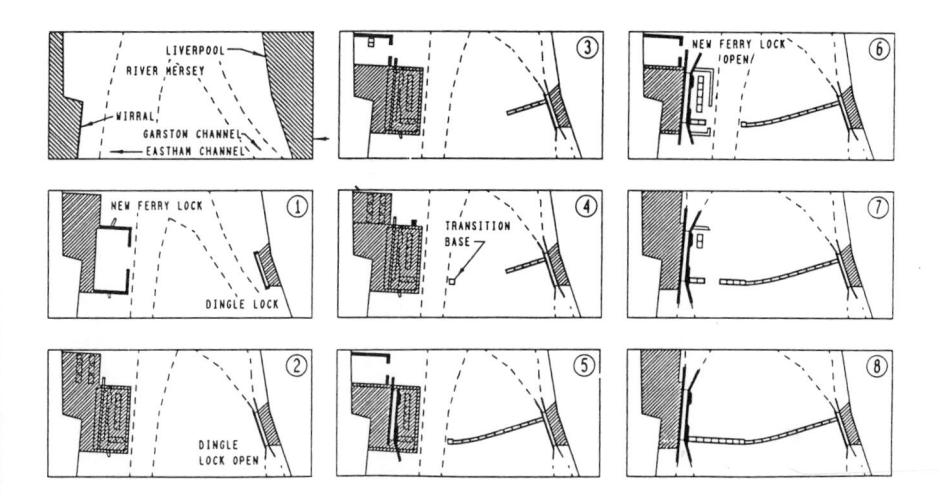

Fig. 9. Construction sequence Stages 1 – 8

be cast in advance, to distribute the production of concrete over a longer period and thereby ameliorate maximum production demands.

On completion of the walls and floor, the lock is to be flooded. However, due to loading restrictions on the lock outer wall, this can only take place when the turbine-generator caissons are complete and the main casting basin is also flooded. Immediately following the removal of the coffer-dam walls from the ends of the lock, the radial sector gates will be floated into position in the gate recesses. Following the isolation of the recesses using stop-logs, the gates will be fixed to preinstalled anchorages, this latter operation being carried out in dry conditions.

The solid lead in jetties, which are required to protect shipping from currents when the barrage is in operation, are to be constructed as twin sheetpile walls, using materials recovered from the casting basin cofferdam.

Construction sequence and programme

The barrage construction has been arranged to extend over a period of 59 months from possession of the site to the barrage becoming fully operational, with barrage closure and partial operation being achieved after 54 months.

The construction sequence and programme, as shown respectively on Figs. 9 and 10, have been developed to achieve the following requirements.

TIDAL POWER

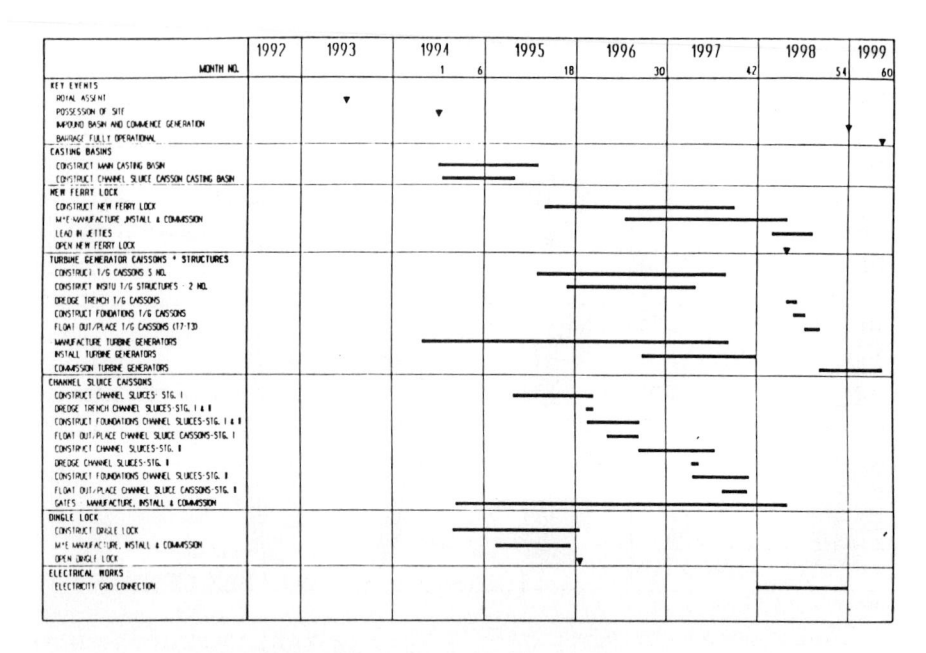

Fig. 10. Mersey barrage Line 3E – Construction programme

(a) Maintain a navigable route to Garston Docks and to the Manchester Ship Canal and the Queen Elizabeth II Dock at Eastham.
(b) Minimise disturbance of existing tidal currents and estuary regime.
(c) Secure the earliest possible date for generation of electricity.
(d) Ensure optimum use of existing mechanical and electrical manufacturing resources and the on-site temporary construction facilities.

Construction will commence concurrently at Dingle Lock and the adjacent reclamation area, close to the Liverpool shore, and at the New Ferry casting basins and reclamation areas, on the Wirral shore. While the channel sluices, turbine-generator structures and New Ferry Lock are being constructed within the casting basins, dredging and caisson foundation works will proceed across the river, generally from east to west. Throughout these operations, the existing navigation channels will be maintained, until shipping can be diverted through the completed Dingle Lock (month 19) and New Ferry Lock (month 46).

Turbine and other equipment manufacture will be arranged to commence at an early stage of the programme, in order that the mechanical and electrical items may be installed in the turbine-generator and sluice structures whilst they are in the casting basins awaiting float-out. Commissioning of these items, where possible, will also be completed at this stage.

Caisson float-out is programmed to occur during the summer months. The two batches of channel sluice caissons will be installed between months 23 - 27 (Stage l) and months 38 - 41 (Stage 2), while the larger turbine-generator caissons will be installed in a single stage between months 49 - 51.

The programme for the design, manufacture, installation and commissioning of the mechanical and electrical plant has been developed in close co-ordination with the civil engineering requirements, in order to achieve the optimum overall construction programme, coupled with minimum construction and project costs. The resulting 5 year construction period is considered achievable with effective project management, close co-ordination, detailed planning and the early integration of design and censtruction requirements. The programme will require the maximised use of off-site manufacture and pre-commissioning of major elements of the works.

Estimated civil construction costs

Estimated costs of construction of the barrage civil engineering works are summarised in Table 3.

The cost estimates for the Mersey Barrage have been prepared on the basis of best, likely and maximum sums. These estimates represent the range of civil construction costs from the maximum sum, including risk and contin-

Table 3: Estimated civil construction costs.

Works Item	Cost £M
Temporary Works - Casting Basins	109
Dredging	18
Foundations and Bed Protection	35
Channel Sluice Structures	47
Turbine Generator Structures	72
New Ferry Lock	70
Dingle Lock	40
Buildings, Roads and Services	1
Accommodation Works	48
Total Estimated Civil Construction Costs	£440M

TIDAL POWER

gent allowances, to the best which excludes these allowances and assumes a subsequant beneficial design and construction development of the scheme.Table 3 summarizes the likely sums, with the corresponding maximum and best sums being £558 million and £395 million respectively.

All estimates are at December 1990 price levels and exclude the cost of design and pre-construction project costs. The total project construction costs, encompassing both the civil engineering and the mechanical and electrical plant components, is estimated to be £861 million.

Conclusions

The Stage III Studies of the Mersey Barrage, completed in September 1991, are considered to have achieved significant advances in the civil engineering aspects of the project. Through the preparation of reliable design and construction proposals, based on comprehensive engineering appraisals, the studies have provided a sound basis for the future project development work required as a pre-cursor to the submission of a Bill to Parliament. Additional work, in the form of an extension to the Stage III Programme, is currently in progress, marking a further step in the development of a tidal energy project on the Mersey.

Acknowledgements

The studies on which this Paper is based were carried out for the Mersey Barrage Company Limited and were partly funded by the Department of Energy. Their permission to publish this Paper is gratefully acknowledged. The Mersey Barrage Company Limited was in turn supported by its Contractor's Group whose assistance is similarly acknowledged. In addition, the support and assistancc provided by Rendel Parkman is appreciated.

References

1. MARINETECH NORTH WEST. *Mersey Barrage Pre-Feasibility Study*, Report for Merseyside County Council, November 1983.
2. MARINETECH NORTH WEST AND RENDEL PARKMAN. *Mersey Barrage - a re-examinatson of the economics for Merseyside County Council*, November 1985. (Unpublished Report).
3. HAWS E.T. et al. The Mersey Barrage. *Proceedings of the Third conference on Tidal Power, ICE, London,* October 1986.
4. DEPARTMENT OF ENERGY. *Tidal Power from the River Mersey, Stage I, Contractor Report*, ETSU TID 4047, 1988.
5. REILLY N. et al. Progress on Civil Engineering and Planning of a Mersey Tidal Project. *Developments in Tidal Power*, Thomas Telford, London 1989

6. MERSEY BARRAGE COMPANY. *Feasibility Studies, Stage II, Final Report, January 1991* (Unpublished Report).
7. SOIL MECHANICS LTD. *Mersey Barrage Feasibility Study Stage III, - Site Investigation, March 1991.* (Unpublished Report).
8. RENDEL PARKMAN. *Mersey Barrage Studies Stage III - Overview Report, - Navigation, September 1991* (Unpublished Report).
9. RENDEL PARKMAN. *Mersey Barrage Studies Stage III - Civil Engineering, Design Overview Report, September 1991.* (Unpublished Report).
10. ALTINK H. et al. The Mersey Barrage - the impact of shipping.*Proceedings of the fourth Conference on Tidal Power, ICE, London, 1992*, Thomas Telford, London.
11. CONROY R. D. *et al.* The Mersey Barrage - Mechanical and Electrical Engineering Developments*Proceedings of the fourth Conference on Tidal Power, ICE, London, 1992.*, Thomas Telford, 1992.
12. POTTS R. AND WILSON E. A. The Mersey Barrage - further assessment of the energy yield. *Proceedings of the fourth Conference on Tidal Power, ICE, London, 1992,* Thomas Telford, London, 1992.

4. The Mersey Barrage - the impact on shipping

H. ALTINK, MSc, Engineering Manager, Hollandsche Beton-en Waterbouw bv, Marine Manager, Mersey Barrage Company Ltd. and G. W. R. HAIGH, BSc, MICE, Assistant Director, Rendel Parkman

This Paper presents the main findings of the shipping and navigation studies carried out in 1990/1991 to assess the impact on shipping of the proposed barrage. Shipping in the estuary has been studied using a traffic model and shiphandling simulations to identify any additional shipping related costs as a result of constructing the barrage and to demonstrate the adequacy of the facilities proposed. Consideration has also been given to the effect of the Barrage on pilotage, towage, dredging and vessel traffic services and a risk assessment has been carried out.

Introduction

The Mersey Barrage Project was described in Papers (refs. 1 and 2) presented to the 1986 and 1989 conferences on tidal power. The Papers related to the initial Stage I and Stage II feasibility studies in which the selection of a barrage location and configuration was presented and discussed and the technical feasibility of the scheme was demonstrated.

In 1990-1991, Stage III feasibility studies have been undertaken. The objectives of this stage of the studies were to promote certainty with regard to construction methods, programme and costs, to carry out a cost benefit analysis and to address the concerns of the shipping and environmental interests.

This Paper outlines the main findings of the shipping and navigation studies which have been undertaken in order to develop the work carried out previously in Stage I and Stage II (refs. 1 - 3). The conclusions from real time ship handling simulations of the situation before, during and after Barrage construction are presented. The development of a shipping traffic model is described. The results of this model have enabled the adequacy of shipping facilities in the proposed scheme to be assessed and the impact of the Barrage on the shipping in the Mersey to be identified. Proposed solutions for reducing this impact are presented and the proposed way forward for future optimisation are indicated.

TIDAL POWER

Shipping interests on Merseyside have been consulted through a Marine Consultative Group set up by the Mersey Barrage Company (MBC) under the independent Chairmanship of Rear Admiral A. F. R. Weir. The assistance of members of the group is gratefully acknowledged although the views expressed in this Paper are those of the authors and are not universally accepted by the members of the group.

Key shipping related features of the project

The proposed tidal energy project is located on the Mersey estuary which accommodates substantial operating port facilities (Fig.1). The barrage alignment which has been the subject of the Stage III studies, is known as Line 3E and runs from Dingle on the Liverpool shore to New Ferry on the Wirral shore, some 1.5 km upstream of the Tranmere Oil Terminal. The proposed layout is shown in Fig. 2.

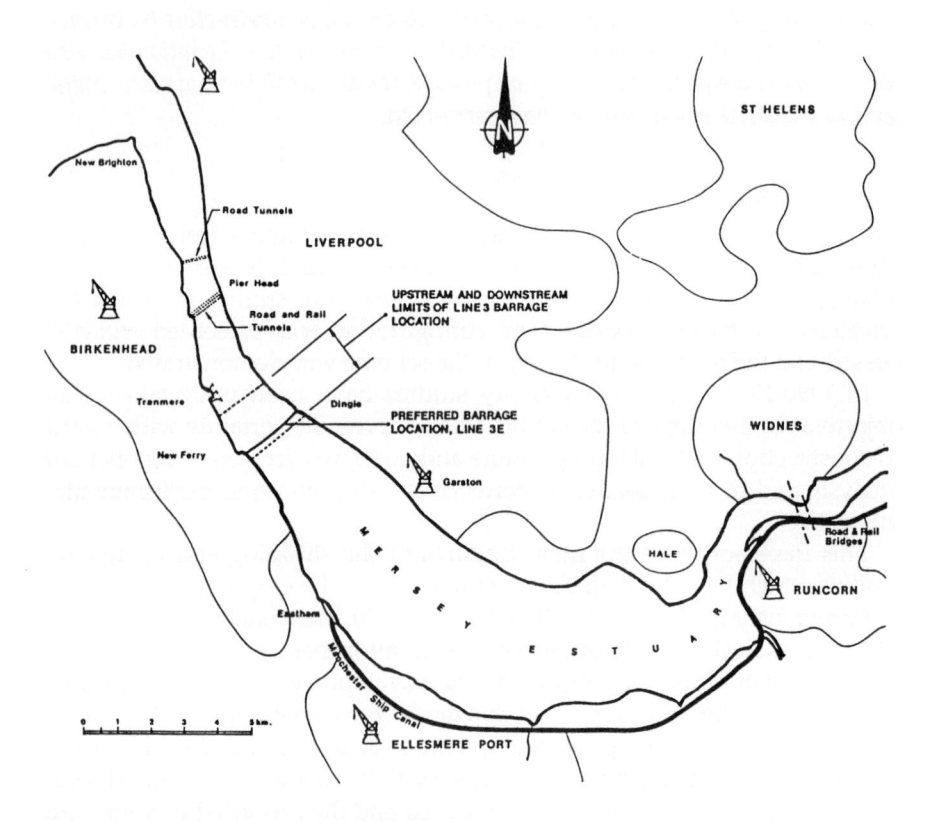

Fig. 1. Plan of the Mersey Estuary

PAPER 4: ALTINK AND HAIGH

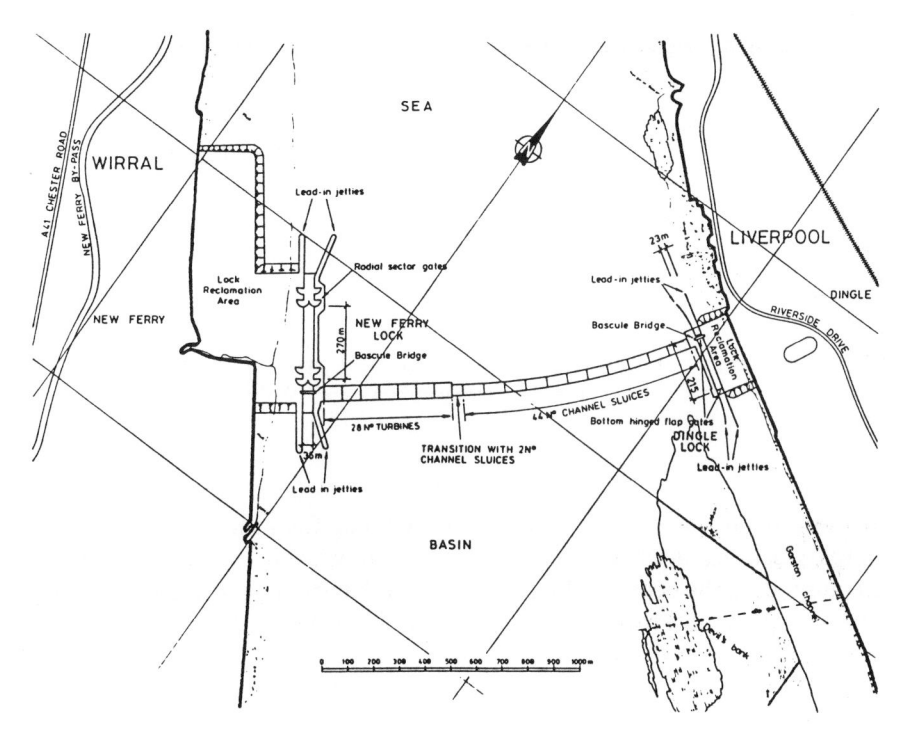

Fig 2. Mersey Barrage Line 3E - general arrangement

A barrage on Line 3E affects the passage of shipping to the upstream port facilities, comprising

(a) Garston Docks on the eastern side of the Estuary, owned and operated by Associated British Ports
(b) The Manchester Ship Canal and the Queen Elizabeth II (QEII) Dock at Eastham on the western side of the Estuary, owned and operated by the Manchester Ship Canal Company (MSCC)
(c) Mersey Wharf at Bromborough.

The Barrage will have effects on shipping traffic to the downstream port facilities because of changes to the tidal cycle, i.e.

(a) Tranmere Oil Terminal on the western side of the Estuary, operated by Shell (UK) Ltd (Shell).
(b) Liverpool and Birkenhead docks, owned and operated by the Mersey Docks and Harbour Company (MDHC).

Two locks are proposed for the Line 3E barrage, one located at the eastern end of the barrage at Dingle to service Garston traffic and one at New Ferry,

TIDAL POWER

Table 1. Shipping traffic, excluding dredgers, bunker barges and tugs, 1989

	No Of Arrivals/ Departures		No Of Shipping Movements	
	Total	Average Day	Total	Average Day
Liverpool	2,642	7	5,284	14
Birkenhead	373	1	746	2
Tranmere	172	0.5	344	1
Garston	684	2	1,364	4
Eastham/QEII	2,680	7	5,360	15

the western end, to give access to Manchester Ship Canal and QEII traffic. The Dingle Lock has internal dimensions of 215 m x 23 m. It has been sized to accommodate the beam of the largest vessel that can gain access to Garston (152 m x 19 m) and the length of two typical regular traders (90 m x 15 m). The lock sill has been set at 6 m below Ordnance Datum (OD), i.e. -1.07 m Chart Datum (CD) which is 0.35 m lower than the sill of the existing lock at Garston.

New Ferry Lock (270 m x 36 m) has been proportioned to accommodate the largest tanker (39000 tonnes dwt) capable of gaining access to QEII Lock. A sill level of 12.5 m below OD, i.e. -7.57 m CD, has been adopted, which is 2.17 m below the sill level at QEII Lock.

The width of both locks allows for a margin on both sides of 10% of the beam of the largest vessel that gains access to the upriver locks, as recommended in ref. 4.

The ancillary facilities required to assist navigation fall into two categories, those required principally to maximize the throughput of shipping, such as lead-in jetties and lay-by berths and those required to ensure adequate safety for vessels and personnel such as lock fittings, navigation lights and buoyage.

The lead-in jetties which are included in the Stage III base case for Line 3E have been developed from a review of lengths and orientations of such structures at other Mersey dock entrances and also worldwide. The views of the Liverpool Pilots' Association (LPA) also influenced the design process. A combination of solid and open structures has been adopted depending on location, with the jetties furthest from the banks of the estuary angled away from the axis of the lock to form a bell mouth.

Existing navigation conditions
Present and future traffic

The numbers of vessels excluding dredgers, bunker barges and tugs passing Line 3E based on 1989 traffic statistics are shown in Table 1. The total number of vessels to the upstream ports has shown little change since the 1988 figures, a reduction in Garston traffic being offset by an increase in vessel numbers to the Manchester Ship Canal/QEII Dock. For the purposes of the Stage III studies, total traffic figures similar to the 1989 levels were adopted. The vessel type distribution and vessel size distribution for the enclosed ports was based on 1985 and 1989 data, while for Tranmere, the size distribution was based on 1989 data.

Future traffic levels have not been predicted explicitly within the studies but barrage operation has been modelled with the base case traffic levels plus 10% and minus 10%.

Existing restraints to navigation for upriver ports

Movement of shipping traffic to the upriver ports is constrained by tidal levels and currents at lock entrances. At Garston no pumping facilities exist to maintain water levels within the dock, but vessels larger than those for which the lock was designed (76 m) make up the greater part of the traffic. The vessels canal through the lock with both pairs of gates open and consequently, movements are limited to a short period around High Water (HW). Small vessels (>76 m) lock through in the normal way outside these times. Eastham/QEII movements take place four hours either side of HW with certain restrictions on inbound vessels because of cross currents off the lock entrance from three hours to half an hour before HW.

In addition, at Eastham access is not possible when the estuary level is above 9.08 m CD because a single pair of mitre gates is closed in each lock to prevent silt-laden estuarine water from entering the canal.

Existing restraints to navigation for downriver ports

The Alfred Lock entrance to Birkenhead Docks is at right angles to the river and has limited depths outside. Large vessels use this lock between HW -2 h and HW. Tankers arriving at Tranmere time their arrival off the jetty to occur immediately after HW to enable berthing to be undertaken on the first of the ebb. The timing of the arrival also has to take into account underkeel clearance limitations at the Bar. Departing tankers make use of the flood tide to turn and generally leave the berth 2 h before HW. Vessels entering or leaving Liverpool Docks through Gladstone or Langton Locks can do so round the clock, but the largest vessels are constrained by draught and current limitations to movements about HW.

TIDAL POWER

Navigation with barrage

Navigation during barrage construction

During the construction of the barrage the arrangements for the passage of shipping fall into three distinct phases which can be related to the construction stages identified in ref. 5.

Stage 1: 19 months: From possession of the site, the construction and presence of the casting basin will cause Eastham/QEII vessels to use the deep water channel mid-river and then rejoin the existing Eastham Channel. The Garston traffic will need to use a newly dredged channel west of the construction activity at Dingle.

Stage 2 to Stage 5 : 27 months: Once Dingle Lock is open, Garston traffic will use Dingle Lock. Sluice caissons will be placed across the newly dredged channel and it will no longer be available to shipping. Eastham/QEII traffic will continue to use the mid-river channel.

Stage 6 and Stage 7 : 8 months: Once New Ferry Lock is open, traffic will use the lock appropriate to each port while the turbine caissons are placed. The mid-river channel is closed to navigation.

Near barrage currents

Based on the results of the two-dimensional (2D) 25 m and 75 m grid

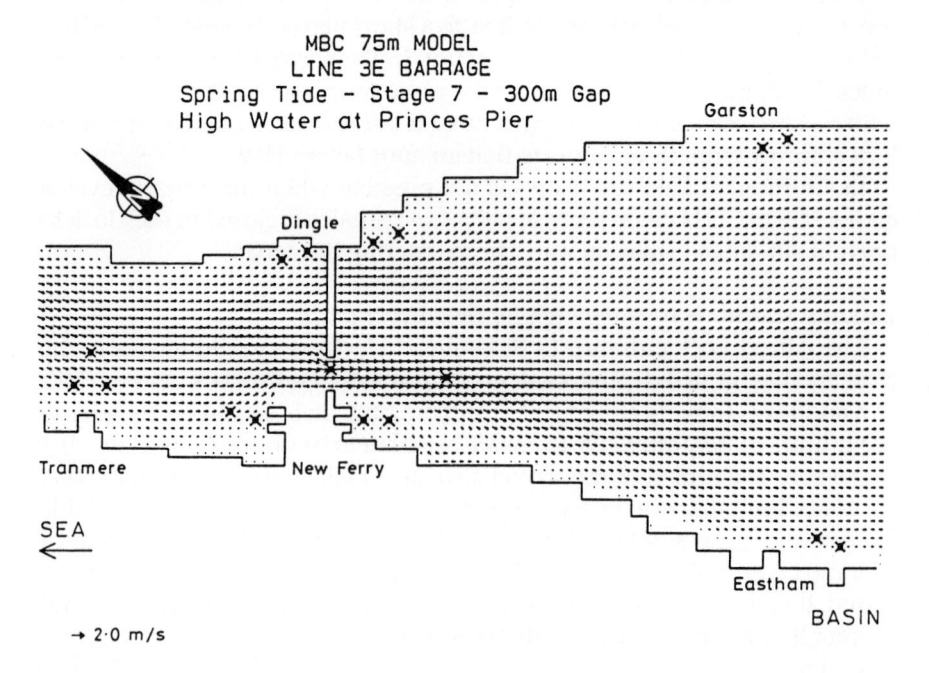

Fig. 3. Typical current vector plot showing location of velocity reference system

hydrodynamic mathematical modelling of the Mersey Estuary carried out by HR Wallingford (HR) and the MBC(ref: 6), velocity time plots for currents for various tidal ranges were prepared for a number of reference locations adjacent to each lock. The reference locations were selected to be of relevance to mariners manoeuvring in the vicinity of existing and proposed facilities at various stages of Barrage construction (including open River) and are shown in Fig. 3.

The data were used with the results of the shiphandling simulations to derive lock access windows for use in the traffic model. The data were also used to establish the effect of the barrage both during the construction stages and during operation on the current velocities off the Tranmere Oil Terminal.

Maximum ebb and flood currents for mean spring and mean neap tides at the reference location in the construction gap at various stages of construction were identified. The velocities during neap tides at the last stage of construction were only 25% higher than the existing open river situation. During spring tides slightly larger increases of up to 41% were calculated.

Shiphandling simulation

The shiphandling simulation was carried out by Maritime Dynamics of Llantrisant, Mid Glamorgan using a real time simulator which allowed the interaction of vessels with the proposed port facilities to be examined in a range of environmental conditions. The system provides a master mariner with a constantly changing display of the visual scene on three monitor screens and numerical data on ship status and environmental conditions (ref 7). Current and water depth data were read directly into the simulator from the output of 25 m grid or 75 m grid mathematical hydraulic models developed by HR (ref. 6).

The objectives of the shiphandling simulation were

(a) to establish limiting environmental and tidal conditions for lock approach at Dingle and New Ferry for a number of typical vessels during Barrage construction and operation
(b) to establish the effect on the tanker operations at Tranmere of the Barrage construction and operation
(c) to identify the limiting conditions for navigation through the 300 m gap in the barrage at the later stages of construction.

The results of the simulation runs were analysed to define the lock entry windows for use in the traffic model. Three vessel sizes were considered

(a) 323,000 dwt VLCC.
(b) 39,000 dwt product tanker typical of a larger vessel to QEII.
(c) 7,500 dwt general cargo vessel with bow thruster typical of the largest regular trader to Garston.

TIDAL POWER

Prior to carrying out any simulation runs, the behaviour of all the ship models was validated. The correlation between the visual scene observed by the mariner and the current vectors experienced by the vessel was also checked.

The two components of current velocity parallel and normal to the ship's centreline during each manoeuvre were recorded. An assessment of the success ratio of the manoeuvres in various current conditions permitted the definition of limiting current conditions. The access windows to the locks were established by comparing the variations in currents at the reference locations with the limiting conditions for the product tanker and the general cargo vessel.

Manoeuvres involving arrival at the departure from Tranmere made use of the VLCC model and up to three tugs and considered both the existing open river situation and also the barrage operational conditions including the effects of flood pumping. This results in the continuation of modest flood tide currents after the time of HW and has an impact on shiphandling in the estuary. Simulation of Tranmere berthing of a VLCC led to the conclusion that berthing of a VLCC at HW +1 h or HW +2 h with flood pumping is possible with adequate tug support and/or appropriate control of pump flow. Further simulation is required to confirm the limiting conditions for different size vessels.

Navigation through the 300 m construction gap by the product tanker and general cargo vessel with head currents was shown to be possible provided one way traffic was adopted. Carrying out the manoeuvre with a modest stern current 1 h either side of HW was also possible, but success was very dependent on the adoption of the correct line of approach.

Traffic model

The construction of a tidal power barrage across the River Mersey will result in many changes to the present day tidal levels and flow patterns. The traffic simulation model was used to simulate shipping movements in the Mersey to and from the upriver ports, Tranmere Oil Jetty and Liverpool and Birkenhead Docks, subject to the effects of the barrage and the limitations of currents and water depths. The model used probability distributions derived from real data to generate traffic during the 10000 h run time of the model. The model was used to simulate the existing (open river) situation, a construction stage and the operational barrage situation. Periods of lock availability at the barrage and the upstream ports of Eastham and Garston were derived from a study of current velocity and tidal height data from HR's hydrodynamic models and from shiphandling simulations (Fig. 4 and Fig. 5).

The voyage times for vessels to navigate from the Bar Light Vessel to the upriver ports or vice versa were identified from the model. The additional

PAPER 4: ALTINK AND HAIGH

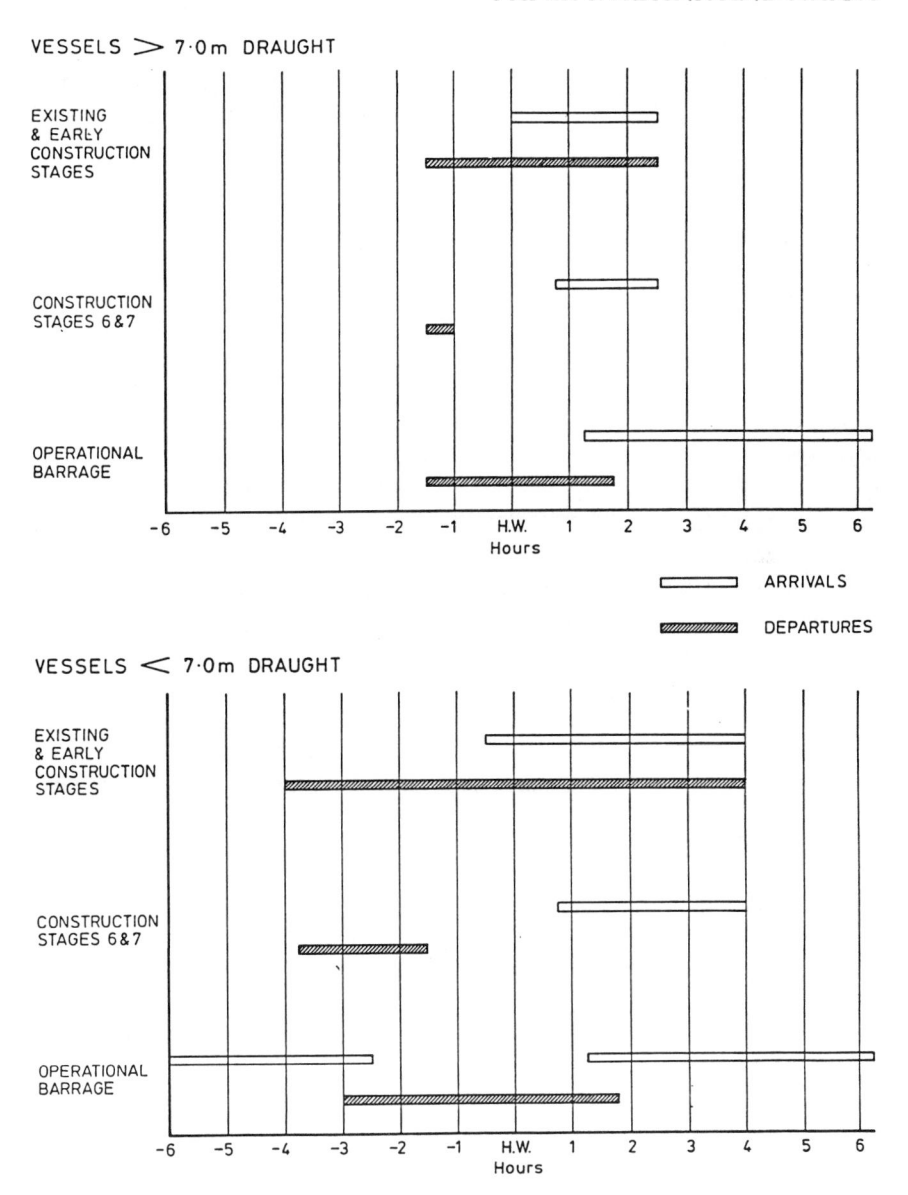

Fig. 4. Times of arrival and departure at Eastham

time required for vessels to pass through the barrage is made up of two parts, the time queueing to gain access to the locks because of traffic congestion and the time required to enter the lock, close the gates, change water levels, open the gates and leave the lock.

TIDAL POWER

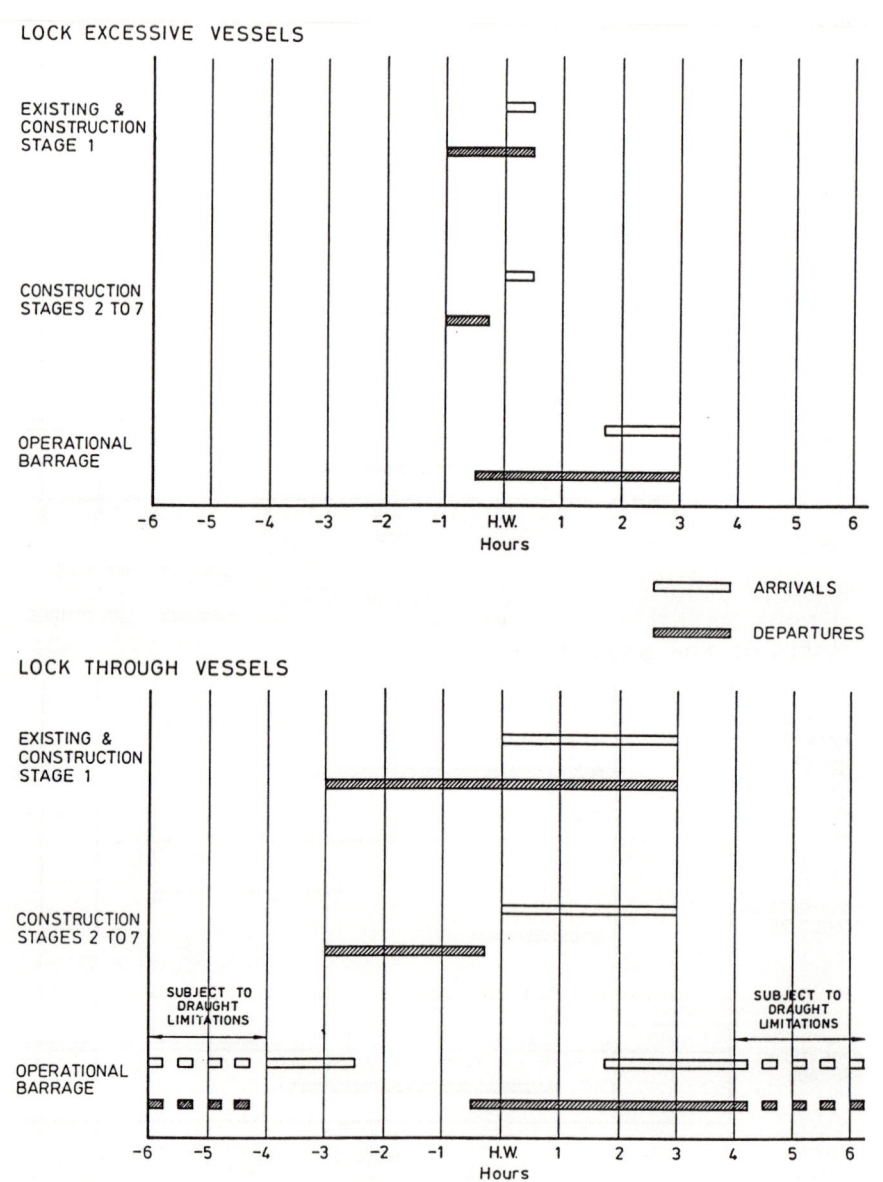

Fig. 5. Times of arrival and departure at Garston

During the course of the study 12 different conditions were assessed to determine the effect of the barrage on shipping in the Mersey and to evaluate the sensitivity to changes in traffic levels, lock entrance times and lock access windows. The conclusions from the study were that the proposed barrage lock sizes are adequate for the operational phase, demonstrated by the steady

Table 2. *Voyage times for spring tides, existing open river and barrage operational*

	Voyage Times (Minutes)	
	Inbound	Outbound
Eastham/QEII		
Average voyage times for open river between the Bar Light Vessel and Eastham/QEII	4 hrs 27 min	4 hrs 07 min
Increase In Voyage Times		
Operational Barrage - Base Case Traffic	50	70
Operational Barrage - Base Case +10%	58	79
Operational Barrage - Base Case -10%	43	64
Lock entry times increased by 20%	93	123
Garston Traffic		
Average voyage times for open river between the Bar Light Vessel and Garston	3 hrs 46 min	3 hrs 41 min
Increases In Voyage Times		
Operational Barrage - Base Case Traffic	42	38
Operational Barrage - Base Case +10%	43	38
Operational Barrage - Base Case -10%	40	37
Lock entry times increased by 20%	63	54

state behaviour of the model, i.e. no steadily growing queues. The typical results for the barrage in operation at spring tides are presented in Table 2, (ref. 8).

Output from the traffic model was used to identify the busiest day for use in a review of towage and pilotage requirements and to assess the number of lay-by berths required. Lay-by berths could be required to accommodate the small number of vessels which are unable to make passage through both the port locks and the Barrage locks on one tide. Analysis of the results of the traffic model showed that two lay-by berths would be required between Dingle Lock and Garston and three between New Ferry Lock and Eastham to provide, during the final stages of construction, a temporary holding area for vessels. The ultimate objective is to eliminate the need for lay-by berths by improvements to the lead-in arrangements which will increase the available periods of access to the proposed locks.

TIDAL POWER

Study results
Operational phase
The proposed locks provide access to the upriver ports for longer periods, but later on each tide than at present. The lock capacity has been shown to be adequate for existing (1989) traffic and only limited use of lay-by berths is required. Voyage times to and from upriver ports will be extended on average by 50 min for Eastham/QEII and 35 min for Garston. Small increases in the number of tugs (2 or 3), pilots (2) and vessel traffic system (VTS) resources (capital and operational) will be required. Maintenance dredging/requirements in the estuary and upriver docks will increase by between 35% and 110% (ref. 6). Further work is necessary concerning the currents induced by flood pumping and the requirements for berthing of VLCCs at Tranmere. The risks to shipping have been evaluated and show little overall change, although there is an increase in risk in the zone between Birkenhead (Woodside) and the proposed Barrage. The risks have been shown to be similar to those accepted for comparable United Kingdom projects. The net shipping related costs during operation have been evaluated as £2.9 million per annum.

Construction phase
The proposed sequence of construction is generally compatible with the needs of shipping navigation in the estuary. Difficulties exist for lock excessive vessels approaching Garston. The throughput capacity of the New Ferry Lock is just adequate, but requires substantial use of lay-by berths. Voyage times to and from Garston will be extended by 4 h 5 min inbound and 50 min outbound. Currents in the 300 m construction gap used by Eastham/QEII traffic up to Stage 6 of construction are higher than at present, but are acceptable. Short term increases for towage and pilotage exist during the construction period. The loss of water levels above the barrage prior to impounding of the basin effects less than 1.5% of vessels travelling to and from Eastham/QEII.

Conclusions
The study has shown that a barrage at Line 3E is capable of being consistent with the requirements for navigation in the Estuary, both during construction and operation. The layout of the barrage on this alignment will be subject to further development in order to reduce the impact of the Project on shipping. In particular

(a) Consideration will be given to forming a channel across Devil's Bank to provide access to and from Garston during construction and operation of the Barrage and relocating Dingle Lock on the western side of theEstuary to form an integrated lock complex

(b) The conflicting requirements for VLCCs berthing and the need for flood pumping will be studied in more detail to establish operational procedures acceptable to both the shipping interests and the MBC.

(c)Revised layouts will be developed for the lead-in jetties to improve the lock access conditions for shipping.

(d)The trends in shipping traffic in the Mersey will be kept under review to ensure that studies reflect changing traffic levels and patterns.

(e)The shiphandling simulations will be expanded to carry out more runs at selected times of the tide to develop confidence in the proposed access windows. A wider range of vessel sizes will also be considered.

Acknowledgement

The studies in this Paper have been carried out for the Mersey Barrage Company Limited, partly financed by the Department of Energy. Permission to publish this paper is gratefully acknowledged. The studies were carried out with the assistance of Captain C. Collings of Global Maritime.

References

1. HAWS E. T., CARR G. R. and JONES B. I. The Mersey Barrage. *Proceedings of the Second conference on Tidal Power, Institution of Civil Engineers, London, 1986.* Thomas Telford, London, 1986.
2. REILLY N. and JONES B. I. Progress on civil engineering and planning of a Mersey tidal project. *Proceedings of the Institution of Civil Engineers Third Conference on Tidal Power*, Thomas Telford, London, 1989.
3. RENDEL PARKMAN LIMITED. *Mersey Barrage Feasibility Study. Navigation And Lock Study, May 1988* (Unpublished)
4. PIANC. *Final report of the International Commission for the Study of Locks.*
5. MORGAN C. D. I., JONES B. I. and PINKNEY M. W. The Mersey Barrage - Civil Engineering Aspects. *Proceedings of the Fourth Conference on Tidal Power, Institution of Civil Engineers, London, 1992.* Thomas Telford, London, 1992.
6. ALTINK H., MANN K. and ODD N. V. M. The Mersey Barrage - hydraulic and sedimentation studies. *Proceedings of the Fourth Conference on Tidal Power, Institution of Civil Engineers, London, 1992.* Thomas Telford, London, 1992.
7. MARITIME DYNAMICS. *Mersey Barrage Feasibility Study 7, Stage III, Ship Manoeuvring Simulation, July 1991* (Unpublished).
8. RENDEL PARKMAN LIMITED.*Mersey Barrage Studies 8 ,Stage III, Traffic Simulation Studies, October 1991* (Unpublished).

Discussion on Papers 3 and 4

H. MITCHELL, Inland Shipping Group
With regard to the infrastructure required for shipping movements and operation, have the traffic forecasts been based on current traffic only or has there been consideration of possible changes in shipping trends and in particular the potential for increased use of waterborne transport, inland and coastal, within and around the UK?

G. W. R. HAIGH, Rendel Parkman Ltd
Changes in the levels of traffic have been considered by evaluating the effect of the barrage on existing traffic numbers plus or minus a certain percentage. The percentages reported in the Paper (plus or minus 10%) were selected following a study of future shipping traffic levels by a firm of consultant economists. The barrage layout currently being considered is thought to enhance the potential shipping throughput compared with the layout shown in the Paper. The new arrangement of double locks improves operational flexibility and provides additional security, while enhancing the ability of the barrage to respond to major changes in shipping numbers, traffic patterns and vessel sizes.

E. JENKINS, Shell UK
Bearing in mind that the cost of the Mersey Barrage will be in the region of £1 billion and that only a relatively small amount has so far been committed in development, could Mr Phillips comment on the confidence he expressed in his costing remarks and the accuracy of his cost estimate?

B. I. JONES, Rendel Parkman Ltd
One of the major objectives of the Stage III Study Programme was to improve the confidence and, as far as possible, remove uncertainty from the project cost estimates.
To achieve this, the Mersey Barrage Company and its Contractors' Group established a comprehensive team in Liverpool which, in addition to civil, mechanical and electrical engineers, planners and estimators, included representatives from the hydraulics, shipping and environmental consultants. These arrangements allowed excellent liaison between the various disciplines involved in or influencing the design and cost estimates for the project

and were particularly beneficial in co-ordinating the design and construction requirements. The resulting design, construction propsals and programme could therefore be prepared to a level of detail significantly higher than would normally be the case at this stage of a project, and it is consequently believed that the estimates available reasonably encompass the likely range of construction costs.

N. SONDHI, M.S.L. Engineering Ltd

In the Stage II studies for the Mersey Barrage, it was proposed to have 20 Venturi sluices. Stages III and IIIA propose to use high level channel sluices and therefore to maintain flow capacity, a total of 46 sluices are required.

Where has the space for these additional sluices come from? Also, have the number of turbines remained the same as for Stage II? Finally, has consideration been given to caissons comprising sluices over turbines? In this manner, a greater number of sluices and turbines may be accommodated in the same length of barrage.

B. I. JONES, Rendel Parkman Ltd

Mr Sondhi may recall that the earlier layouts for the Mersey Barrage included a length of rockfill/sand closure band, in addition to the basic operating elements of sluices, turbines and locks. In the Stage III, Line 3E design the channel sluices extend over the area previously occupied by the closure bunds, with the bund, in a truncated form, essentially providing the sluice foundation.

Regarding the possible combination of sluices and turbines within an integrated structure, the requirement that all structures above the turbine water passageway accomodate either operating control and transmission equipment or permanent ballast would appear to preclude such an arrangement in the case of the Mersey Barrage. The Authors accept that a combined turbine and sluice structure may be economic in smaller projects, where the available width of the estuary may inhibit other arrangements, but on the Mersey, where adequate estuary width is available, there would perhaps be little merit in locating sluices over turbines, only then to have the problem of completing the closure of the barrage with alternative structures.

E. T. HAWS, Consultant

I should like to comment on the very substantial effort made in planning and estimating for the Mersey Project in Stage III of the studies. A team of 50-60, including senior engineers, planners and estimators operated together at Liverpool for a year. The consultant designers were also in close proximity with excellent liaison. The result was a degree of preparation and costing accuracy unknown in my experience at this stage of a project. Perhaps Mr Phillips would elaborate a little more on this aspect.

J. ELLIS, British Steel Technical

Can Mr Jones or Mr Phillips please indicate what they see at this stage as the major factors controlling the selection of steel or concrete for caisson construction.

B. I. JONES, Rendel Parkman Ltd

Steel caisson construction remains an option for the Mersey Barrage and further design and evaluation work will no doubt be required in this area as the project advances. Overall, the selection of steel or reinforced concrete will be based on cost considerations, with the option showing best performance overall (allowing for maintenance, programme and resultant implications for total financial commitment) being adopted.

C. G. RABBITTS, Associated British Ports

It is gratifying to note that MBC is at last paying serious attention to the needs of the ports. Much of this may be attributable to the chairmanship of Admiral Weir. But much remains to be done to convince port areas that the barrage will not be damaging to port operations. Who will pay for the cost of lock operation and for maintenance dredging of the proposed new channel? At the last Tidal Conference, I was forced to point out that pumping flexibility would be limited by the need to have published tide tables for at least 12 months ahead. Has this been taken into account?

G. W. R HAIGH, Rendel Parkman Ltd

It is the Authors' opinion that the scope and depth of investigation of shipping related issues have been consistent with the objectives of each stage of the feasibility study programme. The importance placed on these issues is illustrated by the fact that MBC set up the Marine Consultative Group over 3 years ago. The Authors would like to acknowledge the role played by Admiral Weir as Chairman of the Group since the end of 1990, in developing constructive dialogue between MBC, RP and the shipping interests.

The Authors acknowledged in their Paper that at the end of the Stage III studies a number of shipping related issues were unresolved. The ongoing studies have resulted in the redefinition of the barrage layout and the adoption of moderated operating regimes, with beneficial effects on the impact of the barrage on shipping.

The MBC has always been committed to the principle that the barrage should have a neutral impact on shipping. Voyage times to upriver ports for some vessels will be increased by the prescence of the barrage. However, the increased periods of access will lead to a reduction in voyage times for other vessels and greater operational flexibility. If the net effect is shown to be a cost to the shipping interests, the MBC may consider making appropriate payments. However, MBC cannot undertake to make payments against any particular item, or indeed any payments at all, until the barrage proposals have been finalised, definitive costs and benefits evaluated and formal discussions with interested parties concluded.

The need to have tide tables published a considerable period in advance is being addressed in the ongoing studies and has been the subject of extensive discussions at the Marine Consultative Group. The proposal under investigation at present involves the adoption of two or more operating regimes, one with the period of flood pumping dictated purely by the commercial operation of the barrage, the other with a curtailed period of pumping to allow current sensitive manoeuvres involving VLCCs at Tranmere to be carried out. The objectives of the studies are to demonstrate that ship manoeuvring to and from the upriver ports can be carried out at similar times under both regimes. Vessel owners, shipping agents and port operators will be able to plan their activities around the revised predicted minimum high tide tables in the basin on each tide. These will be the minimum high tide levels to be provided by any operating regime. If full flood pumping is able to take place because the pumping cost/generation value ratio is favourable, and no large tankers are due to arrive at Tranmere, the levels in the basin will be higher than those guaranteed. If only partial or no flood pumping takes place, the levels achieved upriver will be equal to or greater than the predicted minimum tide level. By adopting this method of defining tide levels, the flexibility required for effective use of flood pumping can be achieved, while providing for the level of predictability required by the shipping interests.

D. G. OGILVIE, *Manchester Ship Canal Co. Ltd*

What type of gates and mechanism are proposed for New Ferry locks? This choice obviously affects maintenance, spares and replacement costs.

B. I. JONES, *Rendel Parkman Ltd*

For New Ferry Lock, radial sector gates are proposed, with the drives being by hydraulic winch and wire rope. This type of gate has been selected largely as a result of its ability to carry hydraulic loading from both sides, a requirement imposed by the barrage operating cycle. These gates are considered relatively robust in operation and maintenance will be facilitated by the ability to withdraw a gate into its chamber and, following appropriate stop-logging and de-watering of the chamber, and to work on the gate in dry conditions, without impeding the ongoing use of the lock.

D. M. D. WOOTTON, *Scott Bertlin*

What is the design life of the caissons and what provisions have been made to prevent the ingress of chlorides and the resulting deterioration of the concrete of the caissons?

B. I. JONES, Rendel Parkman Ltd

The barrage caissons are being designed for the 120 year operating life of the project.

The matter of ingress of chlorides and resulting deterioration of reinforced concrete sections was considered in depth during the StageIII Studies. In general terms, it is proposed to secure the life of the caissons through the adoption of adequate cover to reinforcement and other built-in steel items and the specification of high quality concrete mixes, utilising replacement of cement with pulverised fuel ash. Cement contents, aggregate size and water/cement ratios will be controlled to suit the specific requirements of the various concrete elements, taking into account anticipated exposure conditions.

5. The Mersey Barrage - mechanical and electrical development

R. D. CONROY, Chief Materials Engineer, NEI Parsons Limited, H. R. GIBSON, Development Engineer, NEI Parsons Limited, and D. PHILLIPS, BSc(Hons), MICE, MIStructE, FIHT, MHKIE, Chief Engineer, Tarmac Construction Limited - Major Projects Division

Since the conference on Tidal Power in November 1989, significant progress has been achieved on the Mersey Barrage Project, under the Stage II and Stage III studies. The Stage III study determined that the Line 3E barrage alignment was the most appropriate location with the governing barrage design parameters and configuration being similarly defined. This Paper describes the development of the Mersey Barrage mechanical and electrical engineering works and summarises the current design, manufacturing and installation proposals. Several areas of specific interest are discussed in detail and the barrage mechanical and electrical costs and programme are summarised.

Introduction

At the Institution of Civil Engineers' Third Conference on Tidal Power in November 1989 a preliminary indication was given of the mechanical and electrical (M & E) aspects of a Mersey Tidal Project (ref.1). From the completed Stage II studies, July 1991, the following conclusions were reached.

- *Barrage location-* within the general Line 3 area between New Ferry and Dingle (ref. 2).
- *Barrage operation-* Ebb generation with flood pumping.
- *Power unit-* Turbine driven generator via speed increasing gearbox.
- *Sluices-* Submerged venturi type.
- *Grid connection-* Connection into the Manweb 132 kV system on both the Wirral and Liverpool side of the Estuary.

It was also concluded that the Barrage should house twenty-eight power units and more recent studies confirmed these conclusions.

A geared power unit (Fig. 1) was determined as the most appropriate based on its suitability to the Mersey hydraulic regime, efficiency, capital cost, manufacture, installation and use of proven equipment. To maximize

TIDAL POWER

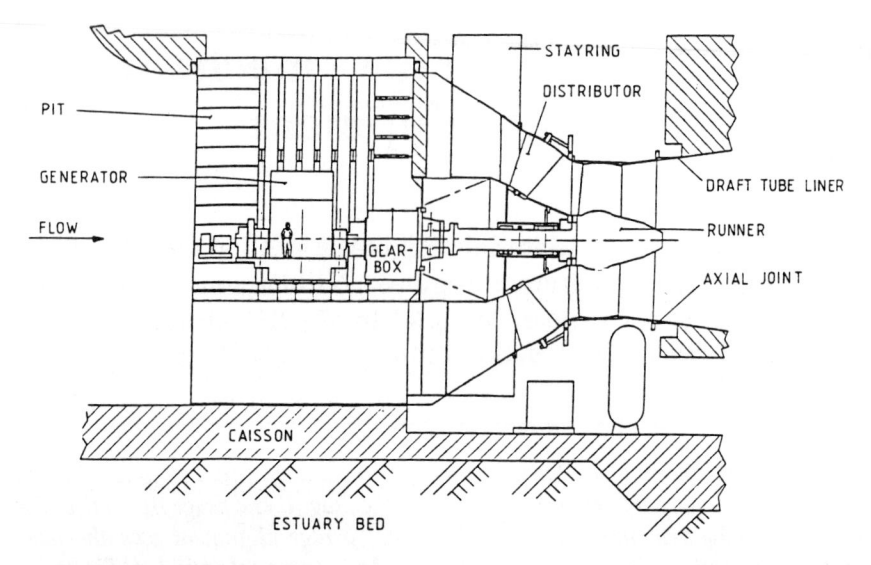

Fig 1. Power unit layout

energy yield, turbines with variable pitch distributor and runner blades were proposed.

In the Stage II study the electrical gathering system was developed to suit connection into the Manweb system with connections made to existing substations at Lister Drive (Liverpool) and Birkenhead (Wirral). During further refinement of the scheme in the Stage III study, it became apparent that a more economic solution is to connect into the National Grid Company (NGC) 275 kV system. This proposal is discussed in more detail below.

During the early part of the Stage III study the twenty submerged venturi sluices were replaced by 46 high level sluices requiring a new type of gate and lighter structure (ref. 2).

The Stage III feasibility studies further addressed hydraulic, environmental, shipping, energy yield, associated civil and M&E engineering aspects together with the project economics and financing. Thus, the barrage location, operation parameters, construction and cost details were refined.

Project engineering and management

The more detailed Stage III study demanded a more interactive arrangement between contributing specialist disciplines and led to the establishment of project teams from the Mersey Barrage Company (MBC) and the Mersey Barrage Contractor Group, consisting of Costain, Tarmac, Cementation, Hollandsche Beton-En Waterbouw bv, Northern Engineering Industries and Alfred McAlpine, in a single project office in Liverpool. However, due to the

Table 1. M&E engineering specialist companies

Company	Responsibility
NEI Parsons	Project coordination.
Sulzer Escher Wyss Zurich	Licensor to NEI Parsons; Turbine hydraulic design and performance.
Sulzer Escher Wyss Ravensburg	Turbine design; Contribution to manufacture, installation and corrosion.
Elin Union	Licensor to NEI Parsons; Generator design; Contribution to manufacture planning.
NEI Allen Gears	Design and manufacture planning of speed increasing gearboxes.
NEI Power Projects	Project management and electrical gathering and distribution.
NEI Reyrolle Technology	Electrical components load flow and fault studies.
NEI Control Systems	Control system.
NEI Peebles	Transformers.
International Research and Development	Contribution to energy yield, hydraulic and sedimentation studies; Fabrication technology; Corrosion management.
NEI Thompson- Sir William Arrol	Design of lock and sluice gates.
NEI Wellman Booth	Gantry cranes.
Balfour Kilpatrick	Cabling; Common services.
Cammell Laird Shipbuilders	Fabrication and assembly of pit and stayring and gates; Storage.
British Steel	Assistance with material supply and fabrication technology.
Global Cathodic Protection	Corrosion protection study.
Manweb	132 kV grid analysis.
National Grid Company	275 kV grid analysis.

TIDAL POWER

scope of the M&E engineering required for a Project of this complexity and diversity, it was not possible to have all the expertise in one location. Therefore, sub-groups of specialist engineering companies were responsible for specific tasks. A list of these companies is given in Table 1.

At the start of the Stage III study detailed reviews were made with relevant study teams, of the operation parameters arising from the Stage II study. An example of the consequences of this review was the geometry of the turbine water passage and, in particular, the length of the draft tube as it influences costs, caisson stability and hydraulic efficiency. An optimised solution was determined for hydraulic efficiency and power unit caisson design. The base case parameters used in Stage III are set out in Table 2.

Table 2. Line 3E - operating parameters

Barrage location	Line 3E between New Ferry and Dingle	
Operating regime	Ebb generation with flood pumping	
Power Unit		
Number	28	
Turbine runner diameter (D)	8 m	
Turbine centreline submergence	-11 m OD	
Number blades per runner	4	
Turbine type	Kaplan driving generator via gearbox	
Turbine shaft speed	50 rpm	
Draft tube length	4 D	
Generator rating	25 MW	
Step-up gear ratio	10:1 (nominal)	
Channel sluices		
Number	46	
Channel width	17 m	
Invert level	4 m OD	
Gate type	Rising sector gate	
Locks		
New Ferry	Width	36 m
	Invert level	-12.5 m OD
	Gate type	Radial sector
Dingle	Width	23 m
	Invert level	-6 m OD
	Gate type	Bottom hinged flap

Features of the power unit design

A horizontally disposed power unit train (Fig. 1) enables the design of the optimum turbine water passage necessary for tidal schemes, in this case with operating heads in the range of approximately 7 m to 1.2 m. An open topped chamber or pit upstream from the turbine stage and positioned symmetrically in the inlet water passage houses the gearbox and generator. The pit structure is usually of steel or reinforced concrete. For this application a steel structure was selected, since for an optimised water passage design, it allows more power units to be installed in a fixed length of barrage and this suits the geometry of the estuary, with its relatively narrow neck at the barrage position.

The pit and stayring (Fig. 2) which supports the distributor blade assembly and turbine runner, is connected to the pit as a single fabrication. The combined mass of the pit and stayring is approximately 400 tonnes. The inner and outer rings of the stayring are connected by two hollow triangular sectioned vertical beams which span across the water passage. They give structural rigidity to the stayring and provide a fairing to the pit structure, which helps to promote smooth flow into the distributor blade annulus.

The pit and stayring is cast into concrete during caisson construction, with concrete also placed into the hollow reinforced triangular beams to give additional structural stiffness. The steel pit support is cast into reinforced concrete to give a rigid connection to the caisson. The torques and forces from

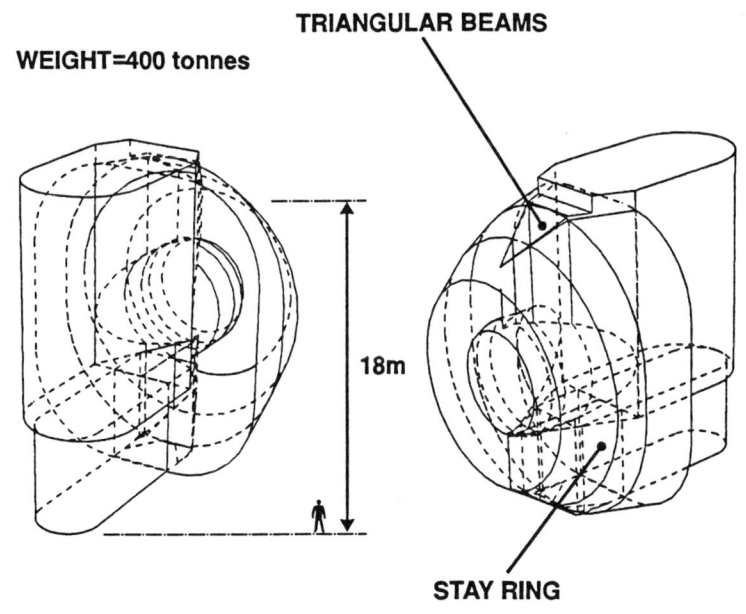

Fig. 2. Pit stayring fabrication

TIDAL POWER

the power unit are transmitted to the caisson through the concrete support plinth and triangular beams.

The torques and forces in Table 3 are considered maximum values set for the design of the power unit caisson.

Table 3. Design torques and forces

Location	Fault	Value
Generator - Moment	MNm	+3.5
Inner ring of stayring - Moment	MNm	+12.0
Outer coned ring of distributor - Moment	MNm	+5.7
Pit/gearbox flange - Axial force	MN	6.3
Stayring/distributor outer flange - Axial force	MN	5.0
Stayring/distributor inner flange - Axial force	MN	3.0

The power unit design has addressed the requirements of its manufacture and installation to provide an overall optimum solution for the barrage project. Schemes have been prepared for the ancillary equipment which include a common lubricating oil and cooling water system for each power unit.

Turbine

The power unit selected for Stage II was a turbine gearbox generator combination, the turbine being of the Kaplan type with variable pitch distributor and runner blades. Similar units of this type, size and power rating are operating successfully in run of the river stations. The intention is to use the variable pitch distributor blades to control the water flow and close off the water passage, obviating the need for a control gate downstream from the turbine runner. However, variable pitch and fixed blade combinations are still being assessed with regard to cost and energy yield predictions. Not withstanding turbine geometry, the preference for a geared power unit still remains valid.

Using model turbine performance results from an experimental programme of work (ref. 3), it was shown that there was a marginal increase in energy yield by reducing the number of runner blades from four to three. Consequently, a three bladed runner is now recommended for the Mersey Project. A three bladed runner simplifies the blade actuating mechanism contained within the runner hub, allowing the diameter of the hub to be reduced, increasing the water throughput for a given diameter of runner with a consequential increase in energy yield.

Mechanical features of design including blade operating mechanisms, water seals, etc. are generally independent of whether the turbine is for a bulb or geared power unit or whether it is for a run of the river or tidal application and are adequately described elsewhere (refs. 4 - 6) and are, therefore, not included in this paper.

During caisson transportation, immersion, ballasting and initial differential head conditions it is probable that the power unit caissons will deform and a means will be required for re-establishing final power unit alignment. The mating flanges of the turbine runner ring and draft tube liner, incorporate a seal to provide for axial expansion. It may also be used to enable corrective measures to be taken in the event of caisson differential vertical deflection, to re-establish alignment and the clearance between the runner ring and tips of the runner blades. Further details on caisson deflections are given in ref. 2. The steel draft tube liner extends until the water velocity, corresponding to maximum throughput, is 6.7 m. At this point the change from steel to a concrete structure must provide a gradual smooth transition.

Gearbox

The two stage speed increasing epicyclic gearbox (Fig. 3), is supported and located by a vertical flange mating with a vertical flange on the pit structure. The low speed shaft assembly includes a combined guide and thrust bearing adjacent to the turbine shaft flange connection at one end and a flanged face at the other, which carries the planetary gears of the low speed planetary train. The high speed stage is of a star gear train and includes a continuously lubricated flexible gear tooth type coupling, for interfacing directly with the generator shaft. The annulus rings, meshing with the planet wheels, are flexibly supported, to assist equal load sharing among all planet wheels.

During normal operation, automatic synchronization will be used and a checking and control process will ensure that transient torque is well within the acceptable limits of twice full load torque. However, the gearbox is capable of withstanding infrequent transient overloads of up to six or seven

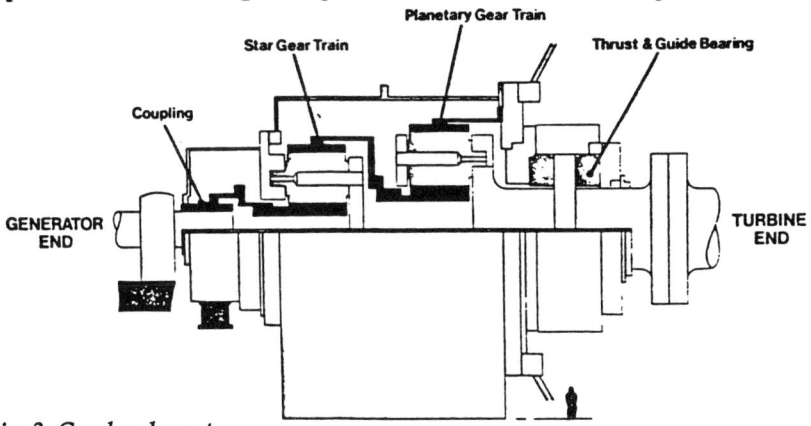

Fig. 3. Gearbox layout

TIDAL POWER

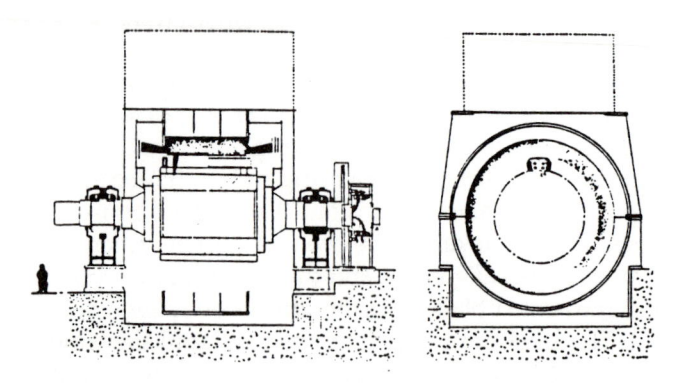

Fig. 4. Generator layout

times full load torque, caused by faulty synchronization onto the grid. The gearbox will also accept a once only transient overload not exceeding ten times full load torque at the gear output shaft, as a result of a short circuit occurring.

Generator

The twelve pole, 500 rpm gear driven generator will have a rating of 25 MW at an over excited power factor of 0.85 and a terminal voltage of 11 kV. The power requirement of the generator, when operated with reverse rotation as a motor for pumping duty, is approximately 10 MW at a power factor greater than 0.85 under and over excited. The maximum runaway speed is 1715 rpm which is below the first critical speed of rotation. The minimum inertia constant of the generator shaft is 1.5 MWsec/MW and is conservatively estimated to be that necessary to ensure power system transient stability for a severe short circuit fault on the electrical connection between power station and the local system. It also assists the reduction of speed variations that may arise as a result of extreme wave activity at the barrage resulting in potential pressure surges on the turbine runner. The gearbox and turbine runner has only a secondary influence on the inertia of the rotating system. The generator design is the well proven salient pole construction, the main features are seen in Fig. 4. The rotor, supported on two pedestal journal bearings, comprises four main components; a cylindrical rotor rim, two stub shafts and laminated salient poles. The stub shafts are bolted to the forged steel rim and centralised by spigots. Torque is transmitted between the rim and shafts by half and half driving dowels. The field winding is fabricated from rectangular sectioned copper strips.

Two shaft mounted fans with radial vanes provide ventilation for operation in both directions of shaft rotation. The slip rings for the supply of

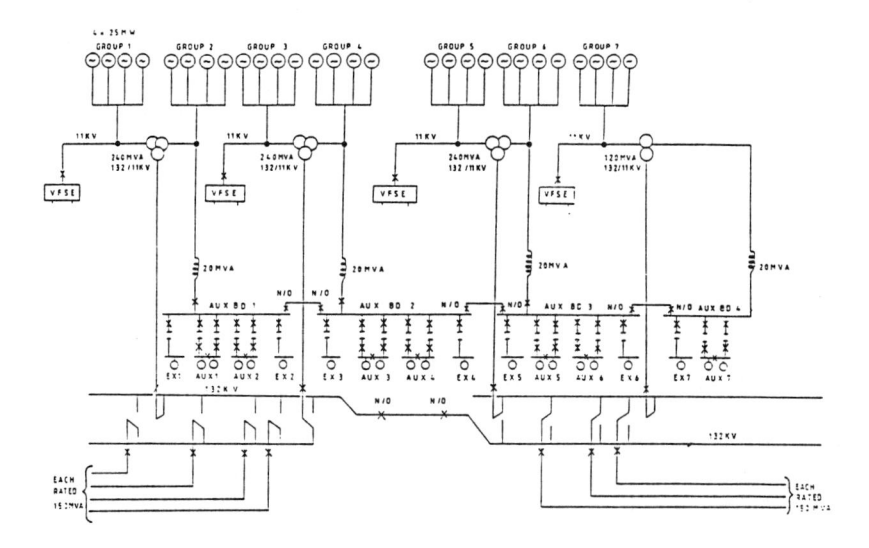

Fig. 5. Main electrical connections

excitation to the field winding is located at the outboard end of the rotor, together with a disc brake to assist run down of the power unit.

The stator core is built from electrical grade sheet steel laminations 0.5 mm thick, located on dovetail shape keys at the core back. The core is clamped between steel end plates by means of insulated bolts passing through the core and then welded at the core outside diameter to the rectangular steel bars, to which the locating keys are bolted. The stator winding is a single path, three phase wave winding, comprising bars made from rectangular copper strips with a locked transposition to reduce eddy current losses. The stator is held within a fabricated steel frame on top of which are mounted water/air coolers.

Electrical gathering and distribution system

The barrage output can be connected directly to the NGC 275 kV or the Manweb 132 kV grid systems. In a previous study, during the privatisation of the electricity supply industry (ESI) it was considered that marginal economic benefits would accrue by connecting the barrage output into the 132 kV system, (ref. 1). Therefore, one of the aims of the Stage III study was to develop and refine the Stage II estimate of costs, on the basis of the 132 kV system and simultaneously review and determine the technical feasibility and economics of connecting the barrage output into the 275 kV system.

TIDAL POWER

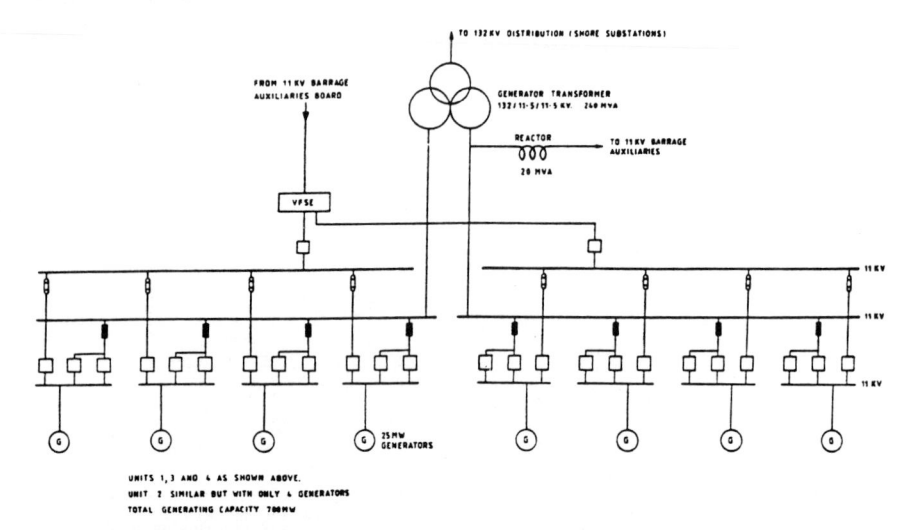

Fig. 6. Generator group

132 kV System

The arrangement shown in Fig. 5 comprises twenty-eight generators, each rated at 25 MW with a terminal voltage of 11 kV giving a nominal output of 700 MW. The generators will be arranged as four generating groups with three of the groups comprising eight generators as shown in Fig. 6 and the remaining group comprising four generators. The output of these units will be fed at 11 kV via phase isolated busbars to the generator transformers where it will be stepped up to 132 kV and then fed to the shore based 132 kV substations into the Manweb 132 kV system as shown in Fig. 7.

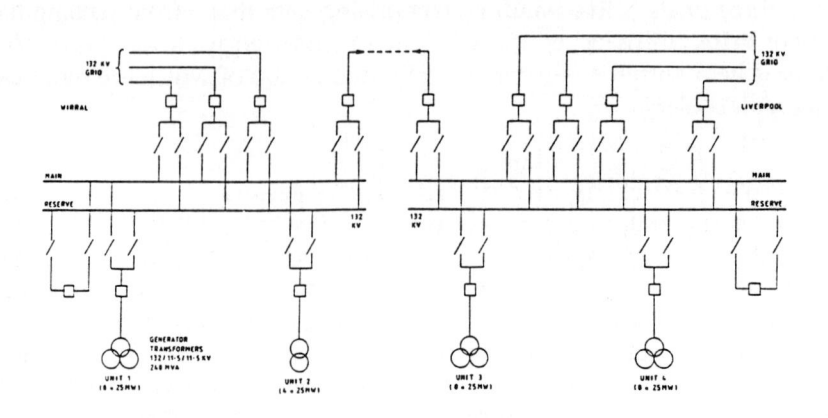

Fig. 7. 132 kV connections

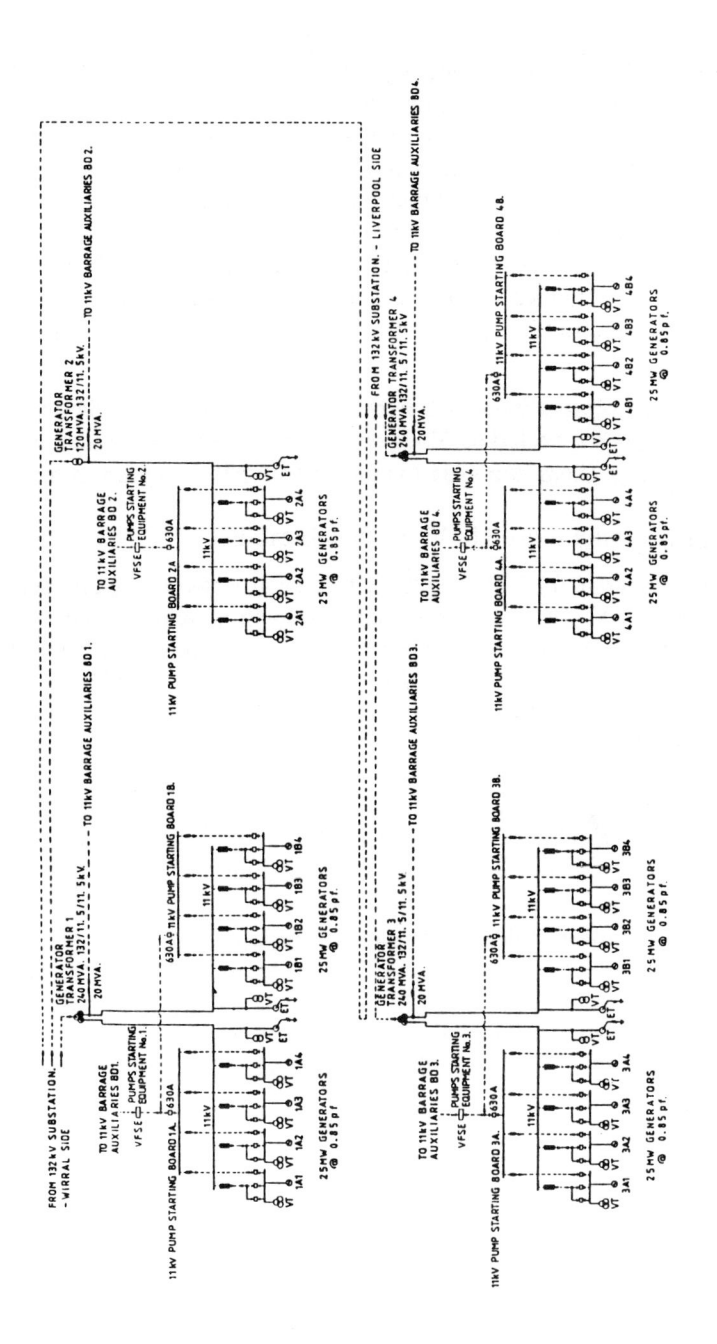

Fig. 8. Main single line diagram - Generator groups

TIDAL POWER

11 kV system

There will be 28 generator switchboards (Fig. 8) each having three 11 kV vacuum circuit breakers, one for connection during generation, one for phase reversal of the generator to reverse its rotation for the pumping mode and one for connection to the variable frequency starting equipment (VFSE) which will be used to run up the power units when used for pumping.

Four 11 kV Barrage auxiliary switchboards will be provided for each generating group. The incoming supply to these switchboards will be derived from one of the 11 kV windings of the associated generator transformers via a fault current limiting reactor. These boards will provide supplies for

- the power unit auxiliary and Barrage common services via auxiliary transformers connected to 415 V common services switchboards
- the generator excitation systems via 11 kV/290 V excitation transformers
- the four channel sluice 11 kV/433 V substations.

On occasions when the 132 kV system is not connected to the Barrage and consequently 11 kV supplies are not available to provide power for the power unit auxiliaries, common services and channel sluices, an alternative 11 kV supply will be provided from the secure supplies switchboard sited in the Birkenhead substation. The switchboard will derive its supply from two incoming Manweb feeders.

Auxiliary systems and services

Each power unit caisson will have a 415 V common services board to supply the power unit auxiliaries and the caisson common services. To overcome the problems of voltage drop associated with long cable runs for supplies to the channel sluices, four package type 11 kV/415 V substations will be established on the sluices. Normal supplies to both locks will be derived from the local Manweb 11 kV system each side of the barrage. To ensure that the locks are operative at all times standby diesel generators will be connected to each of the lock 415 V service boards to cover for any loss of mains supply.

Plant layout

It was recognised that the layout of electrical plant and cabling will have considerable influence on the cost of the electrical gathering and distribution system, as well as the design and cost of the power unit structures and caissons. The layout of the plant has, therefore, been designed to provide sufficient access for installation and maintenance and to ensure that the majority of cabling will be unitised within individual caissons with a minimum number of interconnecting cables between caissons.

On the 11 kV and 415 V systems all connections between equipment, with the exception of the phase isolated busbars to the generator transformer low voltage windings, will be by cable. A cableway beneath the electrical gallery will house cables running to the power unit and between equipment housed in the electrical gallery.

The high voltage cables from the generator transformers, which are located in one cell of every other power unit caisson, will be run in a high level cableway traversing the length of the barrage. This cableway will also accommodate the low voltage cables which pass between the caissons.

Barrage connection to the National Grid

The study involving the connection of the 700 MW barrage generation, 400 MW to Lister Drive substation and 300 MW to Birkenhead substation, into the 132 kV distribution system is not fully complete, but in the event was overtaken by a parallel study of connecting the barrage output into the NGC 275 kV system.

The option to connect the barrage output to the NGC 275 kV system has been considered in detail during the Stage III study, when the options, technical feasibility, costs and programmes for a 275 kV connection were assessed by MBC and NGC.

The study was based on the NGC transmission system and generation/demand pattern given in the 1990 ESI Seven Year Statement for the year 1995, updated to include the committed new generation and with the demand estimate extrapolated to the end of the century.

Bulk supplies of electricity for distribution within the Liverpool and Birkenhead area, are taken by Manweb from supergrid points connected into a 275 kV ring. NGC have now approved the replacement of the existing Birkenhead Lister Drive 275 kV cable for commissioning in 1993 and this aspect was included in the study network.

An examination of the transmission system capability indicated that the existing and committed system has adequate capacity at times of peak winter demand, to carry the forecast transfers with adequate margin for the barrage. Two options for the connection of the barrage output were identified. The least cost and preferred option is to extend the planned Birkenhead-Lister Drive replacement cable into the barrage 275 kV substation, by means of 3.4 km of new double circuit 275 kV cable. The connection of the barrage output will not require any grid infrastructure reinforcements for currently forecast plant and demand conditions, however, four 275 kV circuit breakers have to be replaced. A similar replacement will be necessary for the 132 kV barrage connection. An additional advantage is that there will also be a reduction in cost of the barrage gathering and distribution system by connecting into the

TIDAL POWER

275 kV system. The cost estimate given below, for electrical gathering and grid connection, is based on the 275 kV system.

Barrage control system

The Barrage power station will normally function continuously although power will only be generated for two periods each day. During this time, the head of water across the barrage will continuously alter, causing a variation in power output. To maximise this output, it will be necessary to adjust the settings of the turbine distributor and runner blades to optimise the energy yield based on predicted estuary water levels. Economic operation of the barrage will require a totally integrated control system giving sufficient flexibility to follow a defined optimized operating cycle, either automatically or with operator intervention.

The system proposed will comprise a distributed control system (DCS) and a unit control system (UCS - one per power unit). The DCS comprises equipment in the control centre connected to data acquisition and control nodes located along the barrage. The UCS will be linked to the nodes and will carry out control directives communicated to it from the DCS and will supply the DCS with data on the state of plant. It will also allow control of individual power units independently of the control centre.

The complex sequence of operations and the large number of units, requires automatic computer control, as the need for high speed decisive action from the operators should be avoided. The system will, however, provide appropriate information and advice to operators on the occurrence of faults, provide clear displays of the state of plant parameters and allow recording of these for historical and other analysis.

Manufacture of power units and hydraulic passage structures

Previous studies have defined a broad strategy for manufacture particularly the manufacture and supply of major components (ref.1). An objective of the Stage III study has been to define a manufacturing strategy based upon the barrage construction programme and to identify the methodology, resources and planning needed to implement this strategy, ensuring as far as practicable, that there is adequate manufacturing capacity to meet the programme. The salient features of the manufacturing strategy are

- pre-engineering preparation of a manufacturing database of methods tooling and process planning
- maximise the use of existing skills and expertise
- maximise the use of existing facilities
- a design for production policy to maximise cost benefits and
- development of a purchasing plan

The barrage construction programme requires a significant number of power unit and water passage components to be completed prior to their subsequent installation. The study, therefore, had to concentrate on the identification and development of the resources required to manufacture, transport, part assemble and store many large components. The large fabrications that are installed into the water passage include the combined pit and stayring and the steel draft tube liner. These steel plate fabrications have been designed to use modern steelwork practices, automatic and semi-automatic welding methods to maximise productivity and minimize cost. The large size and weight of the fabrication, require the resource of substantial fabricators with suitable waterfront roll-on/roll-off facilities. A manufacturing plan and programme has been developed to produce 12 assemblies by week 25 of the barrage construction programme and a further four assemblies every four weeks thereafter.

The turbine assembly includes very large diameter stationary components that also form part of the outer wall of the water passage (Fig.1). They are large high integrity fabrications that must be machined to close tolerances. Manufacturers capable of machining these components have been identified, together with those who have heavy duty plate rolling and fabrication equipment, including large manipulators and heat treatment facilities. As there is a very close interdependence between fabrication and machining operations, the preferred location of the fabricator is within easy reach of the machining resource.

Machining resources are readily available for many of the rotating and moving turbine components, but large diameter turbine components require special vertical boring machines of 13 m swing, which also have a profiling capability. Specialised machining facilities are also needed for the runner blade profile machining. Suitable facilities have been identified in the United Kingdom and other European Community countries.

The manufacturing strategy of the gearboxes and generators use existing facilities, including subcontractors and planned investment in new machine tools and processes. Manufacture of the gearbox external and internal components have been considered and sufficient facilities are available for all machining, grinding, heat treatment operations and assembly. Testing of the gearboxes will be undertaken at two sites where they will be full speed, no load tested prior to being despatched to site.

The generators could probably be manufactured by a single specialist manufacturer, but as a means of ensuring adequate manufacturing capacity, two have been considered in this study. The design adopts a commonality of plant and processes to ensure the production of components to a common standard of quality and cost. Materials can be purchased in the quantities and quality required to meet the programme. The installation and manufacturing programmes do not allow delivery of major components on a just in

TIDAL POWER

time basis. In addition, it is not possible to assemble fully the large turbine components in the manufacturer's works and deliver direct to the barrage site. It has, therefore, been necessary to identify suitable riverside facilities that enables the assembly and storage of the major turbine assemblies, including the pit stayring, and the storage of other major components, including the gearbox and generator. By the time of the first major M&E deliveries to the barrage construction site, 18 complete sets of power units and a considerable number of sluice gates will have been made and placed in storage.

Plant installation and commissioning

A principal objective within the Stage III study has been to develop a fully integrated civil and M&E site construction and installation programme to enable all site activities to be substantially completed within the casting basins prior to the caisson placement in the estuary. The civil construction plan is to construct 11 channel sluice caissons and one transition caisson, having two channel sluices in one casting basin and two in-situ power unit structures and five power unit caissons in another casting basin (ref. 2). The radial sluice gates, power units and associated electrical equipment will be installed before caisson float out. Two locks will be constructed and the lock gates installed, fully commissioned and made operational to maintain unrestricted passage of river traffic prior to placement of caissons. The casting basins will have suitable facilities to allow off-loading from barges and access to road transportation.

The erection programme enables power generation to commence with minimum delay after placement of the final caisson. Twenty power units will be available for generation when an operational head across the barrage can be developed. Due to caisson stability restraints, the final eight power units will be progressively commissioned to generate power on completion of the final Civil and M&E in-situ barrage works.

Typical sizes of the major M&E components of plant are given in Table 4.

These items will be loaded on to multi-axle trailers and moved onto a roll-on/roll-off barge for transportation to the barrage site, which will have a purpose built landing facility. Careful scheduling of the components to be transported optimizes the available deck space and prevents double handling or storage problems within the casting basin. Remaining items of plant will be scheduled to arrive on site on a just in time basis from the manufacturer. Four 175 t overhead gantry cranes will be required during the installation of the power units, two for the in situ structures and two for the caisson structures. Two of these cranes will be made early in the programme and used during the placement and construction of the concrete draft tube sections. The cranes will then remain on the barrage to service the equipment during operation.

Table 4. Plant components - dimensions and mass

Component	Approx. size (m)	Approx. mass (tonnes)
Pit stayring	17 dia. x 17.5 long	400
Distributor	12.5 dia. x 4 long	220
Runner assembly	8 dia. x 3 long	130
Gearbox	3.5 dia. x 5 long	100
Generator	5 dia. x 5.5 long	120
Generator transformer		150

The combined installation method and programme developed for the Project (Fig. 9) requires the installation and commissioning of the principle M&E plant items to be undertaken in three principal phases which are extensively interfaced with civil construction activities. Phase 1 and Phase 2 of the installation programme take place while the caissons and structures are in the casting basin and Phase 3 includes all post floatout work when the caissons are finally positioned in the Barrage line.

In Phase 1 the pit stayring structure will be transported into the casting basin and positioned in the base of the caissons and structures to the required datum, which will be monitored during subsequent concreting operations. Following the release of the power unit cell by the civil installation team, the pit stayring temporary stiffening structures will be removed and protective

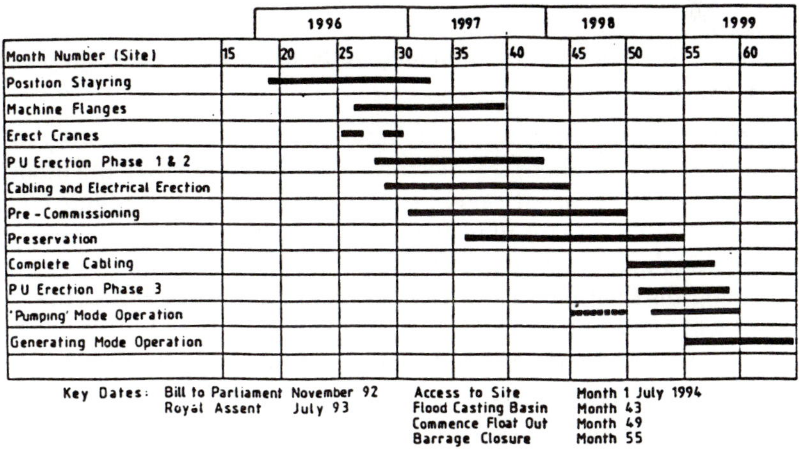

Fig. 9. Installation programme

TIDAL POWER

surface treatments repaired. In-situ machining will then be undertaken on the support flanges for the distributor and gearbox.

Phase 2 includes the installation of the power unit components, to ensure a leak-free water passage and the installation of their auxiliaries, electrical equipment and the caisson services. All the equipment will be lowered into the caisson using the overhead travelling cranes. A specific team will undertake alignment of the power unit train which will be aided by using segmental distance pieces, to the required thickness, between the gearbox and pit vertical support flange.

Each caisson will contain a local control room with a unit control panel and control equipment which will be installed in parallel with the turbine components. Power, control instrumentation and generator 11 kV cabling will be routed and terminated between local plant and the electrical gallery distribution boards, switchgear and control panels.

For two in situ power units structures, work will proceed directly into Phase 3 with the connection of the draft tube liners and the electrical connections to shore. Concreting in of the remaining draft tube liners will not occur until after placement of the five caissons and alignment checks on the power unit trains. A temporary semi-flexible joint will probably be used to accommodate any misalignment which may occur between the runner ring and the draft tube liner during caisson placement.

Pre-commissioning work will be undertaken using power from a diesel generator connected into the 415 V distribution boards. This will allow

- the flushing and operation of the fluid systems to safeguard the preservation and integrity of plant
- instrument loops will be proved from source to the unit control panel
- command signals to/from and plant condition status will be provided to the unit control panel
- the DCS signals will be simulated at the system interface unit in each power unit structure.

For the five power unit caissons Phase 3 of the installation programme starts after their placement. An electrical team will undertake the interconnections between adjacent caissons. The completion of the high voltage electrical connections, will allow the generator transformers to be energized after proving their electrical protection.

To enable the final alignment of the power unit to be progressed, the water passage will be dewatered by fitting stop logs at each end of the passage. Only one cell may be de-watered when an operational head exists and a maximum of two cells at any other time. On completion of the power unit alignment and setting, the steel draft tube liner will be fully concreted in and the water passage, then slowly flooded to check for leaks. The power unit will then be released to the commissioning engineers for final testing prior to its operation in a pumping mode.

After the energizing the generator transformer the 11 kV systems will be commissioned. The plant systems will then be operated from the unit control panel and the system interface unit until the data highway can be completed. The power unit will be run to speed in the pumping mode and the operation of the distributor guide vanes and runner blades will be proved under dynamic conditions.

Following completion of the DCS system, central control of the barrage functions are possible, allowing a head of water across the barrage to enable the power units to be commissioned in the generating mode.

Cost estimate

The cost estimates for the Mersey Barrage have been prepared on the basis of best, likely and maximum sums. These estimates representing the range of M&E construction costs from the higher maximum sum including risk and contingent allowances to the best which excludes these allowances and assumes a subsequent beneficial design and construction development of the scheme.

Table 5 summarizes the likely sums with the corresponding maximum and best sums being £485 million and £395 million respectively. All estimates are at January 1991 prices and exclude the cost of design and pre-manufacturing costs. The initial civil and M&E construction is £861 million (ref. 2).

Table 5. M&E works - cost estimate

		£million
Power unit	(Turbine, Gearbox, Generator, Auxiliaries)	256
Water control	(Sluice gates, Lock gates, Stoplogs, Cranes)	50
Electrical works	(Cabling, Switchgear, Control and instrumentation, Sundry works)	60
General services		17
Grid connection		15
Project management and engineering		23
Total		**421**

Conclusion

Work undertaken in the Stage III study has successfully achieved the objective of developing the M&E engineering of the Mersey Barrage Project to a position, which it is believed, gives the required degree of certainty and

TIDAL POWER

confidence, with regard to construction methods, costs and programme; to allow its inclusion in the renewables section of the Non-Fossil Fuel Obligation.

In earlier studies the favoured option was to distribute the power from the Barrage to both sides of the Estuary into the 132 kV system. However, based on more recent studies by the National Grid Company on behalf of MBC, it has now been decided to connect the Barrage output into the planned Birkenhead - Lister Drive 275 kV cable.

Acknowledgements

The studies on which this Paper is based were carried out for the Mersey Barrage Company Limited and were partly funded by the Department of Energy. Their permission to publish this Paper is gratefully acknowledged. The Mersey Barrage Company Limited was in turn supported by its Contractor's Group and specialist mechanical and electrical engineering companies whose assistance is similarly acknowledged.

References

1. BOLTER J. R. *et al.* Electrical and mechanical engineering aspects of a Mersey Tidal Project. *Developments in tidal energy, Proceedings of the Institution of Civil Engineers Third Conference on Tidal Power, November 1989*, Thomas Telford, London.
2. JONES B. I., MORGAN C. D. I., PHILLIPS D., and PINKNEY M. W. The Mersey Barrage - Civil Engineering Aspects. *Proceedings of the Institution of Civil Engineers Fourth Conference on Tidal Power, March 1992.* Thomas Telford, London
3. POTTS R. and WILSON E. A. The Mersey Barrage - further assessment of energy yield. *Proceedings of the Institution of Civil Engineers Fourth Conference on Tidal Power, March 1992.* Thomas Telford, London
4. HOLLENSTEIN M. and SOLAND W. The Bulb Turbines For Racine Power Station. *Escher Wyss News*, 1,1981/1,1982.
5. FISCHER F. The New bulb turbine at Felsenau Power Station Berne. *Sulzer Technical Review*, 2, 1990.
6. DONMA A., STEWART G. D., and MEISER W. Straflo turbine at Annapolis Royal - The First Tidal Plant In The Bay Of Fundy. *Escher Wyss News* 1, 1981/ 1, 1982.

6. The Mersey Barrage - hydraulic and sedimentation studies

H. ALTINK, MSc, Engineering Manager, Hollandsche Beton-En Waterbouw bv, Marine Manager, Mersey Barrage Company Ltd, K. MANN, MICE, H. R. Wallingford and N. V. M. ODD, BSc, MHydEng, FICE, H. R. Wallingford

Feasibility studies of the proposed tidal power barrage across the River Mersey are in progress. This Paper describes some results of the work on hydraulics and sedimentation carried out in Stage III of these studies. In general, the results confirm earlier studies that a Barrage between New Ferry and Dingle will increase sandy sedimentation seaward of the Barrage in the approach channel and alter significantly the pattern and type of sediments settling landward of the Barrage.

Introduction

This Paper describes the hydraulic and sedimentation studies carried out in the Stage III of the feasibility studies for a barrage across the River Mersey and comprised

- ongoing and new field measurements
- development, refinement and application of mathematical models of flow and sediment transport
- studies on an existing physical hydraulic model of the River Mersey
- a desk appraisal of changes in the bathymetry of the Estuary

The results from the mathematical flow models were also required for

- the estimation of Barrage energy yields
- overall operating costs
- detailed hydraulic flow conditions to be encountered by contractors during successive stages of the construction programme
- shipping movement studies
- environmental studies including water quality and effect on intertidal zones
- accommodation works studies for flood defence and drainage.

The work was undertaken by H.R. Wallingford (HR) together with staff of the Mersey Barrage Company (MBC).

Field studies
In any study concerning the proposed construction of a barrage for tidal power generation or for water storage it is important that field measurements are carried out before the works are constructed for several reasons. In this study the main reason is to provide background information on hydraulic and sedimentation conditions in the Estuary and Liverpool Bay before the Barrage is constructed with which future conditions may be compared, i.e. provide a baseline set of conditions in the existing Estuary. Field data is also required for the calibration and validation of mathematical models and physical hydraulic models, which are used to provide an estimate of the effects of the construction of the Barrage and after completion, on flows and sedimentation.

Silt monitoring
Concentrations of suspended solids of silt have been monitored continuously at three permanent stations in the Estuary, i.e. at Runcorn Bridge, Eastham and Prince's Pier, Liverpool. Measurements, which are made with optical silt meters, were begun in 1989 and are still continuing. The results have shown a definite correlation between silt concentrations and tidal range at all three sites. Silt concentrations measured at Prince's Pier appear not to be affected by wave activity in Liverpool Bay and at Runcorn silt concentrations are increased on ebb tides during high River flows. To supplement the permanent silt monitoring studies a programme of surveys by boat was initiated in June 1989 and was continued until June 1991. This work has given an improved understanding of temporal and spatial silt concentration distributions in the Estuary. The surveys have confirmed existing information (ref. 1) that silt concentrations are higher by a factor of four or five on the Liverpool side of the river compared to the Wirral side on the transect at Prince's Pier. This pattern is most evident during Spring tides, but is also evident at Neaps. The highest silt concentrations have been reported at the Runcorn transect and moving downstream concentrations decrease to a minimum near Stanlow before increasing again in the Narrows at Prince's Pier (Fig. 1). These data have been used in the mathematical model studies which will be described later.

Bathymetric survey
To provide up-to-date bathymetry for the mathematical models and to provide a baseline for the Estuary at 1990, a bathymetric survey of Liverpool Bay and the Estuary was made by Associated British Ports (ABP). It was carried out using hydrographic survey methods and had lines at 900 m intervals in Liverpool Bay and 150 m intervals landward of Prince's Pier. In some of the higher reaches of the Estuary where Spring tides seldom cover

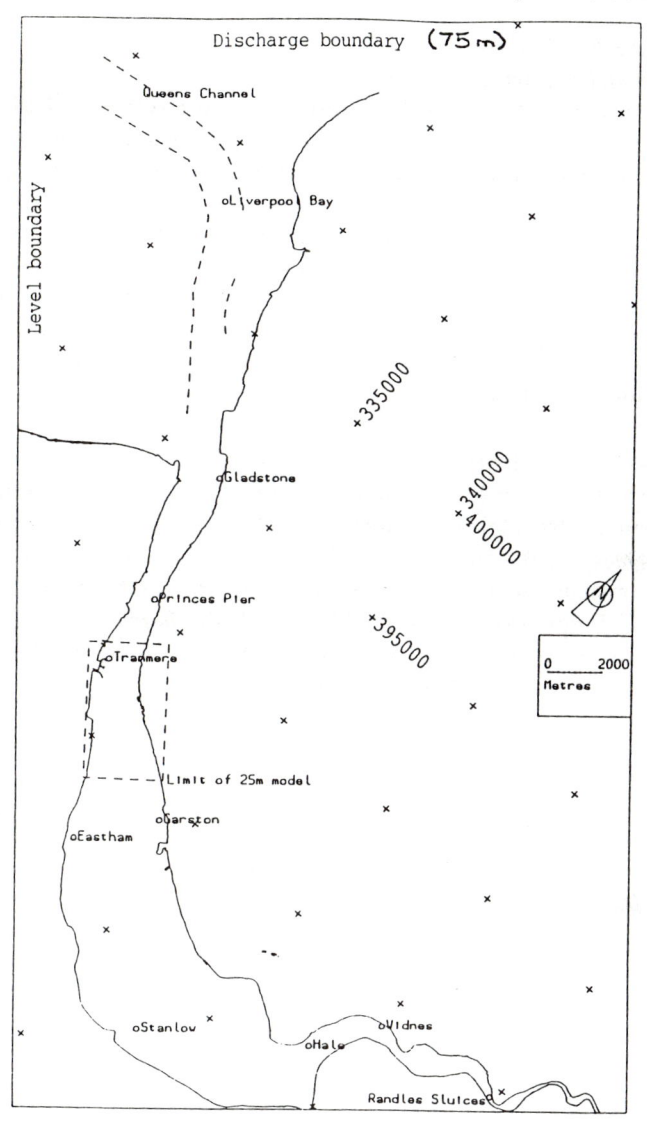

Fig. 1. Location plan of the estuary

the marshes, land-based surveys were carried out and a more comprehensive aerial survey was made. The soundings had an accuracy of +0.25 m.

A comparison of the new survey and the last complete survey of the River Mersey in 1977 and a partial one in 1984, was carried out to determine the changes that had occurred. The results showed only minor changes in bed levels.

TIDAL POWER

Hydraulic modelling
Physical model studies

An existing fixed bed, tidal hydraulic physical model at HR Wallingford was used to provide an estimate of the changes to flow paths and current velocity strengths in the vicinity of the Barrage as a result of the operational Barrage and during the construction stages (ref. 2). Water levels were also measured. The model was also used to test lead-in jetties for the locks on the Barrage and the localised flows through the high level channel sluices with turbine caissons and through the closure gap as successive stages of the Barrage construction were simulated. The model, which is shown in Fig. 2 has been in existence for several years and has been fully calibrated for both velocities and tidal levels. It has scales of 1:600 horizontally and 1:80 vertically.

As a result of the distorted scale of the model the flows in the model were not a geometrical replica of those in nature. Nevertheless, comparative information on the tidal flow patterns and their strength was an invaluable aid in this stage of the feasibility study.

Three construction stages were tested in the model these being construction Stage 4, Stage 5 and Stage 7 with gaps in the Barrage of 750 m, 300 m and 150 m respectively. Stage 8 which is the completed and quasi-operational Barrage scheme was also tested briefly.

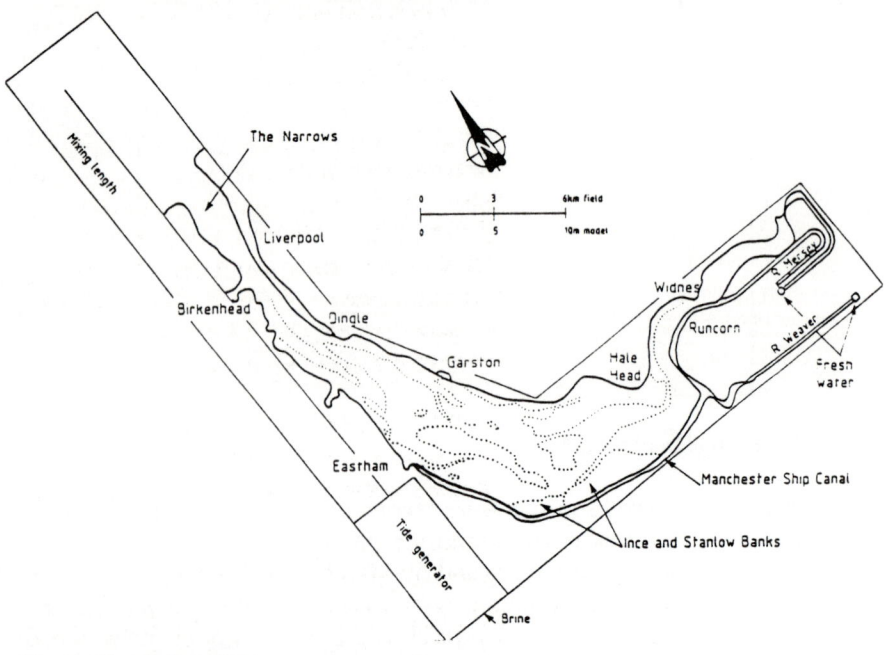

Fig. 2. Physical model

With construction Stage 4, results from the physical model showed that below the Barrage, water levels were virtually unaffected and that currents were generally higher on the ebb tide, whilst those at positions above the Barrage some increases were recorded on the flood tide. At Tranmere, ebb velocities were slightly increased in the centre of the river, but near the Tranmere oil terminal they were slightly reduced. With construction Stage 5 (300 m gap) peak velocities near the works were increased by up to 1 m/s compared to existing conditions. Whilst low water levels on both sides of the works were unaffected, high water levels upriver were slightly reduced due to the constriction caused by the works. With construction Stage 7 (150 m) gap water levels were very similar to those with Stage 5. Current velocities near the Barrage were increased by only 1.3 m/s compared with existing conditions due to the permeable structure of the Barrage and flows across Devil's Bank upriver of the Barrage were increased in the late ebb tide similarly to those experienced with Stage 4 and Stage 5 construction phases.

The completed Barrage was examined only briefly in the physical model. However, the model results showed that the lead-in jetties reduced the flow through the turbine nearest to the lock. Design modifications will be incorporated to overcome this effect.

Mathematical models

The main thrust of the work carried out in this phase of the studies was the construction of a new 2-D mathematical model of the Mersey Estuary

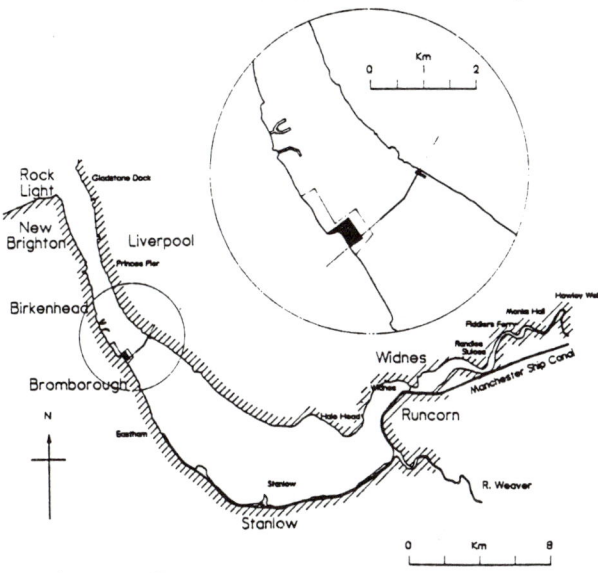

Fig. 3. Construction stage 5

TIDAL POWER

extending to the end of Queen's Channel. The model TIDEFLOW 2D, which had a uniform 75 m grid, was part of HR's TIDEWAY suite of mathematical models and was used in conjunction with MUDFLOW-2D and SAND-FLOW-2D models to determine the effects of the Barrage on tidal levels, flow patterns and mud and sand transport (ref. 3). As well as these models, a 2-D

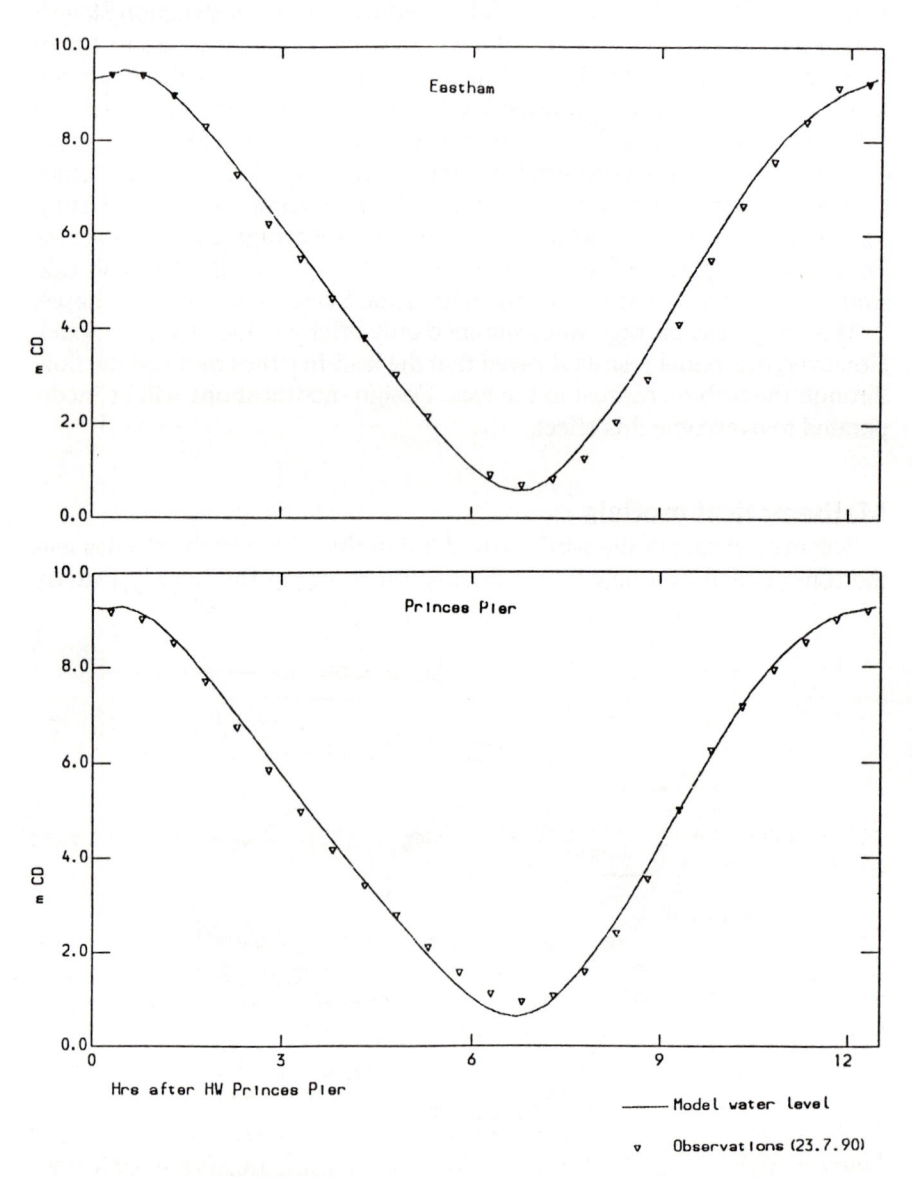

Fig.4. Comparison of observed and modelled mean spring tide levels (MDHC 1990)

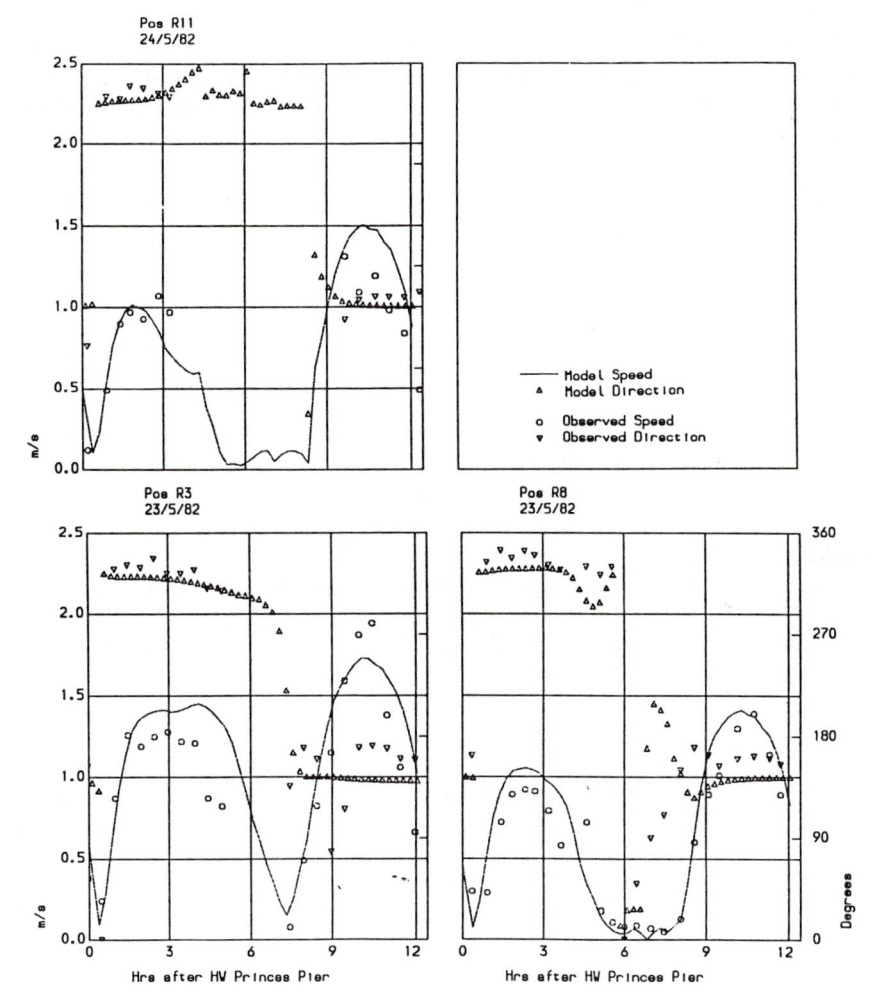

Fig. 5 . Comparison of observed and modelled depth: mean tidal velocities ,mean spring tide (HR Data 1982)

25 m grid model of the area local to the Barrage was set up for use in determining flow patterns and velocities for use in shiphandling studies. This model was also used to examine erosion in the vicinity of the Barrage. The model boundaries are shown in Fig. 1.

As well as the HR models, MBC set up 2-D flow models for use in-house to refine the design parameters of the Barrage and for energy calculations. MBC constructed three models and they were all based on the Bradford University DIVAST model system. This is also a 2-D depth-averaged formulation. A 150 m grid model had boundaries in Liverpool Bay and covered the

TIDAL POWER

whole of the River Mersey to Howley Weir. A 75 m grid model covered exactly the same area, whereas the Crosby model only covered as far seaward as the Crosby Channel. This model utilised a modified tidal curve based on results from the 150 m model and was used for comparing flow patterns for alternative Barrage schemes and not for purposes of examining energy yield or for sediment studies.

Model calibration

The boundary conditions for the HR 75 m grid model were obtained from an existing 2-D two layer model of Liverpool Bay. It was calibrated using a variety of data including observations by West in 1980, HR in 1982 - 1983 and Mersey Docks and Harbour Company/MBC/ABP data in 199Q over a range of tides from Spring to Neaps. Comparisons of observed and modelled tidal levels for Spring tides are shown in Fig. 4. Tidal velocities at three sites in the Middle Estuary near Eastham are shown in Fig. 5.

Model Predictions

The HR 75 m model was used to examine three construction stages and the operating Barrage for Spring tides, the operating Barrage at Mean tides and construction Stages 5 and 7 at Neap tides. The 25 m model was only used to test the operating Barrage at Springs and on a Mean tide only, these results being used for shiphandling studies by Maritime Dynamics Ltd (ref. 4).

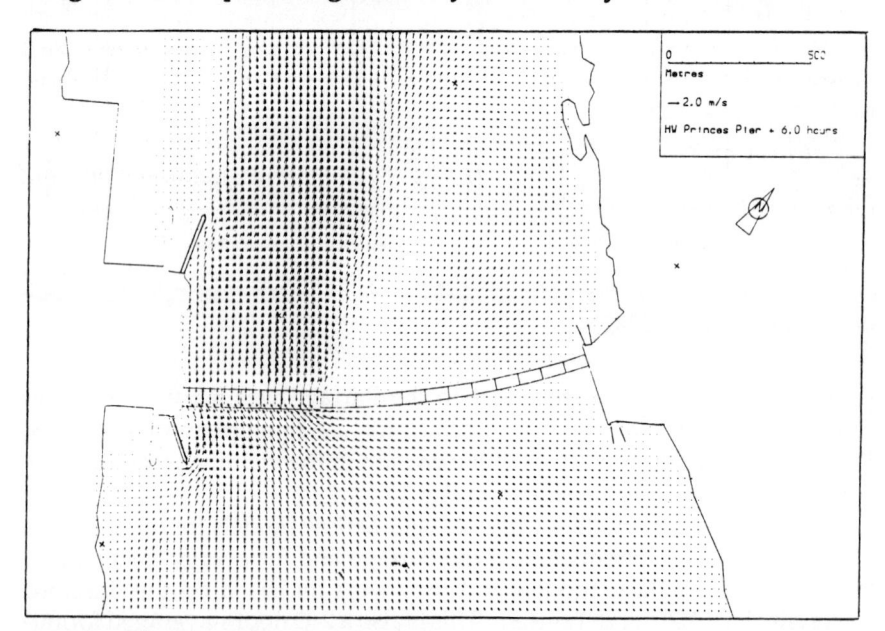

Fig. 6. 25 m Model peak generation velocity vectors : Barrage , mean Spring tide

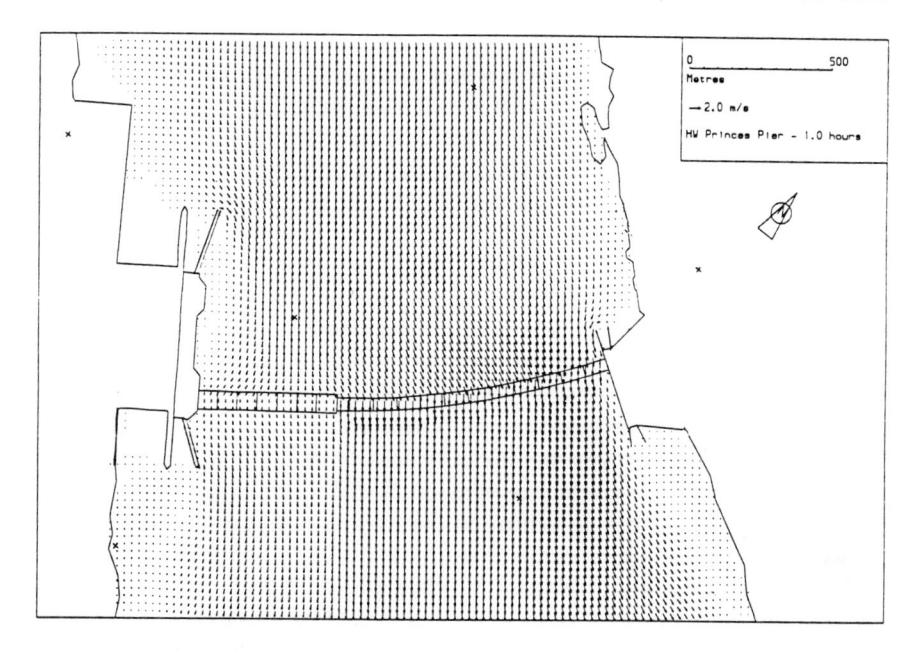

Fig. 7. 25 m Model peak flood velocity vectors - Barrage, mean Spring tide

Below the Barrage, low water levels were raised by up to a metre, but high water levels were unchanged. Above the Barrage low water levels were raised to a level of about 6 m CD (1 m OD). These results agree with those from the physical model.

Flow velocities from the 25 m model are shown in Fig. 6 and Fig. 7. They indicate the flow patterns at the time of peak generation on the ebb tide and flood tide vectors for a mean Spring tide. Although no results are shown here a comparison of the flow patterns and velocities between the HR and the MBC 75 m models showed very good agreement.

Sediment transport modelling

One of the main concerns after the Barrage is built is the effect that it will have on sediment transport within the Mersey Estuary. This applies to both the mud and sand fractions of the sediment and HR used three process models from their TIDEWAY suite MUDFLOW-2D and SAT-2D and SAND-FLOW--2D to simulate the mud and sand transport.

Mud Transport

The MUDFLOW-2D model simulated and predicted conditions in the Mersey Estuary with and without the Barrage. However, this depth-averaged model cannot simulate the gravitational circulation in the Narrows, which is one of the main processes which contains and controls the distribution of suspended mud in the Estuary. It was, therefore, decided to locate the seaward boundary of the mud transport model in the Narrows and to prescribe the concentration of mud on the incoming flood tide based on observations made by MBC and the Water Research Centre (WRC) (refs 5 and 1).

Previous experience at H.R. Wallingford showed that mud models indicate that in some cases it is possible to have no net siltation on Spring tides, but to have a long term build up of material which is mostly due to net changes on Mean tides. In this study both Spring and Mean tides were investigated. It was decided not to simulate Neap tides, because the observed flux of mud through the Narrows on Neaps was only one tenth of that on Spring tides and there was significantly less mud siltation on Neap tides because there is so little mud in suspension. A scaling factor was then applied to the siltation quantities according to the frequency of occurrence of tides, in particular tidal range, in order to estimate the annual siltation rate.

To use these sediment transport models, physical properties of the mud in the Mersey Estuary were required. These were taken from laboratory tests carried out in Stage II feasibility studies (ref. 6) and from field measurements made in the Mersey in 1990 (ref .7). These parameters included the critical shear stress for erosion, i.e. critical shear stress for deposition, the erosion constant, , the settling velocity of the mud flows, and a minimum settling velocity, which was applied to the mud concentrations below 100 ppm.

The magnitude of the gross ebb flux and the net flux of mud trapped in the Estuary is governed by the flow conditions upstream. The model predicted that about 2% of the incoming mud would settle and be trapped in the Estuary which agreed approximately with the net landward fluxes of mud observed by WRC (ref. 1).

The model was run for both the Spring tide and the Mean tide with the operating Barrage. In both cases the boundary mud concentrations were the same as those with the corresponding open river case. The mud fluxes through the Narrows during a Spring and Mean tide are shown in Table 1. It can be seen that the total fluxes with the Barrage were about 65% of those with the open River and that the net flux (and deposits on the bed) at the equilibrium condition are approximately twice those for the open River condition. These were about 2,700 tonnes and 1,500 tonnes for Spring and Mean tides respectively. In terms of annual mud siltation, this amounted to about 1.2 million dry tonnes per year deposited upstream of the Barrage which is about twice as much as for existing conditions.

The distribution of muddy siltation in the Estuary on Spring tides is shown by comparison of Fig. 8 and Fig. 9 which show the areas of deposition are considerably larger with the Barrage than in the open river situation.

Table 1. Simulated suspended mud flux through the narrows - open river and post barrage conditions

	Tidal range (m)	Flood (tonnes /tide)	Ebb (tonnes /tide)	Net (tonnes /tide)	Net/ Ebb (%)
Open river	8.4	-60,100	58,800	-1,300	2.2
	6.5	-35,400	34,600	-800	2.3
Post-barrage	8.4	-40,200	37,500	-2,700	7.2
	6.5	-23,600	22,100	-1,500	6.8

There are wider bands of deposition along the banks of the shipping channels. The increase in the water level upstream of the Barrage will reduce the effects of wave action in the permanently submerged areas. In the Garston and Eastham channels there will be an increase in the amount of muddy siltation.

A sensitivity test was carried out to determine the effect of reduced concentrations of mud entering the Estuary in the Narrows as a result of the Barrage. It is considered that the Barrage is likely to reduce the suspended mud concentrations on the flood tide, but this effect can only be predicted using a fully interactive three dimensional model. In this test the concentrations were half of those used previously. The resulting quantities of mud trapped within the Basin with a reduced concentration was about half that with the original values and about the same as existing conditions. This effect is being addressed in subsequent studies.

Fig. 8. Predicted mud deposition: Open river

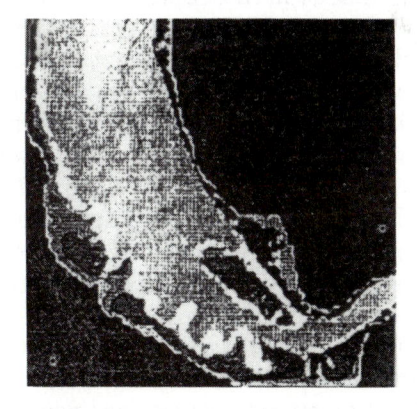

Fig. 9. Predicted mud deposition: with barrage

TIDAL POWER

Sand transport

Despite the fact that the Mersey Docks and Harbour Company has been dredging sand from the shipping channels in the Mersey for almost a century, relatively little is known about the details of the pattern of sand transport in the Estuary. The relatively strong gravitational circulation in the approach channel in the Narrows is probably the main cause of deposition of large quantities of sand in the Upper Estuary in the period about 1910 to 1960, following the construction of the training walls and the maintenance of a deep channel across the Bar. At present, the rate of siltation in the Estuary appears to be fairly steady and low compared to the first half of the century. Wave action on the shallow areas of Burbo Bight probably increases the flow of sand across the training walls into the Estuary on the flood tide during westerly storms. Analysis of both these processes was excluded from the present studies.

Two standard types of 2-D sand transport models were used for the studies. Firstly, a model, SAT-2D, which assumed that the flow is saturated with suspended sand at all times and that there was no limitation of the availability of sand on the bed surface was applied to the Upper Estuary where there is a relatively uniform cover of sand. The second sand transport model, SANDFLOW-2D, took into account the unsteadiness of the flow and the time taken for sand to move in and out of suspension and the presence of inerodable strata. This model was applied to the seaward approaches of the Estuary.

Observations of suspended sand flux in the Estuary carried out in 1990 (ref. 8) were used to provide an empirical function to define the saturated equilibrium rate of sand transport as a function of the local instantaneous tidal velocity in the Upper Estuary.

Without the effects of gravitational circulation and wave action, the models predicted a net seaward transport of sand along the canalised approach channel and the seaward movement in the shallow channels in Burbo Bight of existing conditions. This pattern matches the pattern of net movement of water near the bed measured in a physical model without salinity effects described by Price and Kendrick (ref. 9).

Calibration of the sand transport models was achieved by summing the amount of sandy accretion in each dredged area in the entrance channel for a single Spring tide which were then used to give the approximate equivalent annual rate of maintenance dredging by scaling with reference to the recorded rate of dredging in Queen's Channel East. The model appeared to reproduce approximately the correct distribution of sandy accretion in the four areas of maintenance dredging as shown in Table 2 and Fig. 10.

Table 2. *Simulated sand maintenance dredging rates for existing conditions.*

	Hopper tonnes/a
Queen's Channel East	900,000
Askew Spit	200,000
Crosby Shoal	200,000
Brazil Elbow	20,000

The same method was applied to the analysis of existing patterns of sandy accretion in the Eastham and Garston approach channels. These results indicate that the models can predict trends in the approximate location and magnitude of sandy accretion with a Barrage in operation. However, the models will need to be improved significantly in the final phase of feasibility study.

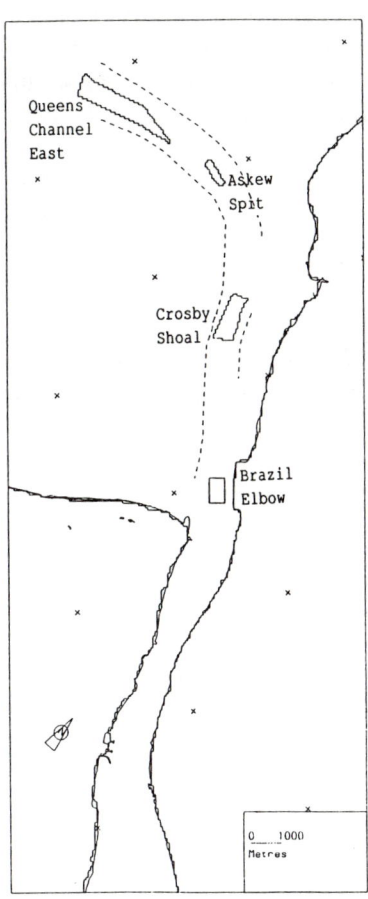

Fig. 10. *Main areas of dredging in Liverpool Bay*

Predictions of sand transport with the barrage

With the Barrage in place, the model (SANDFLOW-2D) predicted a similar pattern to the existing accretion pattern in the entrance channel, but with increased accretion in Queen's Channel East, Askew Spit and in Crosby Channel. It predicted an increase of deposition amounting to about 50% in these areas.

Using the SAT-2D model in the sandy areas in the Upper Estuary indicated that the Barrage would reduce sand siltation in the Eastham and Garston Channels by between 15% and 30%.

Effect of construction stage 5 (280 m gap) on sand transport

The effect of the year-long construction Stage 5 was investigated using the results from the 75 m flow model to drive SAT-2D. The model predicted a slight increase in sandy accretion in Queen's Channel East, Askew Spit and Crosby Channel because of the reductions in tidal flows as a result of the constriction during Barrage construction. No significant change was predicted for the Eastham Channel, but a slight increase in Garston Channel.

Conclusions

The studies carried out in Stage III of the Mersey Barrage feasibility study have enabled better predictions to be made of future flow patterns, tidal levels and mud and sand siltation, both during construction of the Barrage and when it is operational. The results confirmed earlier studies that a Barrage between New Ferry and Dingle will increase sandy sedimentation in the approach channel and change the pattern and type of sediments settling landward of the Barrage. This work has been carried out using new field and laboratory data, physical model studies and 2-D mathematical models of flows and sediment transport. These studies are being continued and a 3-D mathematical model developed to assess more accurately the possible effects of the Barrage on the hydraulics and sedimentation in the Mersey Estuary.

Acknowledgements

The authors wish to thank the Mersey Barrage Company Limited, the Energy Technology Support Unit of Department of Energy and HR Wallingford for permission to use the results of research on which this Paper is based.

References

1. COLE J. A. and HEAD D. C. *Mersey Metals Flux Study* (NDE 9337 SLD). WRC Environment, Medmenham.
2. HR WALLINGFORD. *Mersey Barrage Studies - Stage III Hydraulic And Sedimentation Studies. Report On Physical Model Study*. HR Report EX 2180, January 1991.
3. HR WALLINGFORD. *Mersey Barrage Feasibility Studies - Stage III 2-D Mathematical Modelling Of Tidal Flows And Sedimentation*. HR Report EX 2303, March 1991.
4. ALTINK H. and HAIGH G. W. R. The Mersey Barrage - The Impact On Shipping. *Proceedings of the Institution of Civil Engineers Fourth Conference on Tidal Power, March 1992*. Thomas Telford, London.
5. HR WALLINGFORD. *Mersey Barrage Feasibility Studies - Stage II. Continuous Silt Monitoring In The River Mersey.May 1989 To June 1990*.HR Report EX 2212, October 1990.
6. HR WALLINGFORD. *Mersey Barrage Feasibility Studies -Stage II. Mud Properties*.HR ReportEX 1988, September 1988.
7. HR WALLINGFORD. *Estuarine Sediments, Near-Bed Processes. Verification Of A Deposition Algorithm In The Mersey Estuary*. HR Report SR 251, December 1990.
8. HR WALLINGFORD. *Mersey Barrage Feasibility Study Stage II. Hydraulic And Sedimentation Study. Sand Flux Measurement*. HR Report EX 2225, April 1991.
9. PRICE W. A. and KENDRICK M. P. Field Measurement Investigation Into The Reasons For Siltation In The Mersey Estuary. *Proc. Instn Civ. Engrs*, Vol 24, April 1963.

7. The Mersey Barrage– preparations for an environmental assessment

J. V. TOWNER, Bsc(Hons), PhD, Technical Director, Environmental Resources Ltd. and E. A. WILSON, MA, MICE, Engineering Manager, Mersey Barrage Company Ltd

Environmental studies have formed a significant component of the Stage III work programme. It has consisted of a programme of field studies initiated to ensure that consistent, long time series of key biological data required for a full Environmental Assessment are collected. Also, the environmental specialists have provided continuous assessment and input to the design and proposed operational regime of the preferred Barrage. Results of the fieldwork programme have enhanced knowledge about the Estuary to a significant degree. However, it is also clear that base line conditions are subject to substantial variability. Understanding and accounting for this variability is important when considering the prediction of the impact of the Barrage. The need for strong interactions between the engineering and environmental specialists in the project team to achieve a successful project definition at feasibility study stage is emphasised.

Introduction

Environmental studies have formed an integral component of the overall feasibility studies for the Mersey Barrage scheme. A previous Paper at the Third Institution of Civil Engineers' Conference on Tidal Power (ref. 1) summarized the main findings of Stage I of the Environmental Study. This paper provides an update on the current status of the environmental work.

As part of the Stage I study, further work was identified that will be required to carry out an Environmental Assessment (EA) to conform with EC Directive (85/337/EEC) and UK Regulation SI1199 on EA. This further work covers a wide range of aspects including Estuary hydrodynamics, siltation, water quality modelling, estuarine chemistry, estuarine ecosystem analysis and ornithological studies amongst others. Some of this work, particularly those aspects relating to estuary hydrodynamics and siltation, is required as part of the engineering studies. Nevertheless, the outputs of these studies are critical to the assessment of environmental effects; biological processes, particularly benthic algal growth, may also provide important influences on sedimentation and erosion.

Since the Stage I study, the scale and scope of the full EA programme has

TIDAL POWER

been refined. Items relating to the effects on human beings of construction work, impacts on landscape and waste and construction material implications have gained greater prominence. Nevertheless, the major emphasis of the overall programme is still focused on effects upon water quality, the ecosystem and nature conservation within the Estuary.

The Stage III studies

As is outlined in other Papers given at this conference (ref.2), the Stage III studies are concentrating on a single barrage alignment (Line 3E) in the general location of the Line 3 Barrage defined during the Stage I studies. Fig. 1 provides a location plan.

The main objectives of the Stage III environmental studies can be summarized as follows

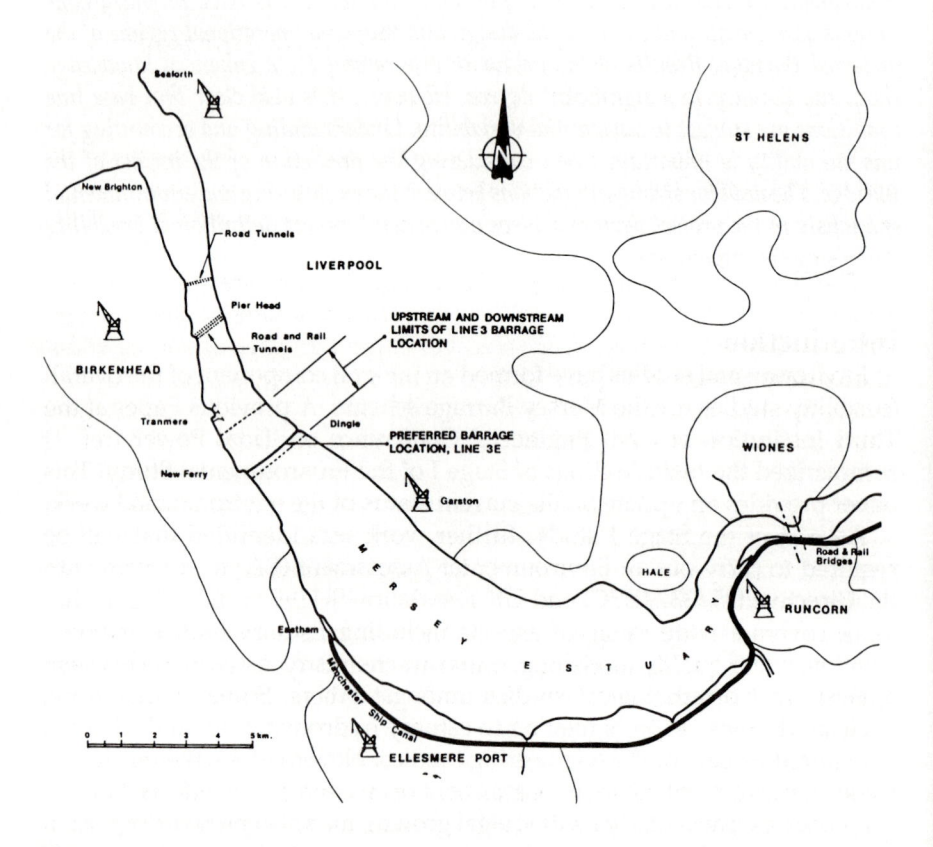

Fig. 1. Mersey Barrage location plan

(a) To update and refine previous findings in relation to environmental effects on the basis of improved hydrodynamic modelling of the Estuary and better defined changes in operating regime (e.g. flood pumping).
(b) To provide input to the overall engineering design and operational regime for the Barrage to minimize or obviate environmental effects.
(c) To carry out essential long-term baseline studies required to more fully understand and describe the current ecosystem and its relationships with prevailing physical conditions.

The baseline studies are not being carried out simply to provide a complete stamp collection of species and communities for a future Environmental Statement. The necessity for this work derives from three interlinked factors

(a) the need to provide necessary data to fill major gaps in knowledge about components, for example, planktonic algae, in the Estuary
(b) the requirements for a consistent and integrated dataset that allows analysis of short term variability and fluctuations in the ecosystem
(c) the enabling of the assessment of long term triends and variations due to, for example, the clean-up of the Estuary.

The main elements of continuous baseline work currently being carried out as part of the Stage III studies are briefly outlined below. The balance of study elements does not reflect the balance of the final EA, which will be of a considerably broader scope.

The phytoplankton and zooplankton of the Estuary are being surveyed on a regular basis (more than twice per month) at a number of stations in the Estuary. The stations have been chosen to reflect the range of salinity within the Estuary, the aim being to characterize the communities and productivity throughout the Estuary. In addition, weekly monitoring has been instituted at an upstream site (Runcorn Bridge) and downstream site (Pier Head) to assess shorter term temporal variations in dominant species and production. Further, fish larvae and planktonic larval forms of otherwise non-planktonic species (for example, certain polychaete worms etc.) have been surveyed. The plankton species of the Mersey have not received significant attention in the past. In particular, the phytoplankton species had not been studied in detail. Since the Barrage may affect the potential for phytoplankton blooms, it is considered important to understand current productivity and species composition in relation to existing physico-chemical conditions in order to assess changes post Barrage.

The salt-marsh of the Mersey is of threefold importance to the Estuary ecosystem. Firstly, salt-marsh acts as a provider of detrital matter that provides nutrients and energy to the ecosystem when primary production may otherwise be low. Secondly, salt-marsh is a major source of bird nutrition to certain species, particularly the duck widgeon; salt-marsh plant seed may also provide a supplementary source of nutrition to other duck species, but is not thought to be a major food item. The third major role of the salt-marsh (particularly that of the southern bank) is an undisturbed roosting area for duck and wader species. Furthermore, salt-marshes constitute important wetland habitats in their own right. So, gaining an understanding of salt-marsh community and production in relation to current tidal regime is of significance. The study components include: intensive vegetation surveys of the Mersey Estuary salt-marsh to examine temporal and spatial succession of species; salt-marsh levelling to assess heights and gradients of the marsh and its communities relative to tides; a detailed floristic survey of salt-marsh and salt-marsh cliff habitats; studies to assess the productivity of salt-marsh in the Mersey. In addition, survey work and productivity studies are being carried out in relation to the benthic diatoms on mudbanks.

The continuing improvement of water quality in the Mersey Estuary has led to the potential for increased importance of the Estuary for marine and estuarine fish, although conditions are still not satisfactory for migratory species such as salmon and sea trout. The likelihood is that over the proposed lifetime of the Barrage, the currently unremarkable and impoverished fish population will improve. For these reasons monthly surveys of fish in the Mersey are being carried out. These are supplemented by larval fish surveys carried out as part of the plankton survey work.

The invertebrates inhabiting the sediments of the Mersey are of key importance as a food resource for birds and fish. Regular monthly surveys of the invertebrate species, communities and biomass in the Mersey have been carried out in conjunction with bird survey work carried out by the British Trust for Ornithology (BTO) - see below. These surveys consider the abundance, spatial distribution and variability of prey organisms, in relation to species and size, and class relationships have been derived between invertebrate communities and habitat factors such as sediment grain size, organic matter content, salinity and shore elevation.

BTO has continued to carry out detailed ornithological surveys of the Estuary. The current Winter survey work constitutes the fourth year of continuous monitoring within the Estuary by BTO. The major objectives of the ornithological fieldwork are: to assess the inherent variability of the distribution patterns of duck and waders in the Estuary; to identify sites that are regularly, intermittently or never used by birds; to investigate relationships between bird densities and food resources, especially invertebrate prey; to study the night-time distribution and diets of teal and pintail in the Estuary. This work has involved day and night-time field observations of birds, radio-tracking studies of Teal and Pintail, analysis of stomach contents of Pintail and correlation of bird densities with invertebrate distributions.

Summary of key findings of Stage III fieldwork

The current programme of fieldwork has been in progress for just over one year and the studies undertaken form the first integrated approach to examining the overall ecosystem of the Mersey. With the exception of ornithological studies data analysis is as yet incomplete on the first year of monitoring. Thus, whilst knowledge of the Mersey Estuary has increased considerably, a complete seasonal cycle for one year alone has not been fully examined. The findings must, therefore, be considered to be of a preliminary nature. Nevertheless, certain significant changes in the biology of the existing Estuary are evidently underway.

Plankton

The zooplankton (planktonic animals) of the Mersey has been found to be typical of a British estuary, with small crustaceans known as copepods dominating. The total number of species identified currently is 62, which is unexceptional. In general, it has been observed that species diversity increases with salinity. This is as expected. The seasonal succession as far as it can be elucidated is also typical of moderately polluted British estuaries. The increased abundance of barnacle and polychaete worm larvae during the spring and early summer is similarly typical of a British estuary. On the basis of the current dataset, it has not proved possible to demonstrate whether the existing zooplankton communities are limited by poor water quality due to current waste water discharges.

The results of phytoplankton (planktonic algae) studies have been of greater significance. In the past, routine monitoring of phytoplankton pigments (chlorophyll a) to estimate biomass have indicated that the phytoplankton production in the Mersey is low. More recent chlorophyll a measurements carried out by the National Rivers Authority during Neap tides have altered this perception, since these measurements have indicated significant phytoplankton biomass.

During current survey work it has been observed that over the year the Mersey Estuary supports a varied and at times abundant phytoplankton assemblage. These results confirm that, the Mersey Estuary has a highly productive phytoplankton. Significant algal blooms have been recorded in the docks adjacent to the Mersey (ref. 3), where turbidity is low and suspended sediment concentrations are below 5 mg. In the Mersey, where suspended sediment concentrations are generally more than an order of magnitude higher than in the docks, peak phytoplankton numbers (70 million cells per litre) similar to those recorded in the comparatively quiescent docks have been attained. The general succession of different phytoplankton groups and the species composition of the diatoms and dinoflagellates have been found to be consistent with those typically found in other estuaries (ref. 4). A marked sequence of successional peaks in the abundance of certain species of phytoplankton assemblage has been noted; in early spring (March - April) a peak of the diatom *Skeletonoma* was observed to be succeeded in May by a peak in numbers of the diatom *Thalassiosira*, while the colonial and buoyant *Phaeocystis pouchetii* peaked in late May and early June. While dinoflagellates have been observed to be presenting the phytoplankton assemblage, they have not been the dominant species during any of the surveys undertaken to date. Potentially toxic (red tide forming) dinoflagellate species have not been observed to be dominant or significant members of the phytoplankton assemblage.

Salt-marsh

The field data collected has revealed that the Mersey Estuary salt-marsh can be broadly categorized into three types.

(a) Accreting and developing marshes in the early stages of succession which are characterized by a relatively species-diverse flora. These marshes are characterized by pioneer species such as *Spartina, Puccinellia* and *Aster*.

(b) Mature *Puccinellia* dominated marshes, which are frequently truncated at the lower edge by steep banks. These marshes tend to have variable grazing pressure. The southern shore Stanlow Bay and Ince marshes are examples of these areas and generally have a well drained substrate. Stanlow and Ince marshes are further differentiated by the lack of livestock grazing on Stanlow marsh.

(c) Marshes which are at the upper limit of tidal influence and are rarely inundated by the tide. These occur upstream of Runcorn.

The differences in the physical and floristic characteristics between the rapidly accreting, developing marshes of the northern shore and the mature marshes of the southern shore are of significance in terms of understanding the effects of a Barrage. The mature marsh is dominated by common salt-marsh grass (*Puccinellia*) and largely free of *Spartina*, which could otherwise reduce the nature conservation interest of the marsh. Further, the marsh edge is eroding and, therefore, the southern marshes do not display the full marsh succession. The northern marshes by contrast are not mature and *Spartina* already forms a major component of the lower marsh flora. These differences in marsh character are likely to determine the susceptibility to invasion by *Spartina* following construction.

Measurements have been made of the above ground biomass and growth rates of vegetation in areas of the northern, developing marsh and the mature marsh of the southern shore. Comparison of these measurements for the northern and southern shores shows that during winter the biomasses of above ground vegetation are approximately equivalent in both groups of marshes. However, during the spring and summer the growth rates of vegetation on the southern, mature marsh are approximately double those of the northern, developing marsh. This reflects the highly productive nature of the *Puccinellia* marsh. Nevertheless, the above ground biomasses (2000 gm^2 - 4,000 gm^2) in summer are high in both groups of marsh and indicate that the salt-marshes are potentially a major source of detrital matter to the Estuary. Rates of decay of *Puccinellia* litter exceed those for *Spartina* on the Mersey (as is found elsewhere) and the invertebrate fauna of the *Puccinellia* marsh is more abundant.

Initial results from the surveys of benthic diatoms in the Estuary indicate that the biomass and production of this component of the ecosystem is high. Preliminary indications suggest that the total primary production of the benthic algae exceeds that of the planktonic algae (phytoplankton). This factor is significant as these benthic diatoms play an important role as primary producers within the ecosystem and hence as a food resource for grazing and detritus feeding invertebrates. Additionally, sediment stability and accretion is greatly influenced by the presence of benthic diatoms. Thus, effects of a Barrage on this group of algae may be important in terms of the overall post-Barrage sediments regime.

Sediment invertebrates

The total number of species of sediment invertebrates recorded in the Mersey is low at 21 and only about half of these species occur in significant abundance. The species found are typical of a polluted, macrotidal estuary. Certain common estuarine species, particularly *Corophium volutator* and *Hydrobia ulvae*, have been reported to be absent or infrequent in the Mersey during previous surveys over the last 10 years (ref. 1). The recent survey work has shown that the crustacean *Corophium* now has a widespread distribution throughout the Estuary and is occasionally abundant, whilst similar survey work in 1987 only found a single individual in samples from the Estuary. The small, snail-like *Hydrobia* was not recorded during the 1987 survey and earlier reports in the 1970s suggested that the species had been in decline in the Estuary previously. The present work has shown that the species is now regularly found in moderate abundance at two locations in the Estuary; at New Ferry, just downstream of the proposed Barrage alignment, and Stanlow Bay, adjacent to the southern shore of the Estuary a few kilometres upstream of the entrance to the Manchester Ship Canal. It is tempting to conclude that the increased populations of both species are a result of improvements in water quality. However, with *Corophium*, at least, it is known that populations are subject to very wide fluctuations for physical and biological reasons other than poor water quality. This aspect of invertebrate population dynamics requires further examination as it may influence bird feeding behaviour and feeding site use.

TIDAL POWER

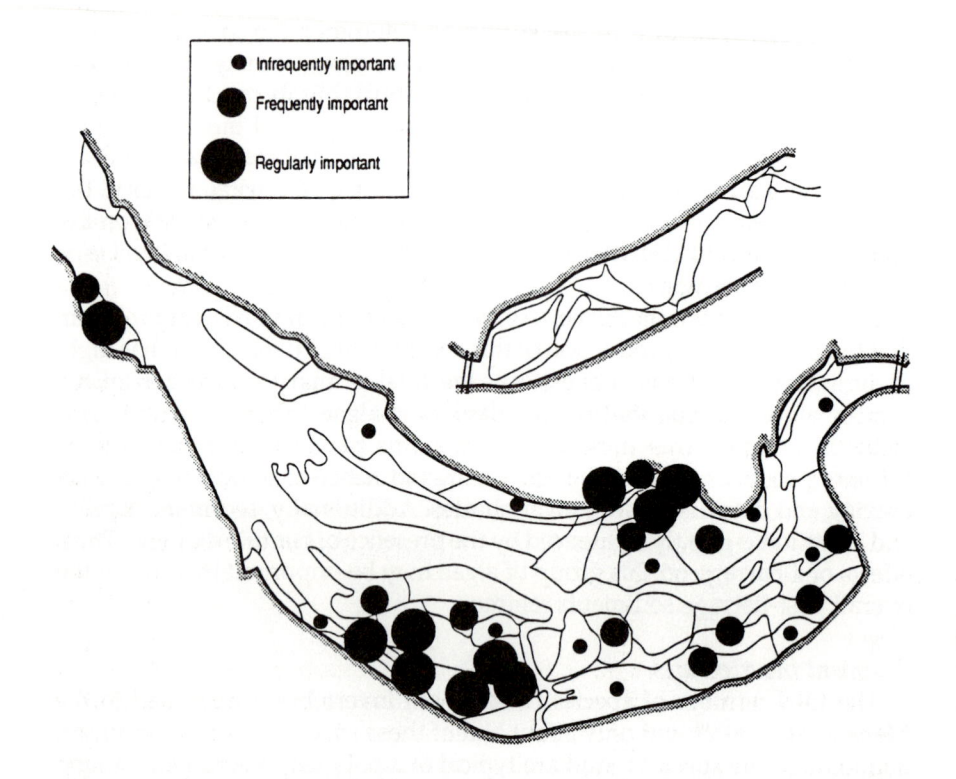

Fig. 2. The relative importance of intertidal areas for all species of invertebrates between November 1990 and June 1991 (Importance is estimated as biomass (kg AFDW) per intertidal area).

The relative importance of various intertidal areas of the Mersey for all species of invertebrates is illustrated in Fig. 2.

It is evident that much of the biomass of invertebrates is concentrated in two major areas; these areas are indicated by the groups of the largest black circles. These are the Oglet Bay area on the northern shore of the Estuary and the Stanlow Bay area on the southern shore of the Estuary. The other area of importance is at New Ferry, just upstream of the proposed Barrage alignment. The elevation of much of the intertidal area supporting these high densities of invertebrates is at or above Mean tide level and the sediments tend to consist of silty sands. The major communities are dominated by oligochaete and polychaete worms, although these sites also support moderate to high densities of molluscs such as *Macoma Balthica* or *Hydrobia* and the crustacean *Corophium*. Lower densities of invertebrates tend to be associated with unconsolidated fine textured sediments in the upper parts of the Inner Estuary, while the coarse sandbanks of the centre of the Estuary support very low abundances of invertebrates or, in some cases, do not contain macro invertebrates at all. These areas of coarse sand tend to dominate the seaward end of the Inner Estuary and are generally present below mid-tide level.

Fish

The fish species recorded during the current fieldwork are typical of a polluted estuary, with flounder and, to a lesser extent, plaice being of greatest importance; the standard method for monitoring estuaries, beam trawls, does bias towards these species, though. The general seasonal trends in fish species observed are also typical, with cod and whiting entering the Estuary in the winter.

On the basis of data collected to date, it has not proved possible to conclusively demonstrate that an improvement in the abundance and diversity of fish species in the Mersey is currently under way, resulting from the improvements in water quality that are occurring. However, circumstantial evidence from sea fishermen and some data collected by the National Rivers Authority does suggest that some improvement is accompanying these water quality improvements.

Birds

Table 1 summarizes data on the national and international importance for waders and ducks during winter for the five year period 1986/1987 to 1990/1991. The Mersey is currently internationally important for shelduck, teal, pintail, dunlin and redshank and nationally important for widgeon, grey plover, black-tailed godwit and curlew. The current position represents a considerable change over the last four to five years, particularly in relation to the significant increase in numbers of the wader species redshank, grey plover and black-tailed godwit. All these species have shown substantial increases in numbers within the five-year period. This has been accompanied by an overall decline in the total numbers of duck. Within this period, the invertebrate fauna has also increased in diversity and there is some indication that abundances of favoured wader prey items such as Corophium and Macoma have also increased during this period. These trends may reflect improvements in water quality in the Estuary. However, increased pressure on other estuarine sites in North West Europe has also been apparent during this period.

In addition to the long-term trends in numbers of different wintering bird species in the Mersey, significant fluctuations in numbers occur over shorter time scales. For example, the peak number of the wader dunlin exceeded 50 000 in the winter 1990/1991 compared with the average over the previous four winters of less than 20 000. Whilst the severe weather of the past winter might partially explain this phenomenon (the Mersey was notably milder

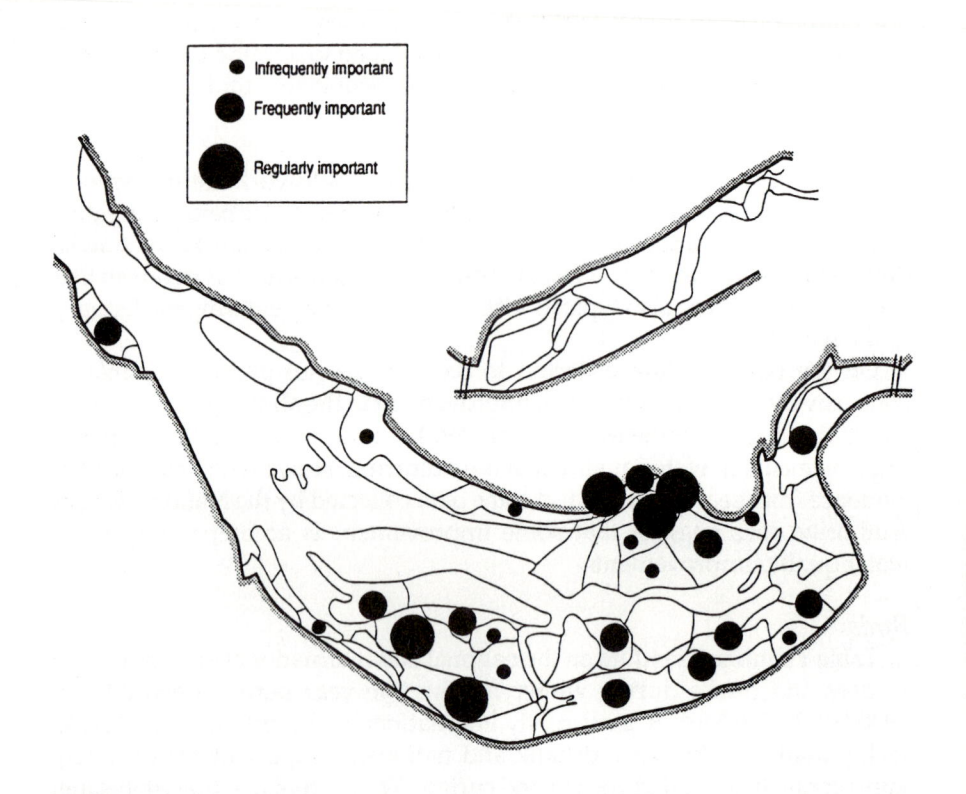

Fig. 3. The relative importance of intertidal areas of feeding birds of all species in the winters 1988/1989 to 1990/1991

than many other estuaries), the peak bird numbers recorded did not coincide with the period of most severe weather recorded by BTO.

The relative importance of intertidal areas for feeding birds of all species in the winters of 1988/1989 to 1990/1991 is illustrated in Fig. 3 (this is reproduced from a map by HTO); it should be noted that this is a generalized overview and individual species preferences for feeding areas differ somewhat. For example, the wader curlew tends to distribute evenly and widely over most of the intertidal area. Nevertheless, it is clear that major bird feeding areas tend to coincide with areas of greatest invertebrate abundance. In addition, the productive salt-marsh of the southern shore of the Estuary supports high numbers of the grazing duck widgeon.

It needs to be emphasized, however, that not only are there inter-species differences in the use of areas, but that use of areas by individual species may change over time. For example, shelduck prefer areas with higher exposure times, while redshank preferences tend to correlate better with sediment

Table 1. *The national and international importance of the Mersey to waterfowl 1986/1987-1990/1991 (Data of Birds of Estuary Enquiry 1991)*

Species	Average Peak Winter Count (Nov - March)	% Of British Population	% Of European Population*
SHELDUCK *Tadorna tadorna*	3,394	4.5	1.4
WIGEON *Anas penelope*	6,716	2.7	0.9
TEAL *Anas crecca*	10,685	10.7	2.7
MALLARD *Anas platyrhynchos*	1,217	0.2	<0.1
PINTAIL *Anas acuta*	5,908	23.6	8.4
RINGED PLOVER *Pluvialis hiaticula*	26	0.1	0.1
GOLDEN PLOVER *Pluvialis apricaria*	727	0.4	0.1
GREY PLOVER *Pluvialis squatarola*	710	3.4	0.5
LAPWING *Vanellus vanellus*	4,852	0.5	0.2
KNOT *Calidris canutus*	316	0.1	0.1
DUNLIN *Calidris alpina*	23,928	5.6	1.7
BLACK-TAILED GODWIT *Limosa limosa*	65	1.3	0.1
CURLEW *Numenius arquata*	1,443	1.6	0.4
REDSHANK *Tringa totanus*	3,824	5.1	2.6

* For wildfowl, percentages are of Western European population, for waders percentages are of East Atlantic flyway population.

type (and by implication reflect prey preferences). Further, there is evidence that changes in sediment quality at some locations towards a finer muddy quality has led to increased use of certain intertidal areas over the last 3 years.

The Mersey ecosystem

The discussion thus far has centred on the nature of the Mersey Estuary at present. Emphasis has been placed on the long term trends and variability of thse Estuary, due to factors such as the clean-up of the Mersey and resulting improvements of water quality. Further, there is considerable evidence of short term, large scale variability which is particularly related to physical factors, such as major events of sediment erosion. Therefore, gaining an understanding of the nature of these trends and interdependencies between physical (for example, tide levels, sediment substrate etc.), chemical and biological variables provides the basis for the prediction of Barrage-induced change. However, an understanding of short term fluctuations and event-driven (for example, wave erosion induced) change is also critical. This is for two key reasons.

(a) The Estuary regime is presently highly dynamic and prone to sudden change, while the post-Barrage regime is likely to be more stable and variability would be expected to occur within narrower ranges.

(b) An understanding of the limits of natural, particularly short term, variation within the system allows improved definition of uncertainty in prediction of post-Barrage change.

As part of the current stage of work an investigation is being carried out of the potential of different systems ecological approaches for predicting post-Barrage conditions. As a first stage of this process, the preliminary energy web shown in Fig. 4 has been drawn up. The objectives of this approach are to integrate the relationships and interdependencies between the various components of the ecosystem. This approach allows an assessment of the effect on the overall ecosystem of changes in the magnitude of sources of energy, such as increased light availability due to post-Barrage reduction in turbidity or loss of detritus to the ecosystem due to reductions in sewage discharges. Similarly, a systems approach can aid the understanding of the effects of reductions in stress energies to the system, due, for example, to reduced tidal currents and scouring or decreased salinity.

A system's ecology approach does not substitute for other methods of prediction and assessment, such as use of statistical relationships between physical and biological factors (for example, sediment quality and invertebrates) but complements them. The precise form of systems approach to be adopted is the subject of continuing review and will be dependent on the quantity and quality of data derived from the field work.

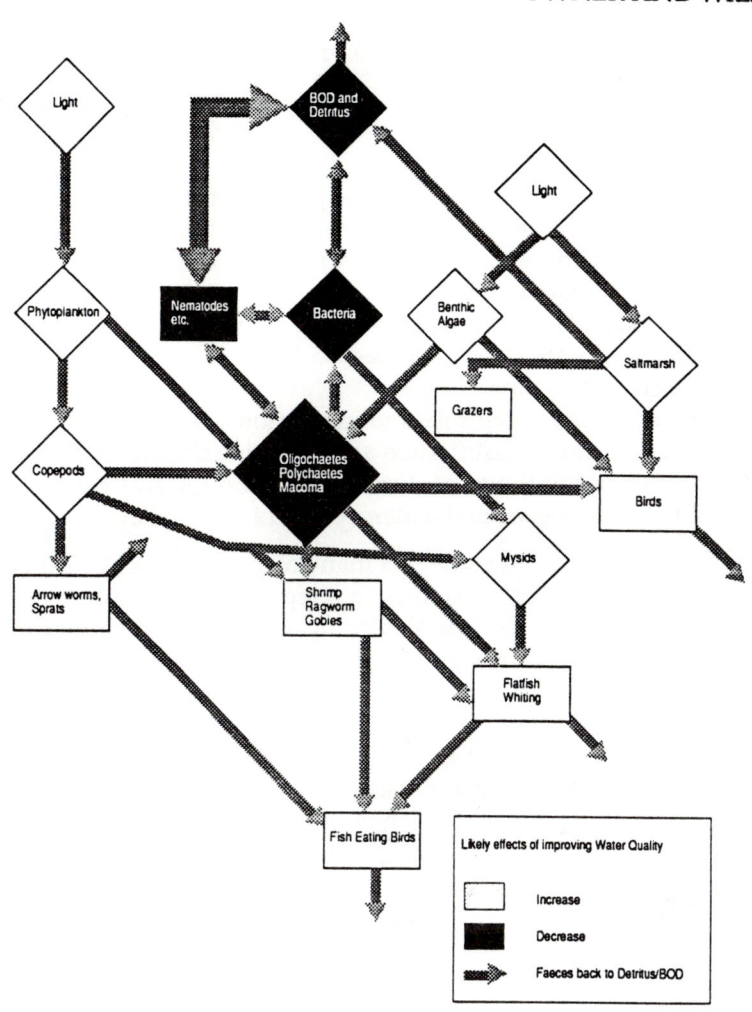

Fig. 4 . Preliminary food web for estuary

Assessment of effects of a barrage
Interaction with engineering and other studies

The prediction of impacts of a Barrage on the environment depends upon gaining an adequate understanding of the physical change induced by the structure, particularly in relation to hydrodynamics and sediment regime. These factors have been assessed using MBC's 2-D DIVAST numerical model of the Estuary and work by HR Wallingford using the TIDEWAY 2-D system (ref. 5).

TIDAL POWER

Key factors of importance that need consideration in assessing the effects of the Barrage include

(a) the degree of reduction in tidal range and changes in High and Low Water levels
(b) changes in the timing and relative periods of ebb and flood flows
(c) the degree of reduction in exposure of intertidal areas available for bird feeding and changes in frequency of inundation and exposure of areas, including salt-marsh
(d) changes in tidal currents and resulting alteration of sediment transport processes and bottom shear stresses
(e) modification of wave climate
(f) changes in suspended solids concentrations
(g) effects on salinity distribution and estuarine circulation
(h) changes in the fluxes of dissolved substances and particulate suspended material between the Mersey and Liverpool Bay.

The prediction and assessment of many of these changes is being carried out as a component part of studies in support of the various engineering studies. Development work is also being undertaken on a water quality model, which combines the hydrodynamics and transport dispersion modules of the MBC DIVAST model with the water quality kinetics algorithms of the US Environmental Protection Agency models QUAL II and WASP. The DIVAST based water quality model is undergoing preliminary evaluation and sensitivity testing at present. The model allows the prediction of salinity distributions, the dispersion of conservative pollutants and *E coli*, as well as oxygen and nutrient concentrations.

Interaction with the engineering project team is important for reasons other than as a source of project information. A key element in the EA process is that the environmental specialists are able to advise the engineering team on environmental implications of different design or operational aspects of the scheme. Examples of this type of interaction with the Mersey Barrage studies include

(a) advice on the implications of differing operational regimes (for example, flood pumping or ebb sluicing on the overall impact on the environment of the scheme
(b) the examination of the environmental implications of alternative construction techniques and options for spoil and other waste disposal
(c) evaluation and advice in relation to the environmental factors influencing choice of components, such as the effects of alternative turbine design and operation options on the potential for damage to fish during turbine passage (ref. 6).
(d) provision of conceptual and preliminary designs for inbuilt mitigation

PAPER 7: TOWNER AND WILSON

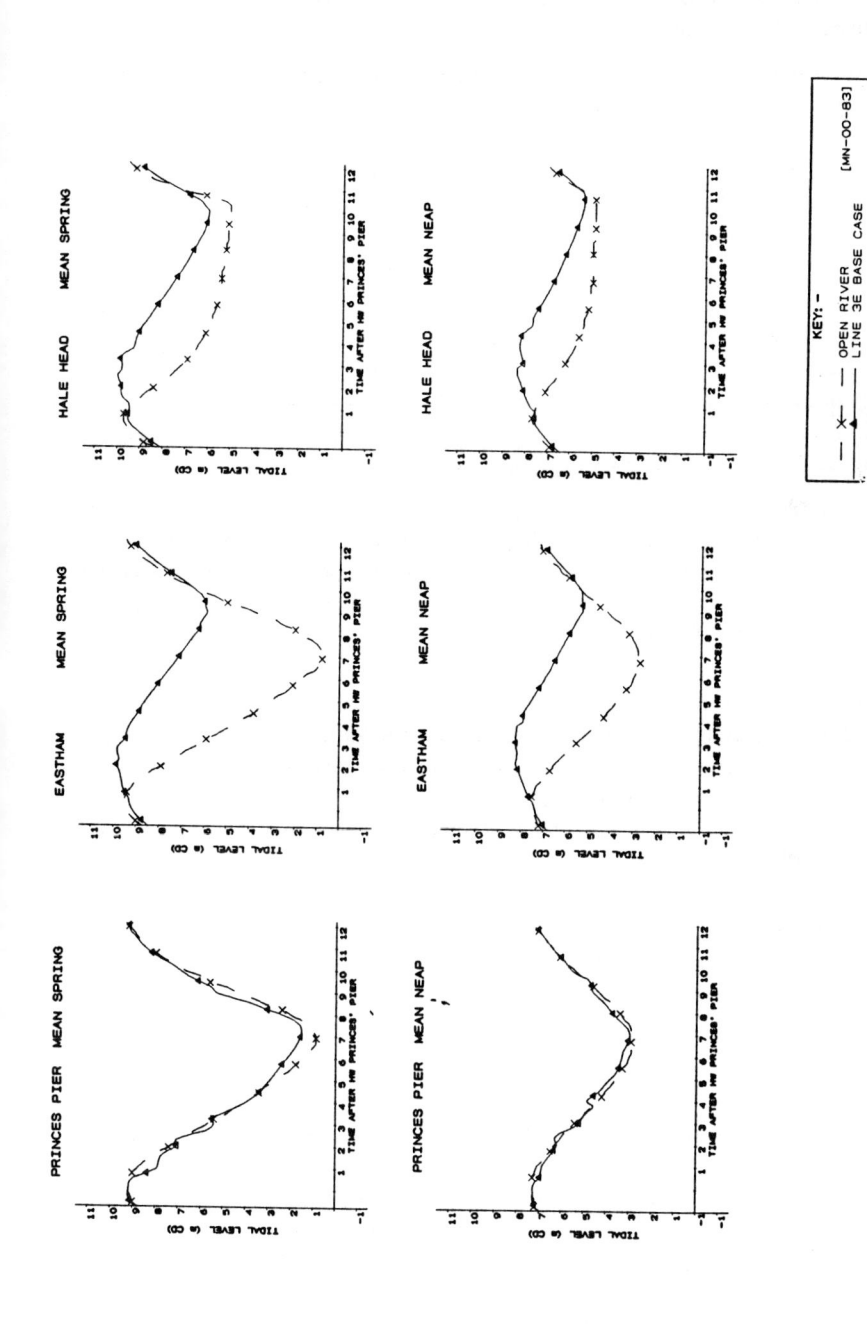

Fig. 5. Water elevations on spring and neap tides

TIDAL POWER

measures, including, for example, fish passes or dredge spoil lagoons designed to support bird life.

Thus, it can be seen that the engineering team plays an important role within the EA process through the building in of mitigation measures at an early stage in design. Clearly the success of this process is dependent on the development of an integrated team between the engineers and environmental specialists. Further, the integration of environmental factors into the design process can be advantageous to the promoter of a scheme. This is because

(*a*) early consideration and inclusion of design or operational features to reduce the degree of impact can minimize later costly design modifications.

(*b*) the engineering design team builds up a knowledge of sand sensitivity to factors (for example, post-Barrage water levels) that influence the degree of environmental impact, which is a significant consideration in the overall promotion of a project

(*c*) it can prove possible to turn engineering problems to environmental advantage, for example, the need to dispose of dredge spoil arising from construction or maintenance can be managed through disposal lagoons designed for the benefit of bird life (ref. 7).

Aspects of the environmental effects of Stage III Barrage Scheme

As noted above, the key first stage in the assessment of the impacts of the Barrage on the environment is to gain an understanding of the changes to estuarine regime resulting from the Barrage. Fig. 5 illustrates the effect of the Barrage on tidal water levels at three stations upstream of the Barrage for Spring and Neap tides; Eastham is the station nearest the Barrage, while Hale Head is the station furthest upstream. The general features of the impact on tidal regime are similar to those noted previously (ref. 1).

However, predicted water levels differ from those derived from the Stage I study for the following reasons

(a) The preferred number of turbines has increased from 21 to 28, their diameter from 7.6 m to 8.0 m and the operating regime now includes flood pumping. These changes have increased the average cubature of a mean tide from 160×10^6 m^3 to 200×10^6 m^3 which results in a tidal curve closer to that of an existing mean tide having a curvature of 250×10^6 m^3.

(*b*) The hydraulic models have been set up and calibrated on the basis of new bathymetric data collected during 1990, while previous predictions were made using a physical model based on bathymetric data collected in 1977, augmented by a part survey in 1984. This has resulted in changes in both open Estuary and post-Barrage water level predictions, particularly in those parts of the Estuary where shallow water effects influence tidal levels.

The latter factor demonstrates the influence on the assessment of impacts of changes in the characteristics of the physical environment of the open Estuary over time. Since predictions of impact need to be made for conditions at some point in the future and possibly for up to 120 years hence, understanding the underlying causes of physical change in their forcing functions is as important as a knowledge of the nature of ecological change. This highlights the need to undertake sensitivity analysis of predictions based on an analysis of alternative future scenarios for post-Barrage and open Estuary physical conditions; this could include consideration, of for example, different rates of sediment accretion.

The overall approach to the assessment of effects of the Mersey Barrage relies on three main elements

(*a*) The use of estuarine intercomparisons of the post-Barrage Mersey Estuary with natural estuaries with similar physical characteristics in relation to tides and sediment regimes (for example, some of the estuaries of the South West coast of the UK).
(*b*) The use of systems approaches which are currently in a simple form, for example, based on simple food web methods (Fig. 4).
(*c*) The use of predictive models and methods, which may be deterministic (for example, the DIVAST based water quality model) or statistical (for example, derivation of multiple regressions between bird numbers and invertebrate densities).

All of these approaches have limitations; for example, whilst the more stable post-Barrage Estuary may bear resemblances to estuaries in the South West, it will still be influenced by its river and marine end-members and the template of its pre-Barrage characteristics such as, pre-Barrage invertebrates communities. Further, consideration must also be given to Mersey specific factors. An example is the existing lack of disturbance of roosting areas as a reason for birds using the Estuary.

In carrying out work on the assessment of post-Barrage conditions, the various methods developed as part of the Energy Technology Support Unit environmental generic studies programme have proved useful. These include the *Spartina* niche model (ref. 8) and the methods developed for the prediction of post barrage densities of shore birds (ref. 9). However, further field data is required and is being collected, to reduce the level of uncertainty in statistical relationships specific to the Mersey.

In terms of general environmental change to the Mersey caused by the Stage I, Line 3 or Stage III, Line 3E Barrage, the magnitude of overall change in either case is predicted to be broadly comparable. The nature of the

TIDAL POWER

resulting new ecosystem would not be expected to be sensitive to relatively minor variations in Barrage scheme characteristics when compared with the substantial gross change caused by the Barrage. A general increase in the productivity of the pelagic planktonic component of the ecosystem would be expected, due to reduced turbidity, while the less aggressive post-Barrage tidal currents and bottom shear stresses would be expected to result in greater invertebrate biomass in the currently less productive sands below current mean tide level. Thus, the post-Barrage Estuary would be expected to be more productive overall mainly due to more efficient utilisation of nutrients and food resources that currently escape the Estuary prior to uptake.

However, in terms of the effects on bird life in the Estuary, which is the principal environmental impact of the Barrage, the availability of invertebrate prey is of key importance. During the current study it has been estimated that about 12% of the total intertidal area supports about 90% of the total invertebrate biomass. A large proportion of the present intertidal area that would become subtidal with the Barrage supports a low productivity fauna. Much, but not all, of the more productive areas are above mid tide level and would continue to be exposed, albeit for shorter periods per tide, on lowest post Barrage tides; It is suggested (ref. 9) that there is no evidence that reduction in foraging time affects shore birds densities. However, a time element in reduced exposure exists due to the effect that certain productive mudbanks that are exposed currently would not be exposed on post-Barrage Spring tides. Therefore, a reduction in the availability of total invertebrate prey to resources to bird species is expected. Further interspecific differences in the degree of effect are expected; for example, curlew are widely distributed over the entire existing intertidal area (possibly feeding on widely dispersed lugworm and large ragworm) and thus, a reduction in numbers approximately proportional to the total reduction in intertidal area (circa 46%) could be expected.

The productivity of the post-Barrage residual intertidal areas is currently high and it is assumed presently that the biomass supported post Barrage would not increase significantly to allow greater bird densities than at present; this conservative assumption is subject to revision if new data collected shows this to be reasonable. For this reason, studies are currently under way to assess potential habitat creation enhancement and management measures to support displaced birds.

Approaches under investigation include recreation of marsh habitats in former salt-marsh reclaimed as low guality agricultural land and the construction and management of dredge spoil lagoons as potential bird feeding areas.

Conclusions

The Stage III studies have increased knowledge about the nature of the existing Estuary and a programme of fieldwork has been initiated that has started to fill some of the major gaps in knowledge about the Estuary.

This work has emphasized, though, that the assessment of the effects of the Barrage needs to be undertaken against a baseline of varying conditions. The variation in the baseline is of three main forms

(a) Steady and gradual changes in regime resulting from the evolution of the Estuary due to natural factors.

(b) Sudden or short term variability, for example, due to meteorological forcing.

(c) Steady medium and long-term changes due to human activities, of which the current clean-up of the Mersey is of key importance.

These changes in the baseline conditions and the operational/design flexibility of the Barrage introduce uncertainties in the assessment of the effects of the Barrage. To reduce these uncertainties, the fieldwork programme is now being extended. When complete, this will allow better definition of the variability in the system and lead to greater certainty and quantification of post-Barrage change as well as to a better understanding of the limitations on predictions due to variability within the system.

A high level of interaction between the engineering and environmental studies has been maintained. Further studies are being undertaken to evaluate alternative operating regimes (for example, reduced flood pumping, ebb sluicing) to minimize ecological effects. Further design work is also being undertaken on management and mitigation measures to reduce impacts; these include salt-marsh creation and spoil lagoon management.

References

1. TOWNER J. V. Environmental Aspects Of A Mersey Tidal Project: Development In Tidal Energy. *Proceedings of the Institution of Civil Engineers Third Conference on Tidal Power, November 1989.* pp 263-274. Thomas Telford, London, 1990.
2. JONES B. I., MORGAN C. D. I., PHILLIPS D. and PINKNEY M. W. The Mersey Barrage - Civil Engineering Aspects. *Proceedings of the Institution of Civil Engineers Fourth Conference on Tidal Power, March 1992.* Thomas Telford, London, 1992.
3. ALLEN J. A. and HAWKINS. S. J. *South Docks Projects*, Report To Merseyside Development Corporation, Final Report, February 1991.
4. McLUSKY D. S. *The Estuarine Ecosystem, second edition.* Blackie, London and Glasgow, 1989.

TIDAL POWER

5. ALTINK H. and MANN K. The Mersey Barrage - Hydraulic And Sedimentation Studies. *Proceedings of the Institution of Civil Engineers Fourth Conference on Tidal Power, March 1992.* Thomas Telford, London, 1992.
6. SOLOMON D. S. *Fish Passage Through Tidal Energy Turbines.* ETSU Report ETSU TID 4056, Department of Energy, 1988.
7. DAVIDSON N. C. and EVANS P. R. Habitat Restoration And Creation: Its Role And Potential In The Conservation Of Waders. *Wader Study Group Bulletin* 49/TWRB Special Public 7, pp 139-145, 1987.
8. GRAY A. J., CLARKE R., WARMAN E. A. and JOHNSON P. J. *Spartina Niche Model.* ETSU Report ETSU TID 4070, Department of Energy, 1989.
9. GOSS-CUSTARD J. D., McGORTY S., PEARSON B., CLARKE R. T., RISPIN W. E., dit DURRELL SEA Le V and ROSE R. J. *The Prediction Of Post Barrage Densities Of Shore Birds: Birds.* ETSU Report ETSU TID 4059, Department of Energy, 1989.

Discussion on Papers 5 – 7

C. P. STRONGMAN, Merz and McLellan

Could the Author provide some performance data for the generating sets, in particular the efficiencies of the turbine speed increasing gear and generator. What is the difference in efficiency when reducing the turbine blades from 4 to 3?

Also what life for the gears has been assumed, i.e. the period after which the gears will have to be replaced?

R. JONES, Countryside Council for Wales

HR's studies show that sedimentation above the barrage could increase by the order of 600 000 tonnes p/a following its construction. Such a loss of sediment from the system could cause a lowering of beach levels and an adverse impact on sea defences seaward of the barrage - has this been considered?

H. MOORHEAD, Severn Tidal Power Group

Mr Mann showed a slide of a bed-frame device for monitoring near-bed sediment and current data in the lower inter-tidal area. It would be useful to know how this device was kept in position on the bed and how the information was recorded, particularly when subjected to wave action. The conditions in Liverpool Bay would appear to be not too dissimilar to those encountered in Bridgewater Bay.

K. MANN, HR Wallingford

The bed-frame was not situated in Liverpool Bay but in the estuary at Eastham and so no wave action was evident.

J. ELLIS, British Steel Technical

British Steel are currently involved in modelling the effect of barrage operating conditions on the erosion of barrage materials by suspended solids.

Will models arising from current or future studies on pre-barrage suspended solids and sedimentation enable or assist one to predict the variation of suspended solids cementation and size in barrage draft tubes and sluices, with height above bed and time into tidal cycle and flow speed?

N. ODD, HR Wallingford

The sediment transport models developed in Stage III did not simulate the degree of detail you seek. There will be relatively short period during and immediately after the construction of the barrage when large quantities of sediment may be put into suspension.

Once the local bathymetry has settled down to the new regime one could expect to find concentrations of several hundred parts per million of a 120-170 micron sand moving in suspension near the bed on spring tides.

I. R. DEANS, Joint Countryside Advisory Service

Do the HR studies assist in projecting possible effects of changes in hydraulics and sediments on mudbanks, sandbanks and salt-marshes, all of which are of significance to the natural environment? Any such changes need to be determined before one can approach an environmental assessment in a considered way.

N. ODD, HR Wallingford

The sediment transport models developed in Stage III predicted the growth of new zones of inter-tidal and sub-tide mud banks at the head of the estuary as illustrated in the Paper. The predictions need to be refined further in Stage IV.

M. KENDRICK, Acting Conservator of the River Mersey

In the pre-barrage runs, the 2D sand transport model, which did not include analysis of the processes of gravitational circulation or wave action, predicted a net seaward transport of sand along the trained approach channel to the Mersey. This was said to match the pattern of net water movement measured near the bed on an earlier physical model (Price and Kendrick) when operated with fresh water only.

This statement is quite correct. It should, however, be added that failure to reproduce the natural salinity distribution on the physical model (which incidentally had a moveable bed) led to poor agreement between the strength and direction of currents in model and field, with the eventual deposition of the model bed sediment in the seaward sump! It was only after reproducing the correct salinity distribution that model water movements became comparable to field values, the residual tidal flow near the bed being reversed from seaward to landward with a consequent change in the net transport of bed sediment.

In the light of this, and in view of the importance of sediment transport studies in the Mersey Barrage project, do the Authors consider it necessary to make any further refinements to their modelling of water movements, and if so, what would they be?

N. ODD, HR Wallingford
The two-dimensional depth-averaged flow model commissioned in Stage III could not simulate the gravitational circulation, which is now being simulated in a 3D model in Stage IIIA. However, the 2D model did predict the effect of the barrage on the primary flood and ebb currents, which directly influence the rate of shoaling at the entrance to the Queen's Channel.

D. H. COWIE, Binnie & Partners
To what accuracy were hydrographic and other surveys used in providing data for the hydraulic model of the Mersey carried out?

K. MANN, HR Wallingford
The specified accuracy of the hydrographic survey was similar to that of previous surveys of large estuaries using echo sounders, i.e. about plus or minus 200mm between adjacent points and plus or minus 500mm overall

M. JEFFERSON, World Energy Council
Dr Towner compared the Mersey with the South Coast in terms of tidal movements. The Mersey, like Morecombe Bay, is an area of great ornithological importance. Has ERL compared the Mersey with other potential tidal barrage sites and attempted to rank them in terms of ornithological importance and environmental acceptability generally? If they have been ranked where would the Mersey come? In Dr Towner's Paper it is suggested, for instance, that the Mersey's curlew population might decline 46% post-barrage and that it is currently assumed there would be no increase in bird densities post-barrage. By such a comparative ranking one might identify 'go' and 'no-go' situations, and allow the 'go' situations to proceed more quickly while stopping fruitless work on the others (e.g. with appropriate mitigation effects, the Severn may well be a relatively attractive proposition and the largest electricity generator).
Reference: ETSU TID 4055, 4057, 4059 and 4076

8. The Mersey Barrage - further assessment of the energy yield

R. POTTS, BSc, FIMechE, Director Engineering Development, International Research and Development Company Ltd and E. A. WILSON, MA, MICE, Engineering Manager, Mersey Barrage Company Ltd

Energy studies have formed a part of the recently completed Mersey Barrage Company (MBC) Stage III feasibility studies. This Paper describes those energy yield studies and the conclusions drawn. Improvements have been made in the accuracy of energy yield prediction. Both zero-dimensional (0-D) and two-dimensional (2-D) models have been used; the former for sensitivity analyses and the latter for base case predictions. The annual energy yield of the base case, Line 3E, Barrage is estimated as most likely to be 1.39 TWh which is a reduction of 6% from the estimate of earlier studies for a similarly configured Barrage.

Introduction

The MBC has recently completed its Stage III studies of a tidal energy barrage across the Mersey, at the Line 3E location. Since the initial assessment (ref. 1) a number of engineering changes, such as the use of three bladed machines, with measured pump characteristics and the use of channel sluices, have been introduced and analysed to assess their impact on the energy yield. Also a number of changes have been introduced into MBC's 0-D and 2-D energy yield computer models to improve their representation of the water levels and ease of use. In particular, an algorithm has been developed, for use with the 0-D computer model, to account for the effect of the approach channel (Queen's and Crosby Channels and the Narrows) on the low water levels near to the barrage during operation. By this means the effectiveness of the 0-D program, to guide the engineering decision making processes, by sensitivity analyses, is much enhanced. A further development is the use of a 2-D hydraulic model as the basis for making a more physically representative determination of the annual energy yield for a 'base case' Barrage configuration.

Development of base case

During the Stage III studies the preferred layout of the base case barrage

TIDAL POWER

has continued to develop. A principal influence on this process has been the site conditions determined by site investigation which effected location and sluice type (ref. 2). Other refinements have been incorporated and the evolution of the parameters which significantly effect energy yield from the initial assessment in 1989 to completion of the Stage II studies in 1990 then to the present is shown below.

	Stage II (1989)	Stage II (1990)	Stage III (1991)
Line designation	3B	3C	3E
Distance upstream of Line 3 in 1985	700 m	1,100 m	1,300 m
Date of bathymetry	1977	1977	1990
Turbine type	Kaplan	Kaplan	Kaplan
Number of turbines installed	26	28	28
Turbine runner diameter	8 m	8 m	8 m
Draft tube length	40 m	40 m	32 m
Number of blades per runner	4	4	3
Runner rotational speed	50 rpm	50 rpm	50 rpm
Generator rotational speed	50 rpm	500 rpm	500 rpm
Generator rating	25 MW	25 MW	25 MW
Sluice type	Venturi	Venturi	Channel
Number of sluices installed	16	20	46

In this Paper, Kaplan is taken to mean a machine with double regulation (variable blade and variable distributor).

Power units
Turbine characteristics

In its earlier work to assess energy yield MBC used modern day, four bladed machine characteristics, supplied by Northern Engineering Industries (NEI) through their licence arrangement with Sulzer Escher Wyss (SEW). Whilst a high level of confidence could be placed in the applicability of the turbine characteristics, there was less confidence in the applicability of the pump characteristics used initially, which were derived rather than measured directly from model tests. This was necessary, at first, because the majority of the pumping operations will take place at very low heads and well removed from the best efficiency point, i.e. operating regimes in which manufacturers do not normally concentrate their efforts when carrying out model tests.

For the Stage III studies NEI-Parsons and International Research and Development developed a strategy for using three bladed turbines on the Mersey. Also, with the support of the Energy Technology Support Unit and

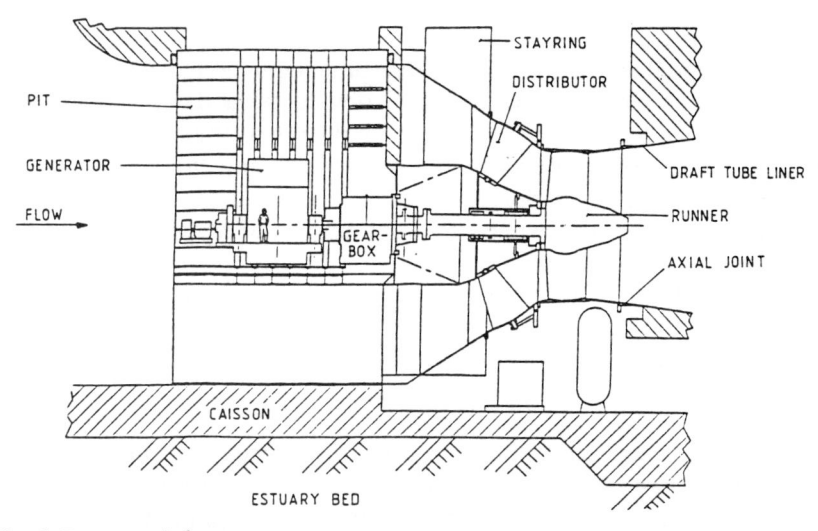

Fig. 1. Power unit layout

in parallel with the Stage III studies, separate arrangements were made through NEI for SEW to undertake pumping performance tests on three and four bladed model turbines, which apart from the number of runner blades, were otherwise the same (ref. 3). The performance of these models as turbines had been established previously by SEW and the characteristics supplied to MBC. It was confirmed from this work that three bladed pump turbines could be used to advantage on the Mersey Barrage.

Generators

From the Stage II studies (ref. 4) is was determined that there will be considerable advantage in terms of cost and to a lesser extent the generator intertia constant to interpose a step-up gearbox between the turbine and the generator and to install both in a pit with open access from above. This permits the use of a high speed generator operating in a dry and more accessible environment. Fig. 1 shows a cross-section of the Mersey Barrage power unit arrangement.

Sluices

Forty-six channel sluices will now be used on the Line 3E barrage rather than the lower number of submerged venturi sluices considered previously. Practical work, undertaken by BHR Group Limited within the Stage III studies (ref. 5) on a model channel sluice, confirmed that its average coefficient of discharge is circa 0.9 as water flows into the basin. There is a dependency between the coefficient of discharge and the water level conditions across the sluice which was taken into account during the model tests.

Computer models

During Stage II studies, as part of their philosophy to have direct control of its computer modelling work, MBC developed a flat surface or 0-D computer model to analyse the energy yield from a single, repeating tide. This model enables the operation of a given barrage configuration in ebb generation and ebb generation plus flood pumping modes of operation to be optimized to produce the maximum energy output or maximum revenue for a prescribed tariff structure.

Work was also carried out to develop a 2-D hydrodynamic model of the estuary that could be used, *inter alia*, for energy evaluation, barrage closure modelling and sedimentation studies.

0-D Model

Zero dimensional modelling is a powerful means of determining the important mechanical and electrical engineering parameters, e.g. turbine runner diameter, number of turbines, generator rating, number of sluices and submergence. The likely benefits from adding flood pumping to the basic ebb generation mode of energy production can also be determined. Although suitable modifications may be incorporated to simulate the estuary and basin, the 0-D model only approximately allows for hydrodynamic effects away from the barrage.

Figure 2 shows typical water level variations on both sides of the barrage during a tidal cycle for the ebb generation plus flood pumping mode of operation. By establishing the actual heads across the machines and sluices, i.e. in particular, allowing for the effects of the velocity head at exit from the draft tube and the subsequent expansion losses which reduce the apparent head represented by water levels at a distance from the barrage, the power

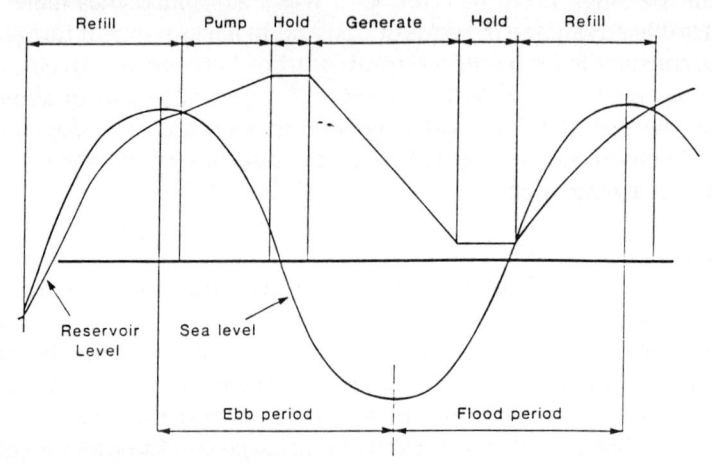

Fig. 2. Typical water level variations - Ebb generation plus flood pumping

and water flow rates throughout the tidal cycle can be defined. The summation of the instantaneous power flows, with allowance for the on barrage losses, defines the energy sent out from the barrage and the energy input required for flood pumping for a given tidal range. A tidal range histogram is then used to determine annual energy output from the barrage.

The 0-D model simulates this operating cycle by four main routines: basin refill; pump (optional); hold; and generate. Account is taken, in particular, of the on barrage losses the effect of the approach channel on flow of water from the barrage during generation and the relationship between water level and basin surface area.

Optimisation is achieved by systematically varying operating strategies (duration of pumping, start generation time and turbine blade angle variations during generation) and then selecting that which gives the greatest net energy yield (or revenue).

The on barrage losses simulated include those losses in the speed increasing gearbox, the generator and the power system between the generator and the grid.

The effect of the approach channel on the flow of water to and from the barrage was simulated by treating the approach channel as an open channel of appropriate section and length and modifying seaward tide levels accordingly (by application of the Manning formula) (ref. 6).

The relationship between water level and basin surface area was revised to that derived from the 1990 bathymetry survey of the Mersey Estuary.

2-D model

To simulate accurately the hydrodynamic effects away from the barrage an alternating difference implicit finite difference 2-D model, DIVAST, was used. This covers the whole area of the Mersey Estuary and Liverpool Bay, (see Fig.3) with the wetted area divided into 20,000 discrete cells, 150 m x 150 m size in plan. This model contains the same barrage operating routines as the 0-D model with the exception that optimisation of the operating path is not included and so the operating path must be defined in the input data.

The DIVAST model, based on the work of Professor Falconer of Bradford University was originally adapted to model the Mersey Estuary and approach channel with a Barrage during Stage II. Also, as part of the Stage II studies, the bathymetry of the Mersey Estuary and parts of Liverpool Bay were resurveyed. Previous surveys were carried out in 1977 with a partial survey in 1984.

In the Stage III studies the 2-D model was updated to include the revised 1990 bathymetry and the area of model coverage was extended further into Liverpool Bay. This was done to minimise boundary effects. Also, grid orientation was rotated to suit the orientation of the Line 3E barrage. To calculate the annual energy yield from the 2-D model single repeating tides

TIDAL POWER

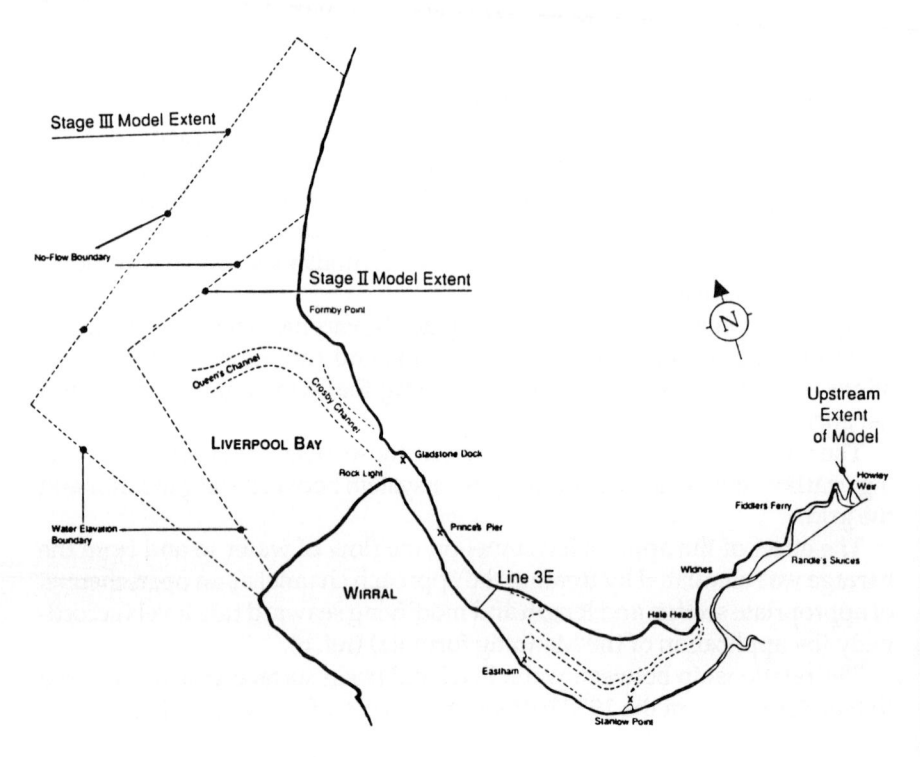

Fig. 3. DIVAST model area

were run for various ranges and these results were then analysed using the same histograms of tidal range occurrence as those used with the 0-D program. The hydrodynamics of the 2-D model were validated using measured water level and velocity field data and the results from a physical model (ref. 7).

Interaction of 0-D and 2-D models

The 2-D and 0-D models, both of which run on in-house desktop computers are now used interactively; optimized operating strategies from the 0-D model are input to the 2-D model and in turn, water level data are fed back to the 0-D model, to allow the approach channel effect algorithm to be tuned up.

For the future it is intended that this manual interaction will be achieved by interfacing the two computer programs directly. This will then provide a very powerful tool for studying, in detail, the operating strategies for the barrage, recognising water level and tariff constraints. Fig. 4 shows the at

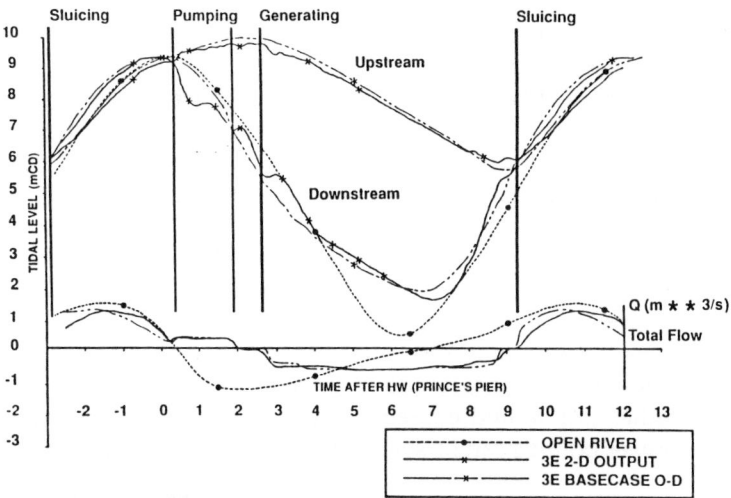

Fig. 4. Line 3E barrage mean spring tide

Barrage water levels obtained from the 0-D model, compared with those from the 2-D model, for the same 8.4 m mean Spring tide at Princes Pier.

Energy yield results
Base case

The Stage III studies assumed a base case set of barrage parameters for the Line 3E location, derived largely from the Stage II studies with the addition of three bladed turbines and channel sluices. As a result of the Stage III studies, including sensitivity analysis on the major parameters, the base case was revised but the principal parameters, as given above remained unchanged.

Energy yield

From the interactive use of the 0-D and 2-D models the net energy yield for the ebb generation plus flood pumping mode of operation was

Tide	Tidal Range Princes Pier(m)	Pumping Energy(MWh)	Gross Energy Output(MWh)	Net Energy Output (MWh)
Low Neap	3.2	327	691	364
Mean Neap	4.5	289	1,149	860
Mean	6.5	229	1,904	1,675
Mean Spring	8.4	258	2,899	2,641
High Spring	10.0	71	3,474	3,404

Using the annual histogram of tidal frequencies for an average year the predicted annual net energy output is 1.31 TWh, with 0.17 TWh required for pumping. The gross energy output is 1.48 TWh.

For a given tide, a prescribed operating regime must first be derived from the 0-D model in order to use the 2-D model to evaluate the energy yield. This is due to the fact that the 2-D model does not optimise the operating path. Because the 0-D model does not fully represent what will happen physically in the estuary, it may be expected that, if the 2-D model were given an optimum operating path allowing for hydrodynamic effects away from the barrage, the energy yield predicted will increase. In this sense the 2-D model predictions may be considered to be slightly sub-optimal. There are a range of other factors which will also tend to modify the net energy output.

An assessment of the magnitude of these factors was carried out in the Stage III studies (ref. 8) and consequently, a most probable estimate with upper and lower bounds can be set for the net annual energy output.

	Net energy output (TWh/a)
Upper bound	1.50
Most probable	1.39
Lower bound	1.31

Comparison with Stage II studies

As a result of the engineering changes, the better representation of the Queen's Channel effect and the interactive use of the 0-D and 2-D models it is interesting to note that there has been no gross change in the energy output from that predicted during the initial assessment in 1989.

	Net Energy Output (TWh/a)		
	Stage II (1989)	Stage II (1990)	Stage III (1991)
Ebb generation only	1.24	1.27	1.20
Ebb generation plus flood pumping	1.43	1.48	1.39

During Stage II the predicted output increased due to the inclusion of a further two turbines in the proposed layout. However, there are several factors which combine to account for the 6% reduction in the annual energy output now estimated compared to that on completion of Stage II

- hydrodynamic effects away from the barrage have been modelled more accurately, in particular, the approach channel
- the preferred line has changed from Line 3H to Line 3E, a move upstream of 600 m
- channel sluices of reduced hydraulic efficiency though greater in number than the venturi sluices have been introduced.

- measured pump characteristics have been introduced to the slight detriment of the previously derived pump characteristic
- the model bathymetry has been updated from that of 1977 to that surveyed in 1990.

Revenue

At present, as the situation following the privatization of the electricity supply industry continues to settle down it is only sensible to make an assessment of the energy related revenue necessary for project viability. This was done on the basis that the Mersey Barrage will be included in the Non-Fossil Fuel Obligation and that, as a result, the cost of the energy for flood pumping will be the same as the value of that sent out from ebb generation. Any change to this unitary tariff will tend to modify the annual energy yield and hence revenue. However, a doubling in the cost of pumping energy throughout the year will only lead to a less than 1% reduction in the net energy yield.

A further effect upon annual energy yield and thus revenue is the 18.6 year tidal cycle. This will cause a sinusoidal variation in the energy yield of +5% about the mean year, with peaks in 1997 and 2016 and troughs in 2007 and 2025.

Hence, in the Stage III studies the revenue was assessed by taking the base case energy yield scenarios given above and applying the 18.6 year tidal cycle variation.

Sensitivity analyses

To check the robustness of the results for the base case, a number of sensitivity analyses were carried out - by varying one parameter at a time from those defining the base case. With the much improved correlation between the 0-D and 2-D models the sensitivity analyses, for convenience, were carried out using the 0-D model.

Machine type

Consideration was given in the Stage III studies to the use of three and four bladed pump turbines of the Kaplan type. The results of this analysis were

	Percentage annual energy output	
	Ebb generation	Ebb generation plus flood pumping
Kaplan, 3 blades	100	100
Kaplan, 4 blades	99.1	98.5

TIDAL POWER

The results of the subsequent sensitivity analyses, reported below, are for a three bladed Kaplan turbine which for the Mersey is expected to show slightly improved performance at slightly reduced cost compared to a four bladed turbine runner.

Number of machines

The results of this analysis (see Fig. 5) showed a progressive increase in annual energy output with the number of machines. Consequently, it will be economic rather than energy yield criteria that determine the number of installed machines.

Turbine runner diameter

It is generally accepted that the largest practicable number of machines should be installed in a tidal energy scheme. With the Line 3E bed levels and

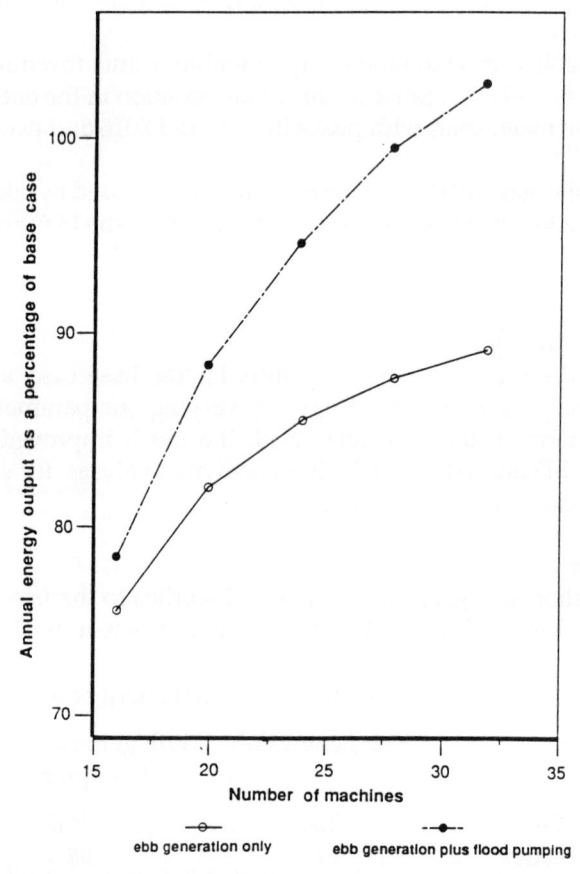

Fig. 5. Sensitivity to number of machines: 8 m diameter machine

the amount of excavation or foundation improved work that may be necessary if more than 28 machines of 8 m diameter were to be installed, consideration was given to the installation within a similar space of a greater number of three bladed Kaplan type machines but with reduced diameter. The results of this analysis were

Runner Diameter (m)	Number of Machines	Percentage Annual Energy Output Ebb Generation Plus Flood Pumping
8.0	28	100
7.0	32	94.7
6.0	40	90.1

Approximate estimates indicate that any reduction in cost arising from the use of smaller machines will not be sufficient to offset the consequent reduction in energy output.

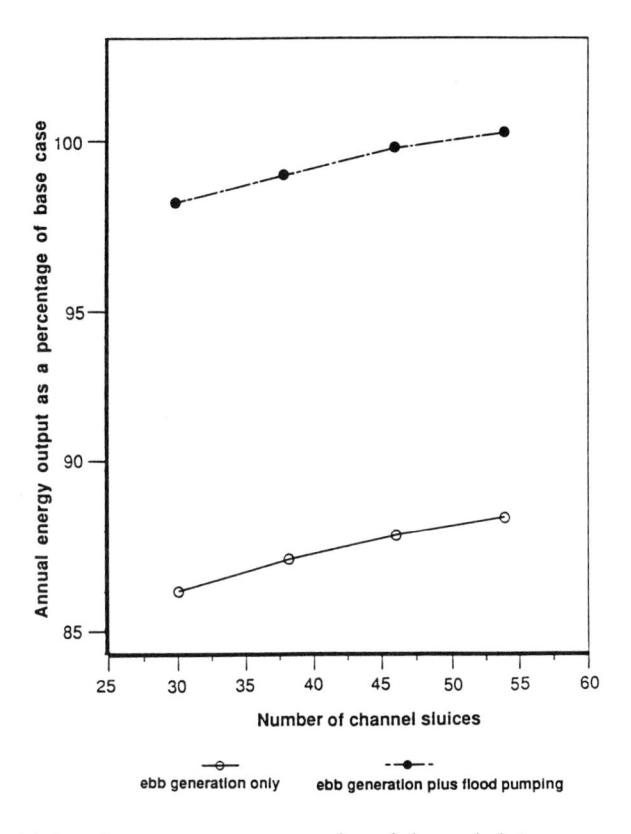

Fig. 6 . Sensitivity of energy output to number of channel sluices

TIDAL POWER

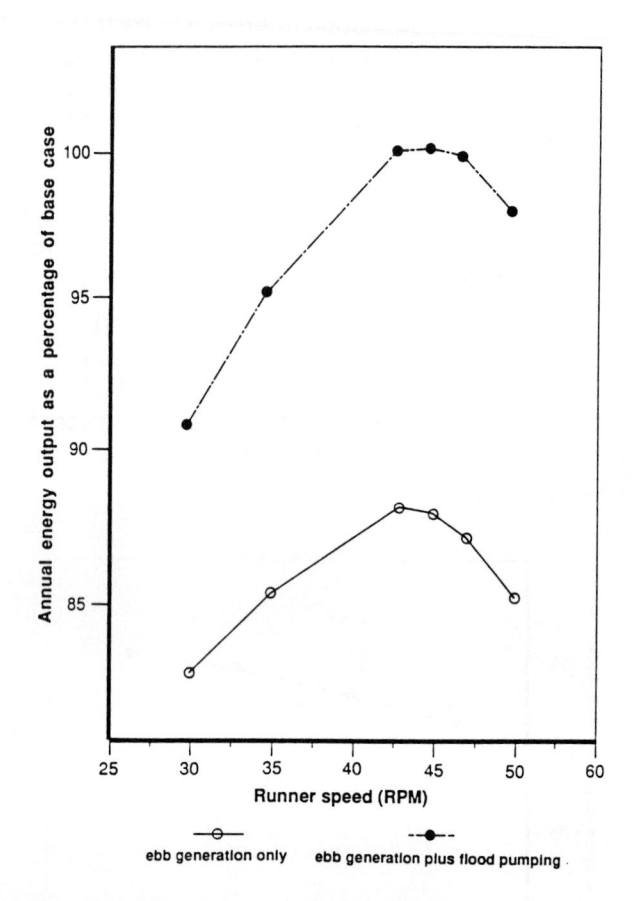

Fig. 7. Sensitivity of energy output to machine speed

Sluice capacity

This analysis (see Fig. 6) showed that changing the number of channel sluices from 46 (with the 28 machines acting as sluices when refilling the basin) gives only a small marginal change in annual energy output. There may be scope to reduce the number of sluices towards 42, for which the small loss in energy output may be offset by a corresponding reduction in cost.

Machine speed

The results of this analysis, (see Fig. 7) for the three bladed Kaplan turbines, showed that the optimum speed for ebb generation plus flood pumping is 50 rpm, the optimum speed for ebb generation only is 48 rpm. As may be seen from Fig. 7 there will only be a minimal effect on energy output as a result of choosing 50 rpm for the ebb generation only speed.

Generator rating

The results of this analysis (see Fig. 8) showed that as the rating reduces from 25 MW towards 20 MW, the annual energy output starts to reduce with the rate of decrease becoming rapid for ratings below 20 MW. Above 25 MW the rate of increase of annual energy output reduces markedly and begins to plateau.

Draft tube length

For a prescribed draft tube geometry, particularly its rate of divergence, the effect of varying the draft tube length is to change the leaving loss. Increasing the rate of divergence to shorten the draft tube is likely to promote flow separation with a concomitant large increase in the hydraulic losses and flow stability in the draft tube. Therefore, for this sensitivity analysis the rate of divergence was held constant with varying draft tube length.

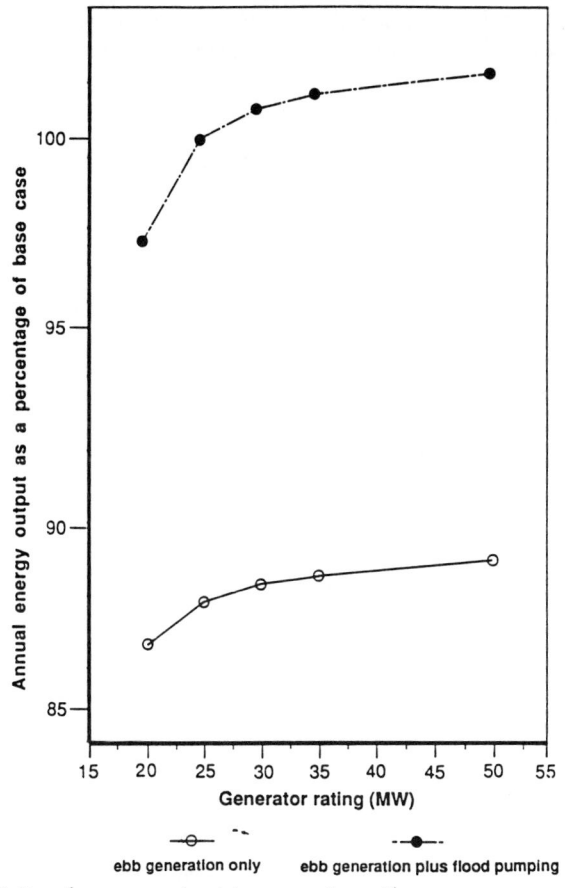

Fig. 8. Sensitivity of energy output to generator rating

TIDAL POWER

The results of this analysis show

Draft tube length (m)	Percentage annual energy output: ebb generation plus flood pumping
28	99.2
32	100.0
36	100.8
44	101.5

The preferred length of 32 m was selected. Below this foundation problems arise whilst above this length the benefit is exceeded by the construction cost increase.

Maximum allowable reservoir level

For flood defence and drainage reasons from time to time there may be a need to restrict the extent of flood pumping to ensure upstream water levels are not excessive. The sensitivity of the energy yield to maximum reservoir level was determined by terminating flood pumping at the maximum level or when the maximum yield for a tide was reached, whichever came first.

Maximum reservoir level (mOD)	Percentage annual energy output: ebb generation plus flood pumping
5.37	100.0
4.87	98.5
4.37	96.2

The highest astronomical tide level at Prince's Pier is in fact 5.37 m OD and a significantly lower level restriction on flood pumping is unlikely, although further studies are required to assess the effects on tidal propagation in the upper reaches of the Mersey beyond Widnes and the consequences of any increases in water levels.

River inflow

Although very small by comparison to tidal flows there is an inflow to the basin from the River Mersey and other sources. Typically, the flow averages 50 m^3/s through the year. The sensitivity to this parameter was shown to be small.

River inflow m^3/s	Percentage Annual energy output: ebb generation plus flood pumping
0	99.5
50	100
100	100.9

Tariff Ratio

To indicate the effect the tariff structure is likely to have on the operating regime of the barrage, a sensitivity analysis of annual energy output to tariff ratio showed

	\multicolumn{4}{c}{Ratio of cost of import (for pumping)/value of output}
	3	2	1	0.5
Percentage import of base case	5.8	26.3	100.0	170.2
Percentage annual energy output	94.3	99.4	100.0	100.8
Percentage annual net revenue	92.3	94.8	100.0	115.6

NB. The annual net revenue was evaluated by using a unit tariff with the cost of import (for pumping) factored as appropriate.

It is apparent that for quite large changes the energy output is inssnaitive to the quantity of power imported for pumping. Thus, the extent of pumping undertaken may well be influenced by economic and environmental as well as financial considerations.

Conclusions

The Stage III studies, for a Line 3E Mersey Barrage, show that the energy predictions made in previous studies are largely unaffected as more details of the hydrodynamic behaviour of the estuary and the use of three bladed machines and channel sluices are introduced into the modelling of the scheme.

Three bladed turbine runners are better suited to the head available on the Mersey, particularly after allowing for the approach channel effect. These machines with measured low head pump characteristics have been used in the Stage III analysis of annual energy output.

For the Line 3E barrage, the most likely net annual energy output is 1.39 TWh with upper and lower bounds of 1.5 TWh and 1.31 TWh respectively. This compares with the 1.48 TWh predicted from the Stage II studies. The reduction in net energy output is in part due to Line 3E being 600 m upstream, compared with Line 3B, but the better representation of the approach channel is the more dominant factor. In this respect, the interactive use of 0-D and the more sophisticated 2-D modelling techniques, with algorithmic refinements in the 0-D model to allow for the approach channel effect, gives greater confidence that the net energy output predicted from the Stage III studies will be achieved.

The Stage III studies indicate a preference for 8 m diameter three bladed Kaplan type turbines, operating in the ebb generation plus flood pumping mode. These machines have an optimum speed of 50 rpm, with generators of 25 MW rating. More detailed study may confirm the potential for a slight reduction in generator rating.

Annual energy outputs have been predicted for varying machine numbers and diameter. However, economic and, perhaps, environmental considerations are likely to dominate any further examination of these options.

The Stage III studies suggest the Line 3E barrage may be slightly oversluiced from a purely net annual energy output point of view; engineering and economic factors are likely to outweigh this tendency.

References

1. WILSON E. A. and POTTS R. Initial assessment of the energy yield and economic aspects of a Mersey Barrage. *Proceedings of the Institution of Civil Engineers Third Conference on Tidal Power, November 1989*, pp. 143 - 154. Thomas Telford, London.
2. JONES B. I. et al. The Mersey Barrage - civil engineering aspects. *Proceedings of the Institution of Civil Engineers Fourth Conference on Tidal Power, March 1992*. Thomas Telford, London.
3. POTTS R. and WATSON D. *The benefit of flood pumping to tidal energy schemes.* Department of Energy Contractor Report BTSU TID 4103, 1992.
4. MERSEY BARRAGE COMPANY. *Tidal power from the river Mersey: a Feasibility Study Stage II Report.* Department of Energy Contractor Report, ETSU TID 4071, 1991.
5. CLARKE A. and READE A. *Mersey Barrage experimental determination of sluice loss/discharge coefficients.* HHR Group Ltd, August 1991 (Unpublished report).
6. WATSON D. *Mersey Barrage Stage III 0-D Energy Modelling.* International Research and Development Ltd, August 1991 (Unpublished report).
7. ALLERY C. *Mersey Barrage Stage III DIVAST Hydrodynamic Modelling.* Mersey Barrage Company, July 1991 (Unpublished report).
8. PORTER J. H. *Mersey Barrage Stage III 2-D Energy Modelling.* Mersey Barrage Company, July 1991 (Unpublished report).

9. The Mersey Barrage – indirect benefits: the economic and regional case

P. J. COCKLE, Independent Consultant, formerly Travers Morgan Economics and J. J. McCORMACK, Mersey Barrage Company

A major cost-benefit exercise, a regional and an Exchequer impact assessment were performed. The cost-benefit net present values (NPV) (1991 prices) of market and non-market streams revealed: capital and operating cost (-£799 million), electricity output (£310 million), CO_2 savings (£627 million), security of energy supply (£18 million), amenity benefit/blight (-£1 million), tourism (£5 million), leisure (£7 million), waders and wildfowl impact (-£8 million) and a road crossing (£132 million). The total mid-range estimate of £370 million confirms the Barrage would increase Britain's welfare. During construction the Barrage would support up to 2,100 Merseyside jobs and 600 in operation. The cumulative gains to the Exchequer are £300 million.

Introduction

The Mersey Barrage Company (MBC) appointed Travers Morgan Economics (TME) in September 1990 to provide an economic assessment of the Mersey Barrage proposal. The study was commissioned in the context of the uncertainty regarding the support that might be available to MBC under the Non-Fossil Fuel Obligation (NFFO).

On a strictly commercial basis the revenues and margins that could be earned from harnessing the tides in the Mersey Estuary to produce electrical power were, on any reasonable projection of energy prices, unlikely to be attractive to the private sector. Nevertheless there were apparent non-market advantages (and some costs) which could make the project worthwhile from a social perspective.If such non-market net benefits existed then assistance, in whatever form, might be worthy of consideration by a number of Ministries. A Ministerial Committee was set up to review the findings of this study.

The brief
The first, in a two phase study, identified the costs and benefits requiring evaluation, determined a broad approach and established whether any

TIDAL POWER

existing public grants applied to the Barrage. The aim of this phase was to gain agreement from the various Ministries on issues of principle or the approach and methodology adopted. The second phase involved the evaluations.

Since the aim of the study was to illuminate discussions between the private and public sector on the private and social net benefits of the scheme, it was considered essential to follow Treasury rules for evaluation of public sector projects. Among other things this meant accepting an 8% discount rate.

Following a review of public grants it was concluded that there was, currently, little other than the NFFO capable of financially assisting the project.

The study

The evaluative exercise was conducted in three parts

(a) an energy and non-energy cost-benefit study
(b) a regional economic impact assessment
(c) an Exchequer impact study.

These were intended to provide three views of the same exercise. The cost-benefit exercise provided the principal basis for deciding the merits of the Barrage, since the regional study simply focused on additional employment and output which arose from investment in Merseyside. Under Treasury rules the presumption of cost-benefit analysis is that resources are fully employed and one is comparing the merits of alternative uses. Though shadow prices and wage rates could have been incorporated in the cost-benefit study to deal with the benefits of employment, it was felt that decision making would be helped most by a simple statement of the extra activity and employment resulting from the Barrage. The Exchequer study was intended to demonstrate that through higher economic activity in the region, there would be less call on the public purse and higher tax revenues. This approach did not seek to distinguish private from public sector net benefits nor whether there were any superior ways of achieving the same net benefits. Since time and space does not permit a detailed description of the Exchequer exercise and its results, let it be recorded that over the period 1990-2005 it was estimated that there could be a cumulative financial saving (more revenue plus less expenditure) of between £257-303 million.

Cost-benefit study

Identification

The process of identifying costs and benefits from the Barrage was carried out in an even handed fashion. Costs were sought along with benefits. The exercise was focused upon areas which, prima facie, were likely to have major cost or benefit implications.

At this stage in the project many of its details were firmly settled but still subject to review. The study was therefore structured so that if new details emerged results could be rapidly adjusted. The study was conducted in 1989 prices, since this was the then price base for the construction and operating costs available to MBC. It was also assumed initially, that annual output would be 1.5 TWh per annum. In the event costs were revised in 1991 and likely output reduced to 1.39 TWh per annum. Thus, towards the end of the study results were re-scaled to reflect these changes.

Construction and operating costs

The costs of construction and operation were established from the engineering studies with the recommendation that 120 years should be considered as the nominal life of the Barrage. In principal, however, this would be a perpetual project and therefore no decomissioning costs would arise. Where possible all transfer payments were excluded from the costs to represent a clear resource measurement of the project. The comprehensive costs were considered under 4 headings

- Development
- Detailed design and preparation
- Site construction
- Operating costs.

Generation benefits

The generation benefits comprise the value of the electricity produced by the Barrage. Though generation benefits did not form an essential part of the study, it was felt that since they were the reason for the project, they should not be omitted from the presentation. It was also essential to validate whether on a reasonable energy price forecast the Mersey Barrage was in fact unviable. As will become apparent a little later, for methodological reasons, the only way it was felt one could determine the social benefits of the Barrage's savings of harmful gases, normally emitted by fossil fuel generation, required a valuation of the Barrage's electrical power output.

Gas emission savings

Fossil fuel generation results in 2 broad types of harmful gas emission

(*a*) acid rain gases (SO_2, NO_x)
(*b*) global warming gases (CO_2).

Approximately three-quarters of UK electricity generation comes from conventional hydrocarbon sources (coal, fuel oil and gas). Of this approxi-

TIDAL POWER

mately 90% is due to coal. In the UK, power plants are responsible for about 70% of SO_2 emissions, 30% of NO_x emissions and 33% of CO_2 emissions (1989 figures).To the extent that the Barrage output displaces consumption of fossil savings in harmful gas emissions. The question is what is this worth?

The theoretical aspects and the work done to date on the costs imposed by contaminated air or higher CO_2 levels was reviewed. The scope for inferring society's valuation of "clean air" generation was somewhat limited. If such valuations were indeed known, the government could construct a policy which assured that a balance was struck between the economic benefits of fossil-fuel generation and that of having "clean air". What they have been forced to do instead, is to join other governments in declaring emission targets and mainly through regulatory measures encourage cleaner electricity generation. This is the case with acid rain gases where Britain has a legally binding commitment through an EC directive on large combustion plants to make cuts in SO from a 1980 base of 20% by 1993, 40% by 1998 and 60% by 2003. In 1984, Britain set out to achieve a 30% reduction in NO_x from 1980 levels by the end of the 1990s. The Government has also announced that Britain is prepared, if other countries take similar action, to set itself the target of returning emissions of CO_2 to 1990 levels by 2005. Thus, a programme of flue gas desulphurisation is in train, gas, a cleaner fuel, can now be used widely for electricity generation and the NFFO has been introduced. While in the main these initiatives go a long way to meeting the acid rain problem they still leave, in the view of many commentators, a great deal to be done on CO.

The approach has been to establish a market price for electricity consistent with the government achieving its CO targets. This was done by considering what level of Carbon Tax would be needed to reach the targets and its consequences for electricity prices. Since the Barrage would be exempt the tax, the difference between the pre-tax price and the post-tax price provides a measure of the valuation of the CO savings from the Barrage.

Security of supply

Energy, labour and capital are crucial resource inputs to most economic activities. Large fluctuations in energy prices have harmed economic output in the western world, causing major misallocation of resources. Cartelised oil prices have not only contributed to two of the major economic recessions since the war but, by setting energy prices at high real levels, have caused labour and capital to be combined in ways which subsequently proved to be inefficient when the cartel power was dissipated and prices fell. In 1974 UK GDP growth fell 1.6% and 0.8% in 1975, while in 1980 and 1981 the falls were 2.2% and 1.1%. Both periods of decline followed on the heels of a large increase in oil prices. Coal and rail strikes in the UK and accidents in the North Sea have shown that energy vulnerability is not an imported problem.

The 1984/85 coal strike knocked 1% off GDP growth. Finally, Britain's oil and gas supplies are finite and will eventually need to be replaced - that output will disappear. UK oil supplies are likely to run out entirely in 40 years and gas in 65 years. Either alternative indigenous sources need to be found or non-energy output must be transferred from domestic consumption to exports to pay for imported oil. Thus, the Barrage by providing an indigenous energy supply and diversifying Britains energy sources is making a contribution to energy security. While this is a benefit it is certainly not easy to measure but needs to be weighed in the favour of the Barrage.

Amenity and blight

Prior to the period of construction certain areas of Liverpool might be subject to planning blight, during construction some areas will suffer the intrusion of the works, but after construction there is the prospect of the impounded lake adding an amenity benefit. If the impounded area of water were thought to provide a more attractive view than the present inter-tidal zones then this would be reflected in property values. Blight would be reflected by a fall in values. Local surveyors provided valuations under certain scenarios.

Leisure and tourism

The impounded lake provides the scope for an increased recreational use of the Mersey. The Mersey is currently used by a range of water enthusiasts for sailing, canoeing, powerboating, rowing, outdoor swimming, waterskiing, wet biking and windsurfing. What was required was an evaluation of what additional use would follow from the development. Surveys were conducted in the area to determine such values.

It is proposed that there should be a visitor centre at the Barrage. Since this would be the first tidal power Barrage in the UK, the second in Europe, it is likely to have appeal as a tourist attraction. A method was sought to evaluate that site as a tourist attraction. A travel cost method was used as part of the valuation.

Flood control

The Barrage can regulate the tidal ranges and so be used as a flood control device. It could thereby reduce the potential costs, direct and indirect, of flooding. The prospects of flooding may be increased by global warming, one consequence of which is to raise sea levels. Varying estimates exist but one respected authority suggested that sea levels might be 650 mm higher by 2050. The benefits of this flood alleviation role were reviewed.

Waders and wildfowl

Since the impounding of a large body of water with a lesser tidal range would reduce the inter tidal area where birds feed, there is a risk of them being adversely affected. Much has been written on the subject of economic valuations of environmental costs and benefits and at the time of doing the study it was not clear what the Whitehall view was on such valuations, some of which are controversial. Since then the Department of Environment has published guidelines on techniques to use. It was felt that as this issue was likely to impose a cost on the project an attempt at quantification should be made but there was uneasiness about the twilight status of some of the techniques that were in existence. Two exercised were, therefore, conducted, one which assessed the cost of providing an alternative habitat and one which sought to identify society's valuation of the site through the contingent valuation method (CVM)

Shipping

The impact of the Barrage on shipping, both during construction and after, was undergoing detailed study in parallel with this work. Rendel Parkman were drawing together this research, part of which included a cost-benefit study which was used in this study.

Crossing

Although the main purpose of the Barrage is to generate electricity, it is feasible to use the structure to provide a crossing over the Mersey. Though a crossing might take various forms — pedestrian way, light rail link or a road — it was decided to investigate, in broad terms, the value of the Barrage as an additional Mersey road crossing. A three lane road with lifting bridges across the locks and a variable message sign system to divert traffic to the tunnels when the locks are open to shipping was assessed. It was necessary to make assumptions about the degree of interuption to road traffic from lock operations. Though there are other crossings of the Mersey (ferries, tunnels, Merseyrail and the Runcorn Bridge) evaluations focussed on the number of trips that would be likely to divert to the Barrage because of time saving and then determine the value of that time saved.

It should be stressed that the crossing represents an option for the Barrage but it is not necessarily part of MBCs plans.

Methodology and results

Below there is a more detailed but selective description of methodologies used and also a summary of the results.

Construction and operating costs. The capital costs of construction were placed at £880 million in August 1989, at then current prices, but revised to £966 million in January 1991 at 1991 prices. The estimate of operating costs associated with the Barrage moved from £11 million per annum to £17.6

million. Completion was planned for year 2000 with the heavy construction costs scheduled in the second half of the 1990s. Though the exercise was conducted originally on the 1989 price base, scaling up to 1991 prices produced a present value of capital and operating costs of -£799 million, using an 8% discount rate and an operating period of 120 years. During the period of operation, turbines would be replaced every 40 years.

Generation benefits. The revised annual output of the Barrage is 1.39 TWh per annum. To compute the present value of electricity revenues forecasts were required of electricity prices going well into the future. A series of model based forecasts was commissioned from the International Energy Service of DRI/McGraw-Hill. First, a broad macro-economic environment was defined and assumptions made about world oil prices based upon judgements on how world supply demand balance would evolve. The horizon was year 2030 and the above developments fell into phases. DRI/McGraw-Hill also made a series of assumptions regarding developments in the Electricity Supply Industry after 1992 when coal contracts come up for renewal. Account was also taken of the new attractions of gas as a fuel source for electricity. They constructed a base price for UK electricity which reflected the capacity and energy supply (fuel cost) pricing mechanism of the pool price. This produced a coal burn generation price of 1.22 p/kWh and a gas generation price of 0.81 p/kWh for 1989. The model was then used to project this price base forward taking into account growth in energy demand and substitution between energy sources. This was the BASE scenario.

Further runs were made to take account of a less rigorous application of the EC Directive on large scale combustion plants (LOW scenario) and faster growth of world oil prices (HIGH) scenario.

Using these price projections the present value of the Barrages' future output at be

LOW Scenario	£269 million
BASE Scenario	£290 million
HIGH Scenario	£297 million

Gas emission savings. All the above price scenarios incorporated prices which assumed substantial measures to reduce SO_2 and NO_x. There were no specific measures to deal with CO emissions. The model was used to determine the level of carbon taxes which would achieve UK government emission targets by the due date. The tax rate that emerged was £25/tonne of carbon incrementing by that amount each year to 2005. Thus by 2005 the tax rate was £325/tonnes of carbon at 1989 prices. The model took account of generation and non-generation fuel uses and their demand and supply sensitivity to price. The present value of Barrage output on these prices was

TIDAL POWER

£944 million. This was based on the HIGH scenario. Deducting the normal generation revenues produces a CO_2 benefit of 627 million.

Waders and wildfowl. The first element of the study looked at the cost of providing an alternative habitat in compensation for the area lost to waders and wildfowl. The legal acceptability of creative conservation is still not sharply defined though there are precedents. Various sites were suggested by Environmental Resources Limited which they regarded as having potential for creative conservation. Each was reviewed and estimates of capital works and annual management charges were assessed. In several instances there were savings on flood prevention since the process of creative conservation required flooding. Though still more information will be needed for a full evaluation, taking a rather generous view of how much land is needed produced a net present value of -£21 million while a tighter criterion produced an estimate of £8 million.

The second element of the study was a CVM exercise in which users and non-users were surveyed and asked to give their value of the site. In the non-user survey (245 usable replies), individuals were selected at random, contacted by letter and telephone and asked how much they would be willing to pay into a trust to preserve the area. A similar user survey focused on bird-watchers and wildfowlers. The exercise was successful in many areas but there were several practical and theoretical problems and it was decided not to incorporate these results into the exercise. First, though one was surprised how seriously people took the challenge of determining their willingness to pay, it was felt that the technique needs to take account of all other hypothetical expenditures that the individual might make (saving the whale, rain forests etc.). Second, despite much goodwill from local bird-watcher and wildfowler organisations it was not possible to draw randomly from these groups and there was a strong element of self-selection and hence bias in the results.

Crossing. A COBA type exercise was conducted in the area. A major road network was defined and traffic survey data obtained from the Merseyside Passenger Transport Authority. Future traffic patterns were generated from the national road traffic forecasts and from local trip end growth. High and low growth scenarios were developed. The assignment of the traffic matrix to the road network was calibrated by reference to actual flows across the Mersey. Barrage tolls were assumed to be equal to tunnel tolls. Rendel Parkman supplied the extra costs of providing the road crossing (£7.55 million) plus additional land costs (£0.45 million) to produce a cost of £8 million. Finally it was assumed that access would be reduced by 30% due to reflect interruptions caused by ships passing through locks. The NPV estimate of the crossing ranged from £92 million to £140 million.

Regional impact study
Aim of exercise

The Barrage represents a large capital investment in an area which despite improvements is structurally depressed. Part of the Barrage expenditure will go on locally supplied materials and labour. These additional incomes will be spent in the area expanding incomes in a local multiplier process. Given the sustained period of Barrage investment, local incomes will receive a lift for some time and as a consequence encourage additional local investment. The aim of this exercise was to quantify this process.

Liverpool model

To do this a model designed by Liverpool Macro-Economic Research Group was used which represents and forecasts the economic activity of the Merseyside region. The model treats Merseyside as if it were a separate economy trading with the rest of the world: part of its economy is the traded sector, part is non-traded. Output of traded sector depends on outside demand while output of the non-traded sector is driven by Merseyside demand. The local content of the Barrage investment and employment was carried through to an increase in non-tradable output and employment, this boosts local incomes and demand and further local output. The output of the Barrage is a direct addition to the traded output of the Mersey economy. Increased output in this sector also raises local demand in a similar multiplier fashion.

Results

Three scenarios were specified with varying local content to the investment. The broad nature of the economic effect of the Barrage is to produce a surge in local GDP and employment which then settles down to a permanently higher level than before the Barrage. At its peak, in 1996, the Barrage adds between 0.6%-0.8% to Merseyside GDP. Peak additional employment, in 1997, is between 1,700 and 2,250. In the long run GDP is some 0.3% and employment around 600 higher. For about 4 years employment is around 2,000 higher than it might otherwise be.

Conclusion
Considerable social net benefits

The Barrage therefore was shown to have substantial net benefits making it a socially desirable investment. These are summarised in Table 1.

Substantial regional impact

The regional impact has a relatively prolonged plateau of activity which

TIDAL POWER

adds between 1,700 to 2,250 jobs and 0.6% - 0.8% to Merseyside GDP. In the longer round it adds 0.3% to Merseyside GDP.

Table 1. Composition of Projects NPV, £M (January 1991 prices, discounted 8%)

Range	Low	Base	High	$CO_2$1	$CO_2$2
Costs[3]	-799	-799	-799	-799	-799
Generation	287	310	317	944	944
Security	12	18*	25	18	25
Amenity	-1	-1	-1	-1	-1
Leisure	5	7*	9	7	9
Tourism	2	5*	8	5	8
Flood Control	1	1	1	1	1
Conservation	-21	-8	-8	-8	-8
Without crossing	-514	-467	-448	167	179
Crossing[4]	92	116*	140	116	140
Amenity	16	16	16	16	16
With crossing	-406	-335	-292	-299	-335

Key

*Mid range

[1] Except for generation benefits all other NPVs are base values

[2] Except for generation benefits all are high values

[3] Capital and operating costs

[4] All values assume a 30% restriction

10. The Mersey Barrage – finance and promotion: the way forward

C. J. ELLIOTT, Barclays de Zoete Wedd and J. J. McCORMACK, Mersey Barrage Company

With long term contracts and the removal of the 1998 cut-off date for the Non-Fossil Fuel Obligation and Levy, large scale renewable energy schemes can be financed.

Introduction

On 18 September 1991 the Secretary of State for Energy said

"Renewables have the potential to contribute simultaneously to meeting our energy needs, to improving security of supply and to reducing environmental pollution. I am proud to have been the first Secretary of State for Energy to make a Non-Fossil Fuel Obligation Renewables Order, and, through this, to help renewables to enter the commercial electricity generating market".

The first Renewables Order was intended to secure 102 MW from 75 small scale and short term projects - nearly 50% of which were existing schemes - by 1995.

On 5 November 1991 Colin Moynihan, Parliamentary Under-Secretary of State for Energy, said

"We believe renewable energy has an important role to play in helping us tackle the serious threat of global warming... no Government in history has ever done so much to promote renewable sources of energy ... and it has done so in two ways... The second way .... has been to create for the first time an effective market place for renewable energy through the establishment of a Non-Fossil Fuel Obligation".

Mr Moynihan went on to announce the second renewables order intended to secure 457 MW of renewable energy in the period 1 January 1992 to 31 December 1998 - from a further 122 short term and small scale projects.

The Mersey Barrage could have added 230 MW to either of these orders, but the NFFO is not being exploited to promote long term or large scale projects as was originally intended - and unless the 1998 cap imposed in March 1990 is removed we would be surprised if there will be any further renewable orders.

TIDAL POWER

The Mersey Barrage Company (MBC) contends that the Non-Fossil Fuel Obligation - as by the Electricity Act - free of the later imposed 1998 cut-off date and utilising the payment mechanisms to advance levy during construction with long term contracts - allows the development of large scale renewable energy projects - such as tidal schemes - and will secure a real reduction in dependence of fossil fuelled generation, real security of supply and diversity of supply.

If the Government

(a) secures the removal of the 31 December 1998 cut-off date for levy and levy contracts as soon as possible

(b) confirms that it will take the necessary action to enable long term NFFO contracts to be placed by the Non-Fossil Fuel Purchasing Agency (NFPA) with contract length related to technologies

(c) presses the Commissioners for the Inland Revenue to ensure that the majority of the capital cost of barrages will be classified as plant and machinery for taxation purposes and will, therefore, attract 25% per annum writing down allowances;

(d) ensures that the real benefits of renewable projects

   (i) no CO_2
   (ii) no SO_2
   (iii) no NO_x
   (iv) no radioactive waste
   (v) security of supply
   (vi) diversity of supply
   (vii) the preservation of finite fossil fuels for more appropriate uses
   (viii) regional regeneration, job creation, tourism and leisure in the case of the larger long term schemes such as tidal
   (ix) long-term supply - over the life of the plant - at prices below those required by fossil or nuclear powered stations over their life time
   (x) for most tidal schemes the possibility of a road crossing

are properly and fully valued and such values taken into consideration in the economic valuation and assessment of renewable schemes

(e) ensures that an appropriate method of evaluation of large scale renewable energy schemes is devised to ensure that a fair evaluation of all schemes is achieved

large scale renewable energy projects will be financable and constructed.

The finance plan

The Mersey Tidal Power Barrage will generate energy at a very low marginal cost for at least 120 years. The main problem faced in the development of the financial structure for the Barrage is that the capital cost, and therefore debt service requirement, is very high in comparison with other generating systems. This problem was exacerbated by the fact that although the operating life of the Barrage is expected to be 120 years, capital repayment would be required over a maximum of 25 years. A significant level of subsidy would therefore be needed during this initial period before the full benefits of the very low operating costs could be enjoyed.

With this background the financial structure was developed in 5 phases with the aim of satisfying the following major objectives

(a) To promote certainty that the funds could be raised to allow the MBC to fulfil the "will secure" test laid down by the Electricity Act for acceptance within the NFFO.

(b) To ensure that the average cost of capital enabled the sale of energy at the most realistic levels and at a level supported by the NFPA.

(c) To provide comfort for lenders and investors that the risks inherent in the Project were adequately covered and that the probability of the returns reflects the risks taken.

(d) To maximise the private sector contributions to the financing of the Barrage.

Development of the finance plan

The financial plan was developed on the basis that the advantages accruing to the Merseyside area, in particular due to the redevelopment opportunities, and to the nation as a whole through the long-term balance of payments, diversification of energy sources and emission control benefits, could not be translated into financial instruments. These benefits are therefore restricted to the economic evaluation.

Phase 1

The initial analysis of the project economics concentrated on gaining assurance that funds could be raised on an acceptable basis. This looked at achieving a robust funding structure that could be raised in the market at a risk/reward ratio acceptable to investors and lenders. This structure utilised the following forms of funding

- Equity
- Long-term fixed rate loan stock
- Long-term index-linked loan stock
- European Investment Bank finance
- Floating rate debt.

TIDAL POWER

The structure was heavily dependent upon refinancing of all the debt prior to the repayment of principal so as to reduce debt service obligations.

The consequence of this analysis was that a unit price of 9 p/kWh and subsequently 8.6 p/kWh was required during the early years of the project. The advantages of this structure were its certainty and the robustness disadvantage was that the unit price of electricity was considered to be too high.

Phase 2

To reduce the unit price of electricity generated, the possibility of utilising the tax losses of the Company during the early years of operation was evaluated. The writing down allowances (tax losses) would be sold at a discount of 15% to shareholders in the Company through a consortium relief structure. The entire capital cost was classified as plant and machinery as opposed to industrial buildings and hence incurred writing down allowances of 25% per annum. There were precedents which supported this aggressive assumption. The sale of all the tax losses generated sufficient revenue to enable indicative cash flows to show unit prices for electricity below 6 p/kWh.

The advantage of this structure was that the unit price was at a more acceptable level. The disadvantages were the uncertainties of important assumptions concerning classification of assets for writing down allowances, the ability of investors to commit to utilise tax capacity more than 10 years in the future, and the risks to the MBC's revenues through potential tax changes.

It was considered that further research along these lines should continue, but before finance could be committed a more comprehensive risk analysis would need to be completed.

Phase 3

Having created a plan which produced low unit rates, it was then necessary to obtain the benefits of the tax losses in a more secure manner. Finance leasing offers this opportunity, although the benefits are generally spread over a longer time period than was assumed in Phase 2. The Barrage itself would be owned by a syndicate of leasing companies (almost certainly subsidiaries of the major United Kingdom banks) which would use the tax allowances associated with those assets themselves and fund the cost of the assets. In return MBC would pay a series of rentals which would repay the principal and interest on the funds invested by the leasing companies over a period of 15-20 years. Since the leasing companies would be able to take advantage of the tax allowances immediately, this would give them a very substantial benefit and much of this would be returned to the MBC in the form of a lower interest rate. The use of lease to fund the entire non-equity requirement was examined and this resulted in energy prices of around 6 p/kWh, assuming that the total capital expenditure was classified as plant and machinery.

A single lease in excess of 1 billion would be the largest ever raised and would have to involve all the major lessors within the United Kingdom. Additionally, the lessors would not take project risk and a banking facility would be necessary to guarantee MBC's payments to the syndicate of lessors. Due to the reducing profitability of the United Kingdom Clearing Banks which crystallized late in the development of the structure, it was considered that a lease of this magnitude would stretch the market to its limits, hence it was considered prudent to continue to develop this scheme and to evaluate alternatives. The advantage of a lease is, however, that it gains MBC the most secure way of benefiting from the tax allowances within the Company.

Phase 4

The Electricity Act allows for Levy to be paid during the construction phase of a project. A discount rate, to reflect long term interest rates, was applied in calculating the value of the advance payments during construction. This analysis assumed that tax losses would be sold. The results were

(*a*) a contract period of 15 years (from start of operation) with a unit price of 9.5 p/kWh; and
(*b*) a contract period of 25 years with a unit price of 7 p/kWh.

Under these two scenarios the quantum of the advance payment remains at approximately the same level, however, the percentage of revenue discounted in each operating year varies quite considerably.

If it is assumed that the NFFO is for 15 years and that the unit price is 9.5 p/kWh, and a significant element of the Levy is available as advance payments, the outturn financing requirement reduces to approximately 700 million. If this is funded 45:45:10 by a lease, a European Investment Bank (EIB) type loan and equity, and it is assumed that half the remaining tax losses available to the MBC are sold in the first 10 years of operation, the structure in a base case capital cost scenario produced acceptable coverage ratios.

Phase 5

The fifth phase of the analysis completed under Stage III was designed to develop a financing structure based on the optimum combination of public and private sector funding. Having reviewed the MBC capital cost scenarios - best, likely and maximum - it was decided that this phase of the financial plan development should be based on the likely scenario.

The MBC's economic consultants produced a long term forecast of the

TIDAL POWER

pool price over the period that the MBC required payments under the NFFO which produced an average of 3.97 p/kWh (December 1990 terms).

It was determined that a structure should be developed that paid a slightly higher rate than the forecast average pool price during operation to give a realistic amount of private sector funding but which maintained a substantial degree of financing from advances. A unit rate of 6.75 p/kWh would be divided between 2 p in advances and 4.75 p during operation. The funding plan would, therefore, be

	£million
Equity	80
Debt	130
Finance Lease	300
Advances	830
Total	1340

This option it is believed offers the best balance between the public and private sector and is, therefore, the "most likely" Stage III financing case. If the outturn figures are re-stated in December 1990 prices they become

	£million
Equity	60
Debt	90
Finance Lease	195
Advances	700
Total	1045

The assumptions underlying this structure are as follows

Capex	£966 millions (*January 1991 prices*)
Construction period	5 years
Operating costs	£17.6 million per annum (*December 1990 prices*)
Inflation	7% per annum - 1991 4% per annum - 1992
NFPA contract period	25 years
Electricity price	6.75 p/kWh (*December 1990 prices*) 2 p/kWh (*paid as advances*) 4.75 p/kWh (*paid during operation*)

Average projected pool price	3/97 p/kWh *(December 1990 prices)*
Lease	15 years from completion
European Investment Bank Loan	20 years from completion
Tax	All capex on the Barrage classified at plant and machinery

It should be noted that the EIB will not give a commitment to any project at this early stage of development, but the broad concept of the Barrage has been discussed with the EIB and it is confidently expected that it will meet their investment criteria.

The important element of this finance structure is the advance payment. The advantages of receiving advance payments include

- removes the risk of interest rate and tax rate fluctuations on a significant amount of the debt
- shortens loan and lease drawdown periods
- saves loan margins and front-end fees
- reduces equity requirement, and
- reduces interest during construction.

A potential problem is the consequence of a shortfall in electricity production in any one year. This financial structure assumes a robust energy yield output, however, if there was a 10% drop in production, this would lead to a 13% drop in revenue if the NFPA required priority because it had provided payment in advance. Even if the lenders and NFPA bore this risk equally, there would still be a liability on the Company to make up the production owed to the NFPA.

Thus, the rights of the NFPA should be

(a) subordinated in each individual year - i.e. a 10% shortfall in planned production of 1.39 TWh would lead to the 4.75 p/kWh being paid on 1,260 GWh with no reduction being made in respect of the electricity already paid for but not delivered; subordinated cumulatively, i.e. in the following year

(b) if all 1.39 TWh were generated this would all be paid for at 4.75 p/kWh. Only if there was additional production would the NFPA receive electricity without having to pay, which would assist in making up the shortfall, and

(c) any remaining liability to the NFPA at the end of the 25 year period would be offset through an extension to the period during which the MBC has to supply a percentage of its power to the NFPA.

The terms of these debt, finance lease and guarantee facilities are those

TIDAL POWER

which it is believed could be arranged in the current market. The base post-tax equity returns of approximately 20% per annum nominal are at a level which should meet the return requirements of investors if they are satisfied with the levels of risk.

The future of the Mersey Barrage is dependent upon the MBC's inclusion within the NFFO and the proposed unit rate and NFFO period, or any alternative with similar effect. The financial markets do not allow the long term benefits of the Barrage to be reflected in investments of similar maturities. The private sector alone cannot fully exploit projects of this nature. The advance payment mechanism allows the project to optimise the combination of public and private sector funding. The effect of the advance payment mechanism is to produce a non-cumulative increase in electricity prices in the year the payment is made. This varies between 0.7% and 2.6% in any one year during the construction period. The Barrage is forecast to produce energy over its operating life which is significantly cheaper than fossil fuel electricity. If UK electricity prices increase in real terms as forecast, it would be worth approximately £3.1 billion by the year 2124. If fossil fuel electricity prices continue to increase at 1.2% per annum in real terms thereafter, then this value would rise to £8.1 billion. It also reduces the need to burn fossil fuels and thus diminishes noxious emissions. The development of the Barrage will also put British contractors at the forefront of the international tidal power market which it is estimated could be worth £150-200 billion over the next 25 years.

Without a positive Government response, it is highly unlikely that further private sector funding will be forthcoming for the development programme. A substantial reduction in unit prices of energy has been achieved during the analysis and the confidence of the private sector will increase with further studies to reconfirm inter alia capital cost, operating costs and the effect of the tidal cycle. This will allow, during the Stage IV Studies, further work on the financial structure to optimise the private sector contribution to the Project and to reduce the levels of support required from the NFFO.

Promotion

The MBC changed considerably in nature during 1990/1991 and the Stage III Studies and will continue to change and evolve during the proposed Stage IV, so that a robust and independent owner and promoter is in place prior to Stage V and the lodging of the necessary Parliamentary Bill. During Stage IV the MBC will develop its Engineering Team under an Engineering Director to monitor and approve the finalisation of the Project Specification and establish an Audit Team to monitor the design and later the construction of the project, to ensure that the Project as designed and constructed complies with the Project Specification and also that budgets and programme are achieved.

The Design and Construct Contract will be a typical turnkey engineering, procurement and construction (EPC) contract, providing the MBC with designs to the agreed specifications and standards, a guaranteed completion date, output and price. The contracts currently in use for the combined cycle gas plants will be used and ensure necessary finance can be raised.

The Design and Construct Contract will also require the QA policy contractor to operate a full Quality Assurance and plan approved by the MBC. Whilst the Contractor's QA Team will be responsible for auditing the QA procedures to ensure compliance with the Quality Plan the MBC Audit Team will review the QA records and ensure compliance with the QA procedures on site.

It will be essential that the MBC Engineering and Audit Teams are of sufficient strength and standing to secure the approval of the MBC funders and so avoid the imposition of unnecessary tiers of inspection and potential conflict between the MBC and its Contractor and the MBC and its funders. Once the Project specification has been finalised the Contractor Group and the Contractor's Designer will separate from the MBC Engineering Team.

Additionally, the MBC will establish an Operating Team for both the locks and the power plant.

Both Engineering, Audit and Operating Teams could be supplied by third parties under contract to the MBC as an alternative to direct employment by the MBC.

Resourcing for the Engineering Team could be from an existing generator or an established engineering consultancy, but real hands on experience would be preferred.

Similarly, the power plant operating team could be resourced from existing generators.

The lock operations will need to be co-ordinated with the Mersey Docks and Harbour Company, the Manchester Ship Canal Company and Associated British Ports - again both direct employment and subcontracting options will have to be considered.

The way forward

The proposed project programme for the Mersey Barrage is 9 years, comprising 4 years for Stage IV(Project Specification and Environmental Assessment), Stage V (Parliamentary approval) and Stage VI (Detailed Design), and 5 years for construction up to generation.

The final development works - Stage IV of the studies should be undertaken as soon as possible to finally determine whether or not the UK will have tidal energy contributions to achieve the desired diversity and security.

TIDAL POWER

The MBC had secured the £8 million required for 50% funding for Stage IV from the private sector and the Government had earlier confirmed that it would continue 50:50 funding into Stage IV.

The MBC, however, asked the Government to confirm that, subject to the outcome of the remaining development works, and, of course, necessary Parliamentary Approval, a long term NFFO type contract would be available to raise the necessary finance.

The Government is still on the fence - it has been there since our December 1989 NFFO Application, and sadly, looks like remaining there.

The key events which must be achieved for Stage IV are as follows.

(a) Initial commitment. After consideration of the Stage III report, agreement by DEn and the Ministerial Committee for the funding of the Stage IV Studies and the remaining development work.

(b) Completion of Stage IV Studies - final definition of project specification, completion of the Environmental Assessment, and the formal consultative process.

(c) Securing by DEn of long term NFFO contracts.

(d) Negotiation of turnkey EPC contract.

(e) Negotiation by MBC of NFPA long term contract and satisfaction of the Section 32 (Electricity Act 1989) "will secure" test.

(f) Promotion of a Hybrid or Private Bill. Agreement by DEn to the promotion of the Project by way of Hybrid Bill or agreement to the passage of a Private Bill, subject to Stage IV confirmation.

(g) Detailed design of the project.

The proposed Project programme divides into five separate stages and two distinct phases.

Phase I comprising

- *Stage IV*. From initial commitment until lodging of the Bill.
- *Stage V*. Promotion of the Bill to Royal Assent.
- *Stage VI*. Detailed Design

Detailed design may commence prior to final Parliamentary approval if a Hybrid Bill procedure is being utilised or if support for the Bill so justifies and enables funds to be generated and released.

Phase III comprising

Stage VII – Construction. Site mobilisation could commence after final Parliamentary approval and the finalisation of the casting basin arrangements and New Ferry Lock configuration (i.e. during the later part of Stage VI).

Stage VIII – Operation. The operation stage would commence with final commissioning and hand-over.

The likely cost for Stage IV is 15 million, Stage V is £9 million and Stage VI is £53 million. All costs are at January 1991 levels.

Bearing in mind the consultation planned in Stage IV, the need for an Environmental Statement and the requirement to satisfy the Office of Electricity Regulation (OFFER), the NFPA and the REC's and Secretary of State in regard to the "will secure" test - the MBC would contend that a Hybrid Bill should be adopted for the promotion of the Project - as with the Channel Tunnel, Dartford Bridge and the Second Severn Crossing.

Adoption of the Hybrid Bill procedure should enable detailed design (Stage VI) to commence during the Parliamentary Stage.

November 1993 seems the most optimistic Parliamentary date which, with a Hybrid Bill, would mean detailed design 1994/1995, construction 1995/1996 and generation at the turn of the century.

Conclusion

Returning to the words of the Secretary of State, where he announced that he was proud to have helped renewables to enter the commercial electricity generating market - it is hoped that the Government is proud enough to secure long term and large scale renewable projects for which the NFFO was originally intended and not just play lip service to the fashionable "green" lobby.

Discussion on Papers 8 – 10

R. WALKER, *Babtie Shaw & Morton*
How does the Treasury standard discount rate of 8% compare with discount rates in other countries of the EC? It is believed that in Germany substantially lower discount rates are used for national construction projects.

It is presumed that the significant CO_2 benefits attributable to a tidal barrage scheme are equally attributable to a conventional hydropower scheme. Could the same benefits be attributable to tidal schemes in developing countries?

In the feasibility design of conventional hydropower schemes, the optimum total installed capacity is established by comparison with a thermal 'equivalent'. How was the total installed capacity of the Mersey Barrage Scheme established?

E. A. WILSON, *Mersey Barrage Company*
In Fig. 5 it was demonstrated that there was a progressive increase in annual energy output with the number of machines and hence increased installed capacity. Estimates were made of the impact of varying this capacity upon costs, both capital and operating, and so the generation cost of varying installed capacities was calculated. A small reduction in installed capacity produced a small change in generation cost which was within the error of the estimate. An increase, however, would result in the need to place the additional turbine caissons on poor foundations as demonstrated by the site investigation and so would lead to a sharply increased generation cost. For environmental reasons generally, it was considered preferable to select as large an installed capacity as possible, so as to minimize the reduction in tidal cubature resulting from operation of the Barrage. Thus, 700 MW (28 machines of 8 m diameter) was chosen as the largest installed capacity not resulting in sharply increased foundation problems. As further investigations and studies are completed economic and environmental considerations may change the preferred installed capacity, though gross deviations are not anticipated.

P. GIBSON, *Severn Tidal Power Group*
I should declare my interest, having recently been given the task of defining the programme for financing studies for the Severn Barrage, which

studies were deferred in the last major barrage studies, consequent on the privatization of the ESI.

I have a number of questions of a general nature which I think will be of interest.

Firstly, I would be interested to know the status and composition of the ministerial committee referred to and the outcome of its review of the cost-benefit analysis.

Secondly, in the presentation of costs and benefit, the benefit arising from elimination of N_{ox} and S_{ox} is not identified. Is it correct to conclude that this is included in the projections of unit power costs?

Thirdly, the benefit of security of supply is assessed at £18 million. How was this derived and did it include any balance of payment effects consequent on ensuring security of supply?

Finally, I would be interested to know how carbon tax projections compare with the proposals for a carbon/energy tax proposed more recently by the European Community?

11. Feasibility of a Conwy barrage

M.E. MATTHEWS, MICE, Project Manager, TH Technology Limited, and R.M. YOUNG, BSc, PhD, MICE, Associate, Binnie & Partners

The Conwy Estuary was identified as one of the most promising small-scale tidal energy sites in the United Kingdom. To further investigate this potential source of renewable energy, a feasibility study was undertaken with the overall aim of establishing the location and arrangement for a barrage, the optimum number of turbines and sluices, the energy output, the principle environmental regional benefits and disbenefits, and the overall cost of the project. The results of this study form the basis of this Paper.

Introduction

A parametric study entitled The U.K. *Potential for Tidal Energy from Small Estuaries*, conducted by Binnie and Partners during 1987, identified the Conwy Estuary as one of the most promising sites for a small scale tidal energy scheme.

In 1989, the Department of Energy through the Energy Technology Support Unit together with Trafalgar House Corporate Development Limited commissioned a preliminary feasibility study to investigate the potential for the development of tidal energy from the Conwy Estuary. The study examined the scheme configuration in terms of the numbers of turbines and sluices, the arrangement of structures to house the hydroelectric plant and equipment, and the effect a barrage would have on the environmental and regional aspects of the Conwy area. Based upon the findings of these studies, estimates were made of the capital cost of the project, the rate of expenditure, annual operating and maintenance costs and the cost of a unit of energy for a range of discount rates.

The Conwy Estuary

The River Conwy lies in the eastern part of the County of Gwynedd in North Wales. The river, with a catchment area of about 590km^2, flows from the mountains of Snowdonia northwards to discharge into the Irish Sea just west of Great Ormes Head, (Fig. 1). The upper reaches of the river are steep but below the town of Betws-y-Coed, about 30 km from the sea, the river flows with a gentle fall through a relatively broad valley about 1 km wide.

TIDAL POWER

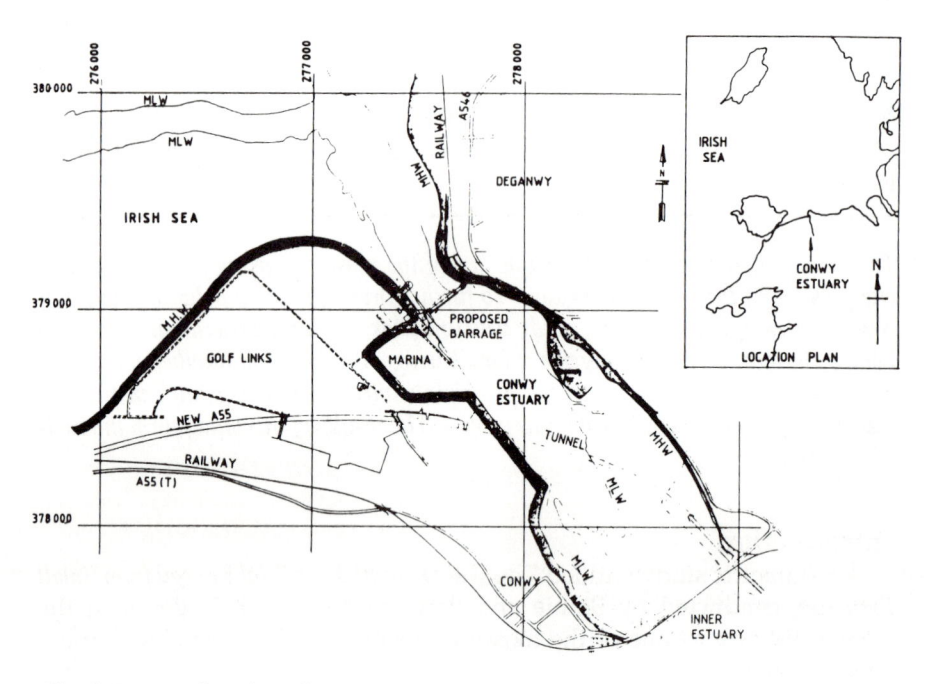

Fig. 1. Barrage location plan

The tidal limit of the estuary is at Trefriw, some 23.5 km from the river mouth. Below Trefriw the estuary broadens gradually as far as the historic town of Conwy with its 13th Century castle where the estuary is restricted to a short channel 150 m wide by the causeway carrying the old A55 road and rail crossing. The lower estuary below the causeway containing extensive sandbanks is about 1800 m long with a uniform width of 750 m.

The estuary terminates at the Deganwy Narrows where the channel width reduces to 200 m. The small town of Deganwy lies on rising ground to the east which contrasts with the flat and sandy ground on the west bank.

Seaward of the Deganwy Narrows are extensive areas of sand banks which are exposed at low tide. The river is drained through the sand banks to the sea by a low water navigation channel about 4 km in length with a depth at mean low tide which falls to about 600 mm.

Recently, the lower estuary has been extensively modified by the construction of an immersed tube tunnel as part of the A55 reconstruction programme. The construction of the tunnel has caused considerable loss of salt-marsh in the estuary, to the south of the Deganwy Narrows for the crossing portals and also for a large tunnel unit casting basin which is now being developed into a marina. Further areas have been lost south of the causeway where spoil from the works is being placed to reclaim land although some of this area will be managed to support plant and bird life.

A positive result from the construction of the tunnel was that during site investigations, the estuary was extensively studied and thus provided a large data base with important sources of information.

Geology and geotechnical conditions

The bed-rock strata of the Conwy Region was laid down during the Ordovician and Silurian periods some 500 million years ago. The geology is complex and comprises mudstone, shale, slates and sandstones overlain by Irish Sea boulder clay together with gravel and cobble deposits. Extensive alluvial sand, gravels, silts and clay material overlie the glacial deposits as a result of sedimentation and transportation of the material by the River Conwy during the period since the end of glaciation.

The geotechnical conditions at the site of the barrage vary considerably from the east to the west bank of the estuary. On the west bank, soil conditions comprise of 13 m of medium sand with some gravel overlying 5 m of boulder clay. Lake deposits underlie the boulder clay to -26 m OD and are described as interbedded firm boulder clay, clayey silt and silty clay with layers of sand. The lake deposits overlie a siltstone bedrock.

A borehole located on the east bank revealed 12m of medium dense sand and gravel under which exist further alluvial deposits consisting of 5 m of soft silty clay with stones and organic matter. These are underlain by stiff boulder clay to -15.8 m OD. A dark grey siltstone is present below this depth.

Tides

Conwy is listed as a secondary port in the Admiralty Tide Tables. Actual tide levels within the estuary have been recorded and the results have been analysed and used subsequently for tidal predictions as follows:

Mean high water Springs (MHWS)	+3.85 m OD
Mean high water Neaps (MHWN)	+2.08 m OD
Mean low water Neaps (MLWN)	-1.75 m OD
Mean low water Springs (MLWS)	-2.67 m OD

The resulting tidal ranges are 6.52 m and 3.83 m at mean Spring and Neap tides respectively.

Location of the barrage

The preferred location for the barrage is a short distance landward of the Deganwy Narrows at the mouth of the estuary and has been selected for the following reasons

TIDAL POWER

- The barrage, although a low-level structure, should be located as far as practicable away from the historic town of Conwy and its castle; the proposed location is about 2km from the castle.
- The barrage should be as short as possible, to reduce both its capital cost and its visual impact.
- The area of the enclosed basin should be as large as possible, to maximise energy production.
- Deep water, which exists at the mouth of the estuary is required to allow the turbines to be adequately submerged.
- Locating the barrage slightly to landward of the narrowest point will ease construction, improve patterns of currents in the basin and reduce visual intrusion for those living along the east shore adjacent to the landfall.

Barrage arrangement

The general arrangement of the barrage is shown in Figs 2 and 3. The principle components of the barrage are: turbine generators and their power house structure; gated sluices which control the refilling of the enclosed basin; and a lock to allow the safe passage of commercial boats and leisure craft past the barrage. A low-level road is provided across the top of the barrage and is intended for maintenance access and pedestrian use only.

Power house structure

It is undesirable that the structure housing the turbines be built in-situ as this would cause undue obstruction of the mouth of the estuary and disturbance to the surrounding population. The power house would therefore be built as two turbine caissons in a suitable dry dock located on the West Coast of the United Kingdom. From the dry dock the caisson would be towed to the Conwy and floated into position before being ballasted on to the prepared bed of the estuary. Electrical and mechanical plant would be installed in the caissons and pre-commissioned at the fabrication dock prior to tow.

The turbine caissons could be built either of reinforced concrete or of steel. Steel caissons would be lighter and therefore easier to tow and float into position, but would require increased maintenance.

The level of the service road across the top of the barrage would be 5.5 m OD with 1 m high wave walls on each side. This will ensure that the profile of the barrage is as low and unobtrusive as possible.

Access for normal operation and maintenance of the electro-mechanical equipment within the turbine caissons would be provided by a continuous internal gallery equipped with a gantry crane, constructed within the caisson beneath the roadway. A small building at the western end of the barrage would provide vertical covered access to this gallery. In addition, for major

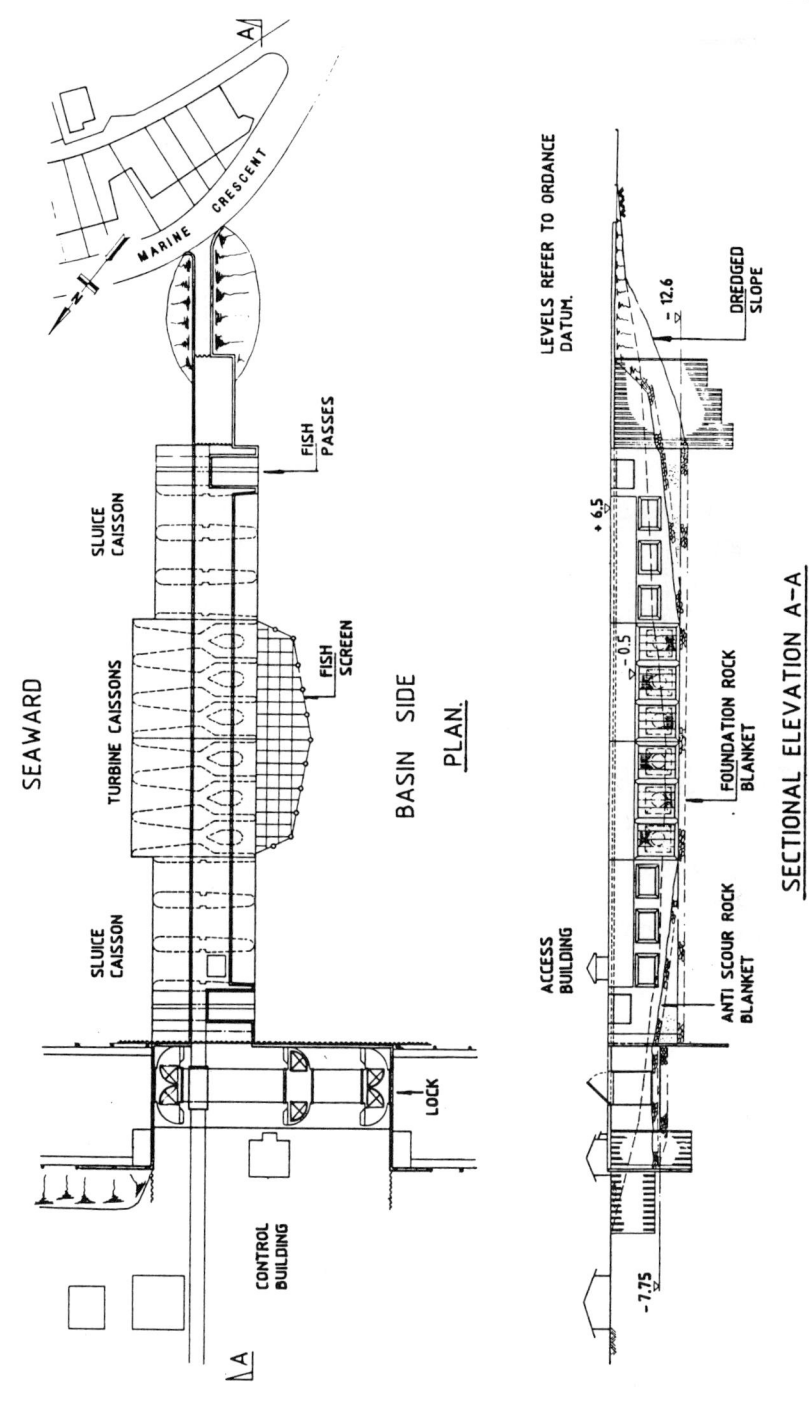

Fig. 2. General arrangement of barrage

TIDAL POWER

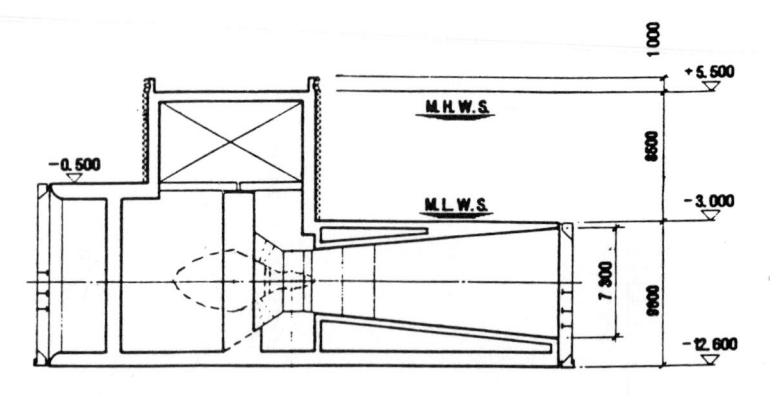

SECTION THROUGH TURBINE CAISSON

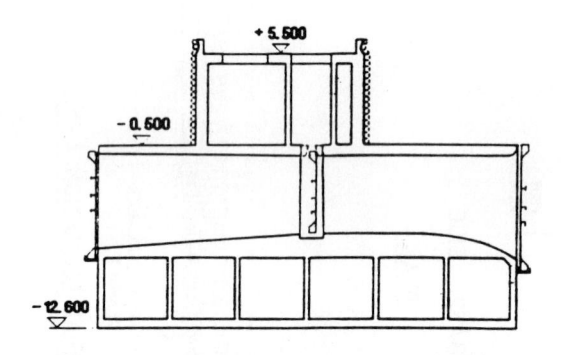

SECTION THROUGH SLUICE CAISSON

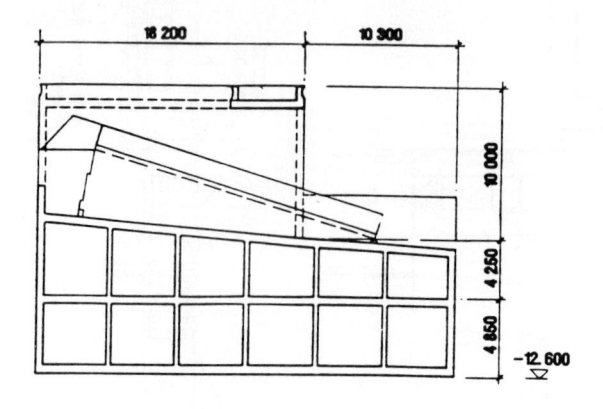

SECTION THROUGH FISH PASS

Fig. 3. Cross-section through caissons

maintenance tasks, there would be removable hatches above each turbine generator in the roadway. Mobile cranes would be used to remove and install equipment as necessary.

Sluice structure

Six large sluices are proposed which would be housed in sets of three in two caissons, one to be located on each side of the turbine caissons. Fish pass gates would be provided in each sluice caisson to allow fish to cross the barrage in both directions. The sluice caissons would also be built in a dry dock, towed to Conwy and ballasted into position.

The sluice gates would be vertical-lift wheeled gates, 8 m wide by 5 m high. Operation would be by electro-hydraulic motors for maximum reliability in the marine conditions. Each gate would be of fabricated steel, suitably protected against corrosion and located in a hydraulically efficient water passage. The water passage would be submerged during normal operation, thus the gates would not project above the top of the barrage when raised, helping to minimise any visual intrusion of the barrage. The gates would be designed to be able to resist water pressure in each direction, to enable the barrage to be used for flood protection in the event of very high tide levels or large river flows.

The special fish pass gates would each be 4 m wide, two sets of passes being installed side by side in a water passage on each end of the sluice caissons. The gates are of patented design, having a fish ladder built into the centre of the seaward face of the gate. The gate itself can be raised and lowered to control the flow down the fish ladder and over the crest of the gate on each side of the ladder. Thus extra flow can be discharged either to assist in the refilling of the basin, or, the evacuation of floods or, more importantly, to attract salmon and sea trout to the base of the ladder.

Navigation Lock

There are at present several hundred boats moored in the Conwy Estuary, including fishing vessels. Plans are underway to convert the new A55 submerged tube tunnel casting basin on the west bank onto a large marina. On the east side, there are plans to create a smaller marina at the disused Deganwy dock. As a result of these possible developments, up to 1400 boats could be based in the estuary.

To accommodate these boats the barrage would include a substantial boat lock to allow movement of craft into and out of the estuary. The lock has a planned size of 60 m length, 9 m width, and with a sill level of -5.75 m OD. This will allow access for boats with drafts up to 2.5 m at any state of the tide.

The lock would be equipped with three pairs of vertical-axis radial gates. One pair would be located at each end of the lock. The third pair would be

TIDAL POWER

located towards the middle of the lock, for use when there was little traffic or when one of the outer pairs required maintenance.

The lock would be built in a dry dock, towed to Conwy and floated into position in the same way as described for the turbine and sluice caissons. Once in position, ancillary works such as lead-in pontoons and anti-scour mattresses would be completed. A bascule bridge would be provided to take the roadway across. Power cables and other services between the shore and the turbine generators and sluice gates would be taken through ducts within the floor and walls of the lock.

The onshore facilities to the west of the barrage would comprise a control building servicing both the barrage and the lock, a small sub-station and a visitor centre. The area would be landscaped and provided with car parks and an access road.

Construction and installation

Construction of the turbine, sluice and lock caissons would proceed simultaneously in a suitable large dry dock on the west coast of the UK. The construction period would be about 18 months. Thus, there would be little activity at the barrage site until the caissons are approaching readiness for float-out.

At the site of the barrage, dredging would be required to deepen the water channel and widen it to form a uniform horizontal surface to receive the caissons. This work would be carried out by a suitable dredger, with the spoil being transported to an approved location for disposal. As the seabed is soft and erodible, dredging would be followed immediately by the placing of anti-scour protection across the barrage site and also for some distance upstream and downstream.

Upon completion of construction, the caissons would be towed to the estuary by suitable tugs, where they will be moored in a waiting area south of the barrage site.

Installation procedure

The lock caisson would be placed in position first, to provide access to vessels prior to the estuary being obstructed by the other construction work. Once in position, the pontoons and other navigation aids would be installed.

The second caisson to be installed would be the east sluice caisson. When in position, a short connecting embankment to the east shore would be completed. The sluice gates would immediately be commissioned and held open to allow the maximum area to pass tidal flows. This would reduce current velocities through the remaining gap in the estuary, which is important both for the installation of subsequent caissons and for craft sailing past the barrage.

The two turbine caissons would be installed as soon as possible after the first sluice caisson. Large diameter piles would be driven into the seabed on the line of the seaward edge of each caisson, to act both as mooring piles and as guides to assist in locating the caissons accurately before they are ballasted down. The ballasting would be achieved initially by flooding the water passages. Thereafter, full stability would be achieved by filling the non-working cavities in each caisson with sand or with high-plasticity concrete. Grout would be injected underneath each caisson to fill any voids and to seal the foundation material against water flows. As soon as possible after installation of the caissons the water passages would be opened to reduce current velocities through the remaining gap. On completion of the caisson installation operation, the mooring piles would be removed. The final caisson to be placed would be the west sluice caisson. The gap width available would be comfortably larger than the caisson to allow some freedom to manoeuvre during this critical stage of construction, which would be undertaken when tidal currents would be at their minimum. Once this caisson was in place, the gap between it and the lock would be closed with two lines of simple stop logs and the space between filled with concrete. Ducts for the various services required would be built into this concrete.

Dredging

At the present time the low water navigation channel seaward of the barrage is inadequate to pass the turbine discharge of about 600 m^3/s, without causing undue head losses at the barrage and consequent loss of output. Thus the cost estimates for the barrage include for this channel to be dredged to an average width of about 100 m and a minimum depth of about 4 m. The dredging would also provide additional draft for the caissons during tow into the estuary.

Hydraulic model of water movements

An important part of the studies has been the development and proving of a one dimensional computer model of the tidal reaches of the Conwy from the point where the low water channel meets the sea at low tide, some 6 km seaward of the mouth, to the tidal limit at Trefriw.

The results of the computer study have been used to predict the energy output of the barrage and form the basis for assessing the effect on the ecology of the estuary.

Energy output

The energy output of the barrage has been based on ebb generation supplemented by running the turbines in reverse to pump additional water

TIDAL POWER

from the sea at high water when sea and basin levels are similar. By pumping this extra water at low head, then discharging it later at relatively high head, a net gain in energy output is achieved. In practice, the operation of the barrage may be modified to optimise income rather than energy, so that time related tariff charges could be taken onto account if appropriate.

The predicted energy outputs of the barrage for spring, neap and mean tides are as shown in Table 1.

Table 1. Net energy output

	Tide range	Without pumping	With pumping
Spring tide	7.08 m	138.8	145.5
Mean tide	5.07 m	67.6	72.9
Neap tide	3.54 m	29.0	31.8
Annual energy output (MWh)		56,800	60,200

Environmental impact

The upper regions of the Conwy have a high rainfall of about 3 m/year. Thus the river is prone to flash flooding and a flood warning system is in operation.

In the last century there was substantial urban development at Conwy and Llandudno. One result is that the sewage from a population of about 49,000 people is discharged to the estuary, from a large number of outfalls. Welsh Water plc are constructing a new collection system for the east bank, and a new system is proposed for the west bank for completion in 1993/4.

Present water quality

The Conwy River's water quality is now Class 1 although some of its tributaries are Class 2. The inner estuary, behind the bridges and their causeway, is classed as fair, while the outer estuary is classed as good.

Dissolved oxygen (DO) concentrations in the estuary are practically always at or above saturation values, due to the large tidal range and constant water movement.

In recent years, algae growth has appeared to peak in the upper tidal reaches and reduces considerably further downstream. This may be due to increased turbidity reducing light penetration and/or salinity killing freshwater species. There are no recent records of algae blooms in the estuary.

Post-barrage water quality

The 1-D computer model described above has been used to assess the effects of a working barrage on salinity and water quality.

Results show that only in the case of dissolved oxygen, the barrage would have an effect on water quality with a modest drop in the middle part of the tidal river. For the other parameters modelled, the pollutant concentrations would either reduce or be unchanged so that the overall effect of the barrage would be beneficial. Salinity would be reduced in the tidal river as a result of the reduction in volume of the flood tide, but would increase near the barrage. This increase is a result of the present low water flows being largely fresh water, particularly during periods of high river flow; the greater volume retained in the basin at low water therefore results in mean salinity increasing.

Sediments

The general reduction in water velocities in the enclosed basin would reduce the movement of non-cohesive sediments (silts, sands and gravels). Using the 1-D model with a sand transport function suggested that the net transport of sand from the sea to the estuary would be reduced to about 15,000 t/year. This, if left to accumulate, would reduce the volume of the basin by 50% in a period of about 200 years.

Plants and animals

The reduction in suspended sediment within the basin, following barrage construction, would improve the appearance of the water. It would also increase light penetration and therefore allow greater growth of phytoplankton. These would then support a food chain leading through filter feeding and bottom feeding organisms to fish and birds.

The loss of the lower part of the intertidal area due to the raising of low water levels would affect the feeding patterns of wading birds. Preliminary conclusions are that the development of *Spartina* and other salt marsh plants would be constrained by the new regime.

The most important freshwater organisms in the estuary are salmon and sea trout which migrate to the sea as molts in the spring and return as adults to spawn in the period June to November. The major influence of the barrage on these species would be the turbines and the changes in water quality already discussed.

Adult salmon and sea trout returning to their river to spawn should be carried through the sluices on the flood tide. Those that are unwilling to do so would be able to traverse the fish passes at any state of the tide. A fish screen would be installed landward of the turbines to prevent adult salmon entering the turbines while waiting for suitable water conditions to ascend the river.

TIDAL POWER

The species and number of birds in the estuary have bee recorded by a number of local ornithologists almost from the inception of the Birds of Estuaries enquiries. The Conwy has at present some 17 species of waders and 14 species of wildfowl which winter regularly. The total population can reach 6000 birds. The redshank numbers represent more than 1% of the West European total for this species, making the Conwy a site of national importance. The Conwy Estuary was evaluated by the NCC in 1977 and found not to merit the status of SSSI.

Mussels have been harvested from beds outside the mouth of the estuary for centuries and are now cleaned in special tanks at the MAFF laboratories just south of Conwy, but this industry is now small.

Regional assessment

There is little industrial development around the Conwy Estuary, the principal source of employment being the service industries, including the holiday industry which is estimated to account for about 20% of the 18,000 jobs in the area. The nearby town of Llandudno is the largest resort in Wales and provides 40% of the total services bed capacity in North Wales.

The current main attraction in the area is the well preserved historic town of Conwy, with its castle and wall, attracting about 200,000 visitors a year.

Following the completion of the A55 coast road, Conwy will be about one hours drive from Chester and 1.5 to 2 hours drive from Liverpool, Manchester and Birmingham. This brings Conwy just within the catchment area for day trips from a very large population. Possibilities for development of Conwy following the relief from through traffic provided by the new road crossing are actively being considered.

In general, people agree that the views of the town and its castle across the estuary are better at high water than at low water. Thus the barrage would improve this aspect of the estuary. The estuary is used for water sports, including sailing, but the scope for this is very limited at low tide. Again the barrage will improve conditions. The barrage itself will be a significant attraction in its own right as a unique source of renewable energy. Based on the numbers visiting the Rance barrage, about 200,000 visitors a year could be expected to visit the barrage if suitable facilities were provided. Thus the barrage would be an additional attraction, especially out of season. Preliminary estimates, produced by a consultant, of the associated economic benefit attributable to the barrage are in the range of £0.5 million per year.

Cost estimates and economic analysis

Cost estimates for the design and construction of the Conwy Barrage have been based upon quantities taken from drawings produced during the feasibility study phase of the project. Design work undertaken during this phase was limited and while the costs contained in the report are considered as reasonably accurate, amendments to the estimates either upwards or downwards must be anticipated as detailed design proceeds. Further studies especially in the field of geotechnical investigations are required in order to confirm the extent of foundation work required for the barrage structure. Budget cost estimates for all major items of plant and equipment were received from suppliers and such costs have been included in the cost estimate.

The cost of the project in 1989 money terms has been estimated as

	Cost
Civil engineering main contract	£7.33 million
Civil engineering sub-contracts	£16.45 million
Plant and equipment main contracts	£3.72 million
Plant and equipment sub-contracts	£19.95 million
Non construction cost	£4.95 million
Total project cost	£52.40 million

Operation and maintenance cost have been estimated as £570,000 per year.

Unit cost of electricity

Table 2 below shows the cost per Kwh of electricity generated by the barrage for a range of discount rates based upon the total project cost and the annual operation and maintenance cost. The cost is shown in 1989 money terms and is for electricity at the barrage boundary.

Table 2. Cost of energy (p/Kwh)

IRR	30 years	Project life 60 years	120 years
2%	4.90	3.82	3.15
5%	6.80	5.89	5.63
8%	9.08	8.44	8.35
12.5%	13.05	12.73	12.73
15%	15.47	15.27	15.27

Project progress

The duration of the project programme from the commencement of additional studies to the generation of tidal power would be 5 years.

Conclusion

The results of the study have shown that using current technology no unexpected engineering problems are likely to be experienced during the construction of steel or concrete caissons to house the turbines and sluice gates. The caissons, as with the navigation lock, may be constructed at a facility remote from Conwy and towed to location where they would be sunk to form the barrage. Foundations to the barrage may require reassessment once a full geotechnical investigation of the site has been undertaken.

Computer model studies of the estuary have shown that, providing the hydraulic characteristics of the Conwy navigation channel are improved by dredging, ebb generation would produce an annual output 56.8 GWh. This output may be increased to 60.2 GWh if pumping is introduced at high water.

A review of the environmental aspects of the study has indicated that a tidal power barrage need not adversely affect the estuary. Conwy is not an important estuary for birds and provision will be made to allow fish to pass through the barrage in both directions and to maintain the existing mussel fishery. Siltation has been examined and found to be slow. However, movement of sand from the outer banks during storms will require further study.

A brief appraisal of the social and industrial impact of the barrage has shown that it will benefit the tourist and leisure industries up to the value of £0.5 million per year.

The cost of electricity at the barrage boundary would range from 3.15 pence per KWh of 2% discount rate to 12.73 pence per KWh at 12.5% discount rate in 1989 money terms.

The overall assessment of the result of the Feasibility Study has shown that the construction of a barrage on the Conwy will have few adverse effects on the estuary and no insurmountable problems may be expected during the construction.

12. Tidal energy from the Wyre

M. E. MATTHEWS. MICE, Project Manager, TH Technology Limited and R. M. YOUNG, BSc, PhD, MICE, Associate, Binnie & Partners

Amongst the estuaries around the United Kingdom, the Wyre has been found to be prominent in possessing the main requirements for a small scale tidal energy scheme. This paper describes the results of a Preliminary Feasibility Study undertaken to assess the viability of the construction of a barrage at the mouth of the Wyre estuary by establishing the form, cost, energy output and the nature and broad scale effect that the project may have on the environment and surrounding region. The study also investigated the potential benefits of utilising the barrage as a road crossing between the town of Fleetwood and Over Wyre.

Introduction

The Wyre Estuary posseses the main requirements for the development of a small scale tidal power scheme in that the tidal range is large and the estuary has a relatively narrow mouth in proportion to the area of its tidal reaches. These physical characteristics would permit the construction of a relatively short barrage with sufficient impounded storage from which to generate a modest but important amount of clean renewable energy. It was further identified that a barrage constructed on the Wyre may also act as a road crossing between Fleetwood and regions to the north of the estuary, thereby avoiding the present diversion via the Shard Toll Bridge, located some 8 km upstream from the mouth of the river.

In September 1990, Lancashire County Council in conjunction with the Department of Energy, NORWEB, Lancashire County Enterprises Limited and the National Rivers Authority, commissioned a preliminary feasibility study to undertake an investigation into the engineering, construction, environmental, regional and economic aspects of a tidal power scheme which may also be utilised as a road crossing to be located at the mouth of the Wyre Estuary. The initial brief for the study required that the barrage location should be limited to sites upstream of the entrance to Fleetwood Dock in order to minimise interference with shipping operations.

Preliminary research indicated that in the past, organisations with interests in and around the Wyre have conducted a number of surveys and studies on the physical and environmental aspects of the Estuary. These

TIDAL POWER

studies were found to be extremely relevant to a tidal power scheme and have greatly assisted in the preparation of the Feasibility report.

Description of the estuary

The estuary of the River Wyre is located on the west coast of England in the County of Lancashire, midway between Blackpool and Morecambe Bay. The river, which rises in the Forest of Bowland has a catchment area of about 45,000 ha, flows generally in a westerly direction psssing through the towns of Garstang, Cotterall and St Michael's before turning north at Thornton to discharge into the Irish Sea at Fleetwood.

The tidal limit of the river is generally at Cartford Bridge some 18 km from its mouth. The river is about 50 m wide at the Cartford Bridge and broadens fairly uniformly as far as Thornton where it is about 700 m wide. It remains at about this width for some 4.5 km to Fleetwood where it narrows to about 500 m immediately before reaching the sea.

Seaward of the river mouth, the coast is fringed by extensive sand banks which are exposed at low water for a distance of about 3 km from the shore. Within the sandbanks, a low water channel drains the river to deep water, taking a direct route just west of north. The channel provides navigation to the mouth of the estuary at all states of the tide.

For the greater part of its length, with the exception of the west bank between Thornton and Fleetwood which is largely urbanized, the river is flanked by agricultural land which is protected by flood banks to a height of +7.3 OD. There is some industry, notably the ICI Chemical and Polymer works at Hillhouse where chemicals are produced from brine obtained from rock salt deposits located to the east of the estuary. A major feature of the estuary is the port of Fleetwood which comprises two docks and a roll-on-roll-off terminal. Access to the docks is restricted to about 2 hours either side of high water although access to the ro-ro terminal is possible at all states of the tide. The estuary supports two sites of Special Scientific Interest, Burrows Marsh and Barnaby Sands Marsh, both of which are important for their areas of salt marsh.

Geology and geotechnical conditions

The underlying hard rock geology consists mudstones of the Triassic era forming a syncline with a north-south axis to the east of the Wyre Estuary. Within the mudstone strata is the Preesall salt which probably consists of collapse breccia. The mudstones are overlain by glacial deposits of boulder clay, marine deposits of very coarse sub-rounded gravel and alluvial silt and sand. To the east of the estuary exists up to 2 m of alluvial silt and sand overlying the glacial deposits. The east shore is bounded by cliffs of upper boulder clay overlying glacial sand and gravel which in turn overlies a strata of lower boulder clay.

In general, the geotechnical conditions at the candidate sites for the barrage comprise of boulder clay which outcrops on the east bank of the estuary and dips to a depth of 14 m beneath the west bank. Overlying the clay, are alluvial and marine deposits of gravels, sands, silts and clays.

Tides

The predicted heights of mean high and low tide level at the Wyre lighthouse have been taken from the Admiralty Tide Tables.

Tide	High Water (mOD)	Low Water (mOD)	Range (m)
Spring	+4.6	-4.0	8.6
Neap	+2.5	-2.0	4.5

Barrage location

During the conceptual engineering phase of the study, three alternative locations for the site of the barrage were identified (Fig.1). These sites were selected on consideration of the physical features of the estuary, constructional and operational requirements and the needs of users of the estuary. Two of the sites, the south and centre sites complied with the original terms of reference for the study in that the location of the barrage should be upstream of the entrance to Fleetwood Docks. The third site identified was located downstream of the Dock entrance and was found to have significant advantages over the other sites without adversely affecting the operation of Fleetwood Docks. An amendment to the contract was issued for the third site (north site) to be incorporated into the scope of the study work.

A review of the merits of each of the sites indicated that the north site possessed the maximum advantages and was adopted as the preferred site for the following reasons

- least length of barrage
- maximum area of enclosed basin and energy output
- minimum dredging requirement
- no changes required to plans for Fleetwood Dock Marina village development
- provides access to Fleetwood Docks at all states of the tide
- traffic to and across the barrage need not pass through the marina village
- most direct road link to existing road network
- close proximity to Fleetwood to encourage visitor spending

TIDAL POWER

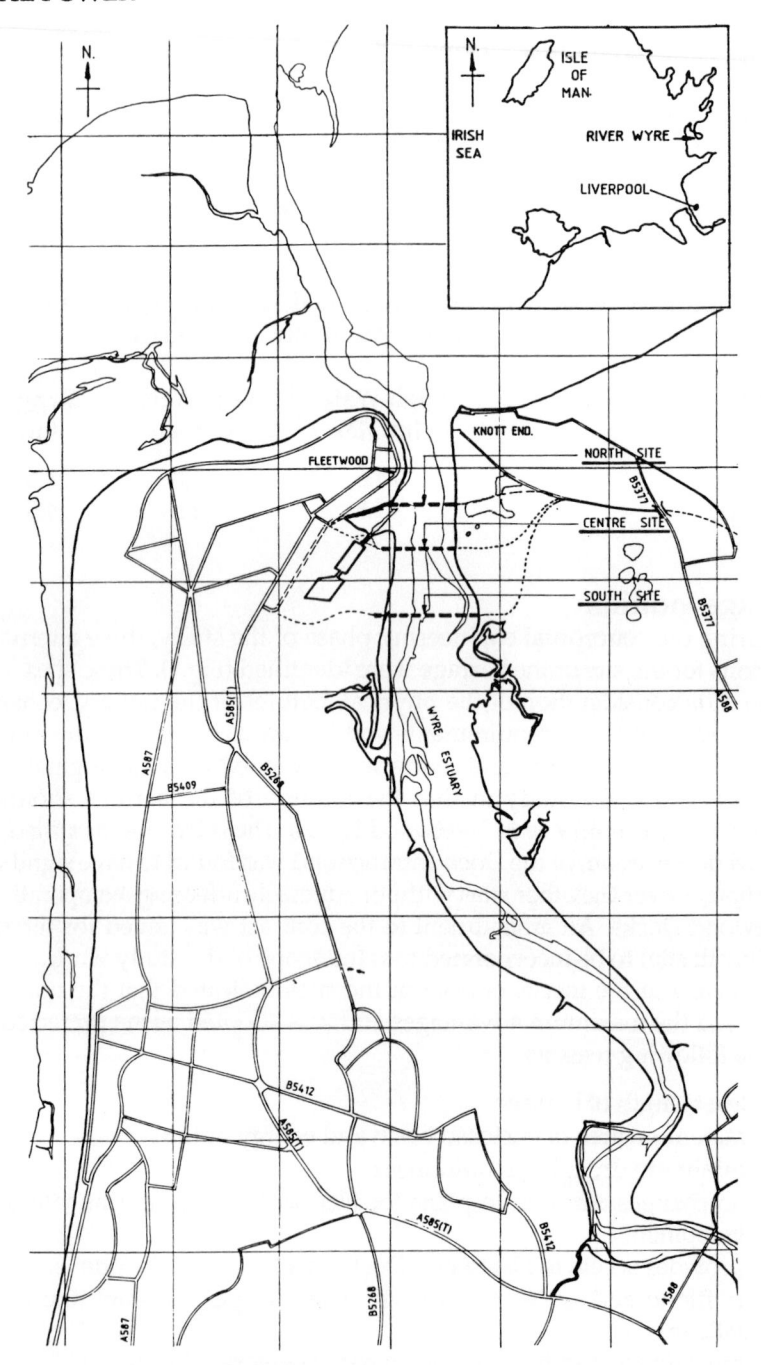

Fig. 1. Plan of the Wyre Estuary

Barrage arrangement

The arrangement of the barrage located at the north site comprises two concrete caissons to house four 6.2 m diameter pit turbines, ten sluices and two fish passes. A simplified plan and elevation is shown on Fig. 2. The caissons would be connected to the east shore by a short length of embankment. To the west of the caissons, a reclamation area would be formed using material excavated from the dredging operation. This area would contain the navigation lock, control building, visitor centre, car parks and a road which would be routed across two bascule bridges, one located at each end of the lock, to provide near continuous flow of traffic. Should the barrage be used as a road crossing, the west link road would be routed along the existing Jubilee Quay in the form of a promenade. The Jubilee Quay would be re-located along the south bank of the reclamation area. The link road to the east would cross the Knott End Golf Club, pass to the north of Hackensall Hall and follow the alignment of the disused Knott End railway line to join the A588 east of Preesall.

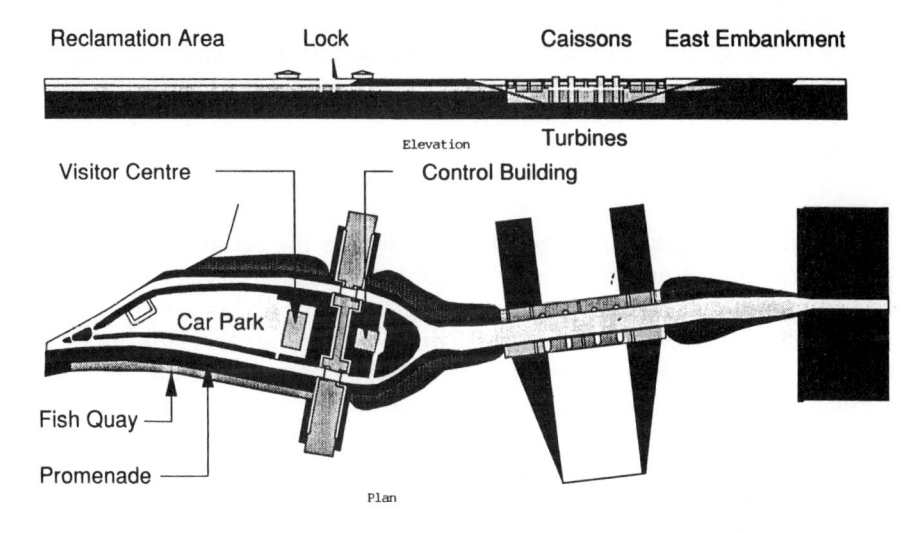

Fig. 2. General arrangement of barrage

Turbine - sluice caisson

The barrage power house containing the hydro-electric plant and mechanical equipment would be constructed in reinforced concrete in the form of two identical but handed caissons at a suitable facility remote from the Wyre. This form of construction would permit the minimum construction period and also allow the installation of the major components of plant including the turbines, sluice gates and fish passes at the construction yard. On

TIDAL POWER

completion, the caissons would be towed to location and ballasted down on to a prepared foundation laid on the bed of the estuary.

The general arrangement of each of the caissons has been developed to provide a structurally efficient configuration in which to house two 6.2 m diameter pit turbines, three 9.8 m wide sluices positioned immediately above the turbines, two 11.0 m wide located to the side of the turbines and a single 4.0 m fish pass. The overall dimensions for each caisson are 78.2 m long by 50.4 m wide with an overall height of 24.2 m. (Fig. 3).

The turbines incorporating the runners, gearbox and generators are located in a pier which extends to the top of the caisson to form the sides of the sluice channels and support the road above. Water-tight hatches set in the top of the piers provide direct access over major items of equipment to permit their removal and replacement by means of a mobile crane.

The sluice gates would be horizontally mounted, electro-hydraulically operated radial steel gates suitably protected against corrosion by a specialist paint system. The gates would be designed to resist water pressure in each direction to permit the barrage to be used for flood protection during large river flows or very high tide levels.

The roadway, which would form the top deck of the caisson, would be constructed either as a 7.3 m wide carriageway should the barrage be used as a road crossing, or 5.0 m wide if access only to the barrage was to be provided.

Navigation lock

A navigation lock would be provided within the reclamation area to permit craft of various types including fishing vessels to pass the barrage to and from the sea. The lock would have a plan size of 74.0 m long by 10.0 m wide with a sill level to seaward of -6.0 m OD and at the basin side of -3.8 m OD. The lock would be equipped with two pairs of hydraulically operated vertical axis, steel radial gates to ensure maximum speed of operation. Steel bascule bridges would be installed at each end of the lock to provide near continuous access for road traffic across the lock.

It is proposed that the lock would be constructed of high modulus sheet piles tied back to piled anchor walls. The floor of the lock would be of reinforced concrete. Lead-in jetties comprised of tubular steel piles and tethered floating pontoons would be installed in line seaward and landward of the lock to provide temporary berthing for vessels waiting to pass the lock and also to act as a safety barrier to prevent craft from entering turbulent water during generation or sluicing operations.

Reclamation area

The main structure of the barrage comprising of the caissons would be linked to the west bank of the estuary by an area of reclaimed land formed

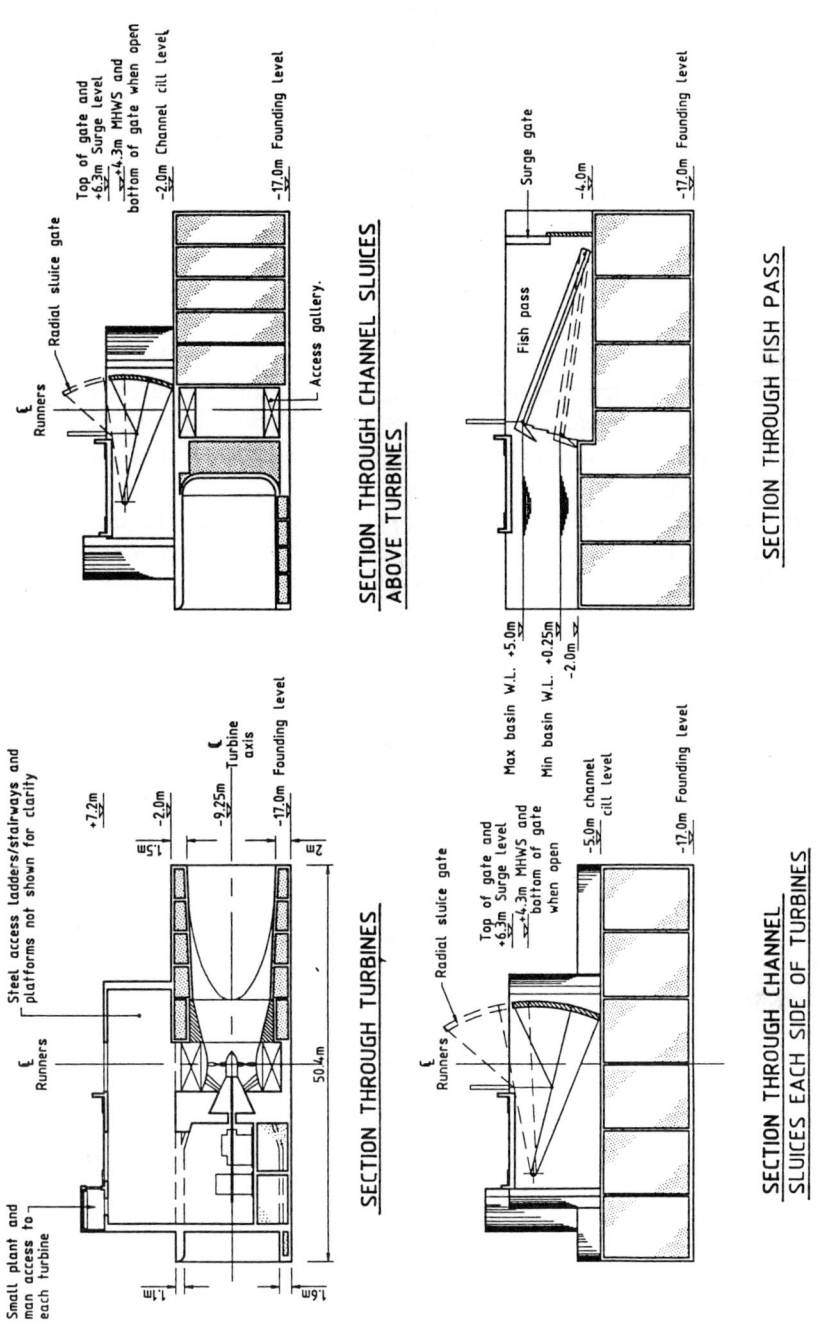

Fig. 3. *Cross-sections through caisson*

TIDAL POWER

with material from the dredging excavation for the caisson foundations. The reclaimed land would be utilised to contain the navigation lock, control building, visitor centre, transformer compound, car parks and carriageways for the road crossing. The facilities within the reclamation area have been arranged to provide a road layout which would permit near continuous flow of traffic across the barrage while making provision for the safety of visitors and the general public. This has been achieved by aligning the approach carriageways to the bascule bridges along the north and south perimeters of the reclamation area. Traffic would be routed to the north road while craft are entering the lock from the south and diverted to the south road where the north bascule bridge is raised. The arrangement of the barrage facility allows for the existing Jubilee Quay, at present occupied by the inshore fishermen, to be demolished and reconstructed to form the west road link in the form of a Promenade. Facilities for the inshore fishermen would be transferred to a new quay to be constructed along the southern side of the reclamation area.

An important feature of the barrage is that it should be of pleasing appearance to attract visitors and it is proposed that the reclamation area would be landscaped and planted with trees, flower beds and lawns.

Construction and installation

Construction work within the estuary would commence with the installation of the sheet piled cofferdams provided to stabilize the channel and to form the outer ends of the reclamation area and embankment. On completion, a contract would be let to dredge granular material above the boulder clay within the lock channel, turbine channel and caisson foundation area. The material excavated would be placed within pre-formed rock bunds to form the embankment and reclamation area. Within the reclamation area, the navigation lock would be constructed, closely followed by the new Jubilee Quay. The navigation lock would be commissioned as early as possible to provide safe passage to boats entering and leaving the estuary during periods that the main channel between cofferdams could be obstructed by dredging or other construction activities.

The construction of the caissons would proceed simultaneously at the selected facility on the coast of UK at a site probably remote from the Wyre. As the caissons neared completion, a further contract would be let for dredging the boulder clay to formation level within the turbine channel, the caisson foundations and the temporary mooring area. The dredged clay would be transported to an approved location at sea for disposal. Crushed rock would be placed over the caisson foundations area to protect the formation from scour and also to provide a blanket upon which to set down the caissons.

Upon completion of the caissons which would include the installation of the hydro-electric plant and mechanical equipment, the west caisson would be towed from the construction yard to the estuary by suitable tugs during a period of spring tides. This period would be selected to take advantage of maximum depth of water in the navigation channel and in order to minimise dredging requirements.

On arrival within the estuary, the caissons would be initially moored against two previously installed mooring dolphins located in a dredged area to the north of the barrage embankment before being warped into position during a period of neap tides and ballasted initially with water on to the prepared foundations. The water ballast would be replaced in the non-working cavities of the caisson by sand previously stored from the dredging operation.

Following initial flooding of the caisson, the sluice gates and turbine draft channels would be immediately opened to allow the maximum area to pass tidal flow. This would reduce current velocities through the remaining gap in the barrage, and would ease installation of the east caisson.

Installation of the east caisson would follow a similar procedure to that of the west caisson.

Grout would be injected beneath the caisson to provide uniform bearing on the foundation of crushed rock and to seal any voids to prevent seepage. Once the caissons are permanently ballasted in position the gap between the cofferdams would be closed with two lines of stop logs and the space between dewatered and filled with concrete.

The final stage of construction would be the erection of buildings, the connection of electrical equipment and services and commissioning of the barrage.

Energy output

The energy available at the barrage was quantified using a flat-estuary model. By investigating the power output for different turbine capacities a suitable size and number of turbines was determined. Similarly, by varying the sluice capacity in the model, the appropriate number of sluices was identified. The power output was confirmed using a 1-D model of the estuary.

The energy output and turbine requirements predicted from the computer model studies based on ebb generation and by operating the turbines or pumps at high water are as follows

Number of turbines	4
Turbine diameter (m)	6.2
Rated output (MW)	15.9
Annual energy output (GWh)	131

Hydraulic studies

A numerical model was used to examine the impact of a barrage on the water regime in the estuary. The studies included predictions of changes in water levels and velocities, water quality and sediment transport.

The results indicate that a significant increase in low water levels would occur landwards of the barrage. The duration and level of high water in this area would both increase. During periods of river flood or tidal surge, the barrage operation may be programmed to provide flood protection and ensure that the existing high spring water level in the estuary would not be exceeded.

The results of water quality tests indicate that no significant detrimental effects are likely to occur due to the presence of the barrage. In general, it is predicted that the increased volume of water in mid-estuary would cause a dilution of any pollutants discharged landwards of the barrage.

Sediment transport tests have revealed that, due to changes in velocity, the presence of a barrage is likely to produce a net seawards movement in the area outside the barrage. The predicted accretion within the basin is expected to be low, with the half-life for the scheme estimated to be in the region of 150-200 years.

Environmental impact
Main features of the existing estuary

The bed sediments in the lower estuary and in the sand banks beyond the estuary mouth are a very fine sand with some mud above the mid-tide mark within the estuary (the proportion increasing with distance inland) and fringed by salt marsh.

The low water channel in the lower estuary was found to be extremely mobile, moving from side to side on a regular basis except where constrained by training works. The low water channel beyond the estuary mouth is dredged regularly to maintain access to the ferry terminal at Fleetwood, as is the channel by and just upstream of the terminal.

Two areas of salt marsh on the eastern bank have been designated as SSSIs as they are ungrazed and contain plant species rare in the north-west.

It is considered by some naturalists that the estuary may act as a refuge in bad weather for birds including flocks of teal, turnstone and possibly other species from the nationally and internationally important Morecambe Bay/Lune Sanctuary sites (although this is as yet unproven) and as a result the designation of the whole of the lower estuary as an SSSI is being considered by NCC.

Both salmon and sea trout are found in the river, although not in large numbers. The estuary is not used for commercial fishing on a regular basis nor does it appear to be of particular importance to any fish populations although local sport fishermen use the estuary in poor weather and some mussels are collected.

A number of sewage treatment plants currently discharge to the estuary, although the largest of these is planned to be diverted by 1995 and secondary treatment will be provided for the remainder by a large sewage treatment plant to be sited on the west bank of the estuary with an outfall direct to the Irish Sea.

ICI Chemicals and Polymers Limited have a plant which currently and in the past has discharged effluent from a process involving the use of mercury to the estuary (although it is planned that this process will be phased out in the near future).

The use of Fleetwood Dock by fishing vessels has fallen dramatically in recent years and the owners are proposing to develop the area as a marina village.

Some areas of agricultural land, particularly on the east bank, are relatively low-lying and these areas drain through outfalls that are fitted with tidal flaps.

Main impacts of the barrage

The change in the tidal regime would alter the pattern of submergence on the foreshore leading to changes in the distribution of salt marsh vegetation and invertebrates.

The extended high water stand landward of the barrage would alter the exposure of the upper foreshore to attack by locally generated waves possibly leading to erosion and recession of mud-flat and salt marsh edges. This effect may be prevented by erosion protection measures.

Reduced turbulence landward of the barrage would probably lead to a reduction in the quantity of sediment in suspension after the barrage was built which may affect the potential for accretion on the upper foreshore during periods of calm weather.

The changes in foreshore submergence and invertebrate distribution may affect birds roosting and feeding in the estuary, although preliminary assessments suggest that acceptable alternative options are likely to exist.

Reverse turbining and pumping are proposed as well as normal turbine operation which may affect fish (including salmon and sea trout) entering or leaving the estuary. It is proposed to make provisions for fish passes and fish diversion devises to minimise this effect.

The reduced flushing within the estuary would lead to small changes in salinity (generally less than 5 parts per thousand) and similar changes to other water quality parsmeters, although in the case of BOD, ammonia and phosphates these changes would be masked by the effect of relocating and improving sewage treatment works discharges.

The changed tidal regime would probably lead to raised ground water

TIDAL POWER

levels near the estuary which may affect the land-fill waste disposal site, the proposed new sewage treatment plant site and drainage form low-lying land. Where necesssry provision would be made to install pumping stations to maintain ground water at existing levels.

Regional and traffic studies

The study has looked at the social and economic impact and the effect on traffic movements by the construction of a tidal barrage across the mouth of the Estuary. The barrage would create an extensive tidal lake which could be used for water sport activity, and the barrage itself would be of considerable tourism interest.

A road could also be built over the top of the barrage which would connect Fleetwood and Over Wyre when previously there was only an intermittent foot passenger ferry, or a long car or bus journey round the estuary via Shard Bridge.

In considering the impact of the impounded lake on tourism and recreation, the study has noted many uncertainties. One of these is the effect of marketing which, if handled successfully, could enhance the number of visitors to the area considerably. Given a reasonable success, however the visitor centre at the barrage would still be the major attraction, perhaps 300,000 visitors per annum including a high proportion of school children. The lake itself would attract water sport enthusiasts, including some activities not now seen in the estuary. It is estimsted that the rate of activity would nearly quadruple from about 6,000 such trips per annum to 23,000 trips.

The effect of constructing the barrage road would be to attract about 5,500 daily vehicle trips across the estuary, diverting traffic from the A588 Shard Bridge and the A586 Gt Ecclestone route.

The provision of a road over the barrage would also provide a beneficial, additional link for traffic to the communities in the Over Wyre.

The Study has shown that in economic terms there would be a strong net contribution from associated developments towards the cost of the barrage. This can be quantified to give about 1 m per annum with a wide range of uncertainty. This is equivalent to a net present value of about £10 million.

Financial analysis
Cost estimates

The total project cost for the construction and installation of a barrage at the north site without public road crossing in 1991 money terms are as follows

Civil engineering	£44.16 million
Plant, equipment and transmission	£34.83 million
Non-construction cost	£10.95 million
Total project cost	£89.94 million

The operation and maintenance cost has been estimated at £880,000 per annum with replacement of the main items of hydro-electrical equipment after a period of 40 years and 80 years at the cost of £30 million on each occasion.

Cost of energy

The cost of energy at the boundary of the barrage site based on the energy generated in a year of average tides at 131 Gwh has been calculated as

IRR

Structure life	2%	5%	8%	10%	12.5%	15%
30 years	3.76	5.25	7.04	8.37	10.15	12.05
60 years	2.96	4.56	6.55	8.00	9.91	11.90
120 years	2.53	4.39	6.49	7.98	9.90	11.90

Project programme

The overall project programme, inclusive of further studies, parliamentary approval, design, construction, installation and commissioning would take six years from the decision to proceed with the next phase of the project to commercial power generation.

Conclusion

The results of the study showed the Wyre Estuary to be physically well suited to a tidal energy scheme and unlikely to have any serious adverse effects on present and proposed developments or current users of the estuary. Engineering studies confirmed that it would be feasible to construct a tidal energy barrage in the estuary which may be used as a road crossing.

The identified environmental impacts are not thought at this stage to be sufficiently serious to preclude further investigation of a barrage scheme.

With active promotion, the barrage would attract significant numbers of tourists and visitors to Fleetwood, thus providing associated additional employment and income to the area.

The overall conclusion of the feasibility study is that a tidal energy barrage located within the estuary could produce a modest amount of clean renewable energy, could be constructed using current technology and would be of considerable benefit to the region.

13. Environmental aspects of small tidal power schemes

R.M. YOUNG, BSc, PhD, MICE, Associate, Binnie & Partners and
M. MATTHEWS, MICE, Project Manager, TH Technology Ltd

Introduction
There has been an increasing realization in recent years that British estuaries are a valuable environmental resource, supporting a wide range of habitats and an abundance of wildlife. There is, however, concern that British estuaries are coming under increasing pressure from a variety of types of development. As a result there has been considerable interest in conservation issues in estuaries with important reports having been published by the Royal Society for the Protection of Birds and the then Nature Conservancy Council.

The construction of a tidal power scheme will have a marked effect on the tidal regime landward of the barrage and a less obvious but still noticeable effect seaward. These changes may have significant effects on the local environment which it is important should be identified and understood before the scheme is built.

Between 1989 and 1991 TH Technology Ltd and Binnie & Partners co-operated in carrying out pre-feasibility studies of tidal power schemes in two small estuaries, the Conwy and the Wyre, described in outline in the two other Papers presented in this session of the conference. The purpose of this Paper is to set out the approach used to assess the impact each scheme would have on the local environment and to describe the results of these preliminary assessments.

Scope of environmental studies
The aims of both the Conwy and the Wyre studies were to determine the energy output and overall cost of each scheme with sufficient accuracy to judge whether it was worthy of more detailed examination and at the same time to discover whether there are any issues which could jeopardize the scheme's implementation. To achieve these aims the scope of the environmental work carried out was

- to collect sufficient information about the estuary to establish its existing condition

TIDAL POWER

- to determine the changes that may be expected to occur if a tidal power scheme is constructed
- to highlight any gaps or other shortcomings in the available data
- to identify work that could be undertaken to ameliorate the effects of the scheme
- to recommend a programme of further studies to be carried out should the project proceed.

The framework for the environmental studies was

Physical environment	- water levels
	- flood evacuation
	- land drainage
	- sediment movement
	- water quality
Biological environment	- plants
	- invertebrates
	- fish
	- birds
Human environment	- navigation
	- industry
	- recreation

This framework was used both to describe the existing condition of the estuaries and to assess the long-term impacts of the scheme proposals. Short-term impacts during construction were also considered but not in such detail unless they were thought to have possible long-term implications. When addressing the human environment only impacts immediately local to the estuary were considered, as studies into the regional impact of the schemes were undertaken separately.

As much use as possible was made of existing material, collected from a large number of organizations, to describe the present condition of each estuary. Additional work was commissioned from external specialists where readily available information was inadequate. In the case of the Conwy the amount of information available was considerable as a large-scale data collection exercise had been carried out in connection with the new A55 road crossing of the estuary. Accordingly, the only work commissioned specifically for the study was an assessment of the effect of the tidal power scheme on salt-marsh development with particular reference to *Spartina anglica* undertaken by Dr A. Gray of the Institute of Terrestrial Ecology. The Wyre had been less well-studied, although some work had been carried out by ICI Ltd in connection with their chemical processing works at Hillhouse and by

Associated British Ports. Additional studies were therefore commissioned as follows

Dr A. Gray	ITE	Saltmarsh
Dr S. McGrorty	ITE	Invertebrates
Dr S. Lockwood	MAFF	Marine fish
Dr M. Elliot	IFE	Freshwater fish
Dr J. Goss-Custard	ITE	Birds

Note: ITE = Institute of Terrestrial Ecology, MAFF = Ministry of Agriculture, Fisheries and Food, IFE = Institute of Freshwater Ecology

On Dr Goss-Custard's recommendation a further special study of ornithological records in the Wyre and neighbouring estuaries was undertaken by the British Trust for Ornithology.

In both studies the key to assessing the effects of the scheme was a one-dimensional numerical model of the estuary extending from the tidal limit to deepwater seaward of the estuary mouth. Each model was built round a hydrodynamic module, which gave water level and discharge information along the estuary for a range of tidal and river flow conditions both with and without the scheme. The information for selected conditions was used as input to modules predicting the transport potential of non cohesive sediments and the movement of pollutants in the estuary taking into account the effects of dispersion and biochemical processes.

The results from the numerical models enabled broad predictions of changes in sediment movement patterns and water quality to be made which were then used in assessing the effects of the schemes on the biological environment in the estuaries. Both the physical and the biological changes were taken into account when assessing the effects of the scheme on human activities and possible feed-back effects were also considered.

Environmental effects of the Conwy tidal power scheme
Existing conditions

The River Conwy has a catchment area of about 590 km^2 and flows from the mountains of Snowdonia to discharge to the Irish Sea just west of Great Ormes Head in North Wales (Fig.1). Under normal conditions the tidal limit is at Trefriw, about 23.5 km from the mouth but extreme tides can influence water levels at Llanrwst some 5 km further upstream. The tidal reaches of the river flow largely through agricultural land lying in a valley about 1 km wide. Below Trefriw the estuary broadens gradually reaching a maximum of 1.5 km wide a short distance upstream from the town of Conwy. At this point it is constricted to a width of about 150 m by a causeway and bridge carrying road and rail links. Further seaward it broadens to a width of about

TIDAL POWER

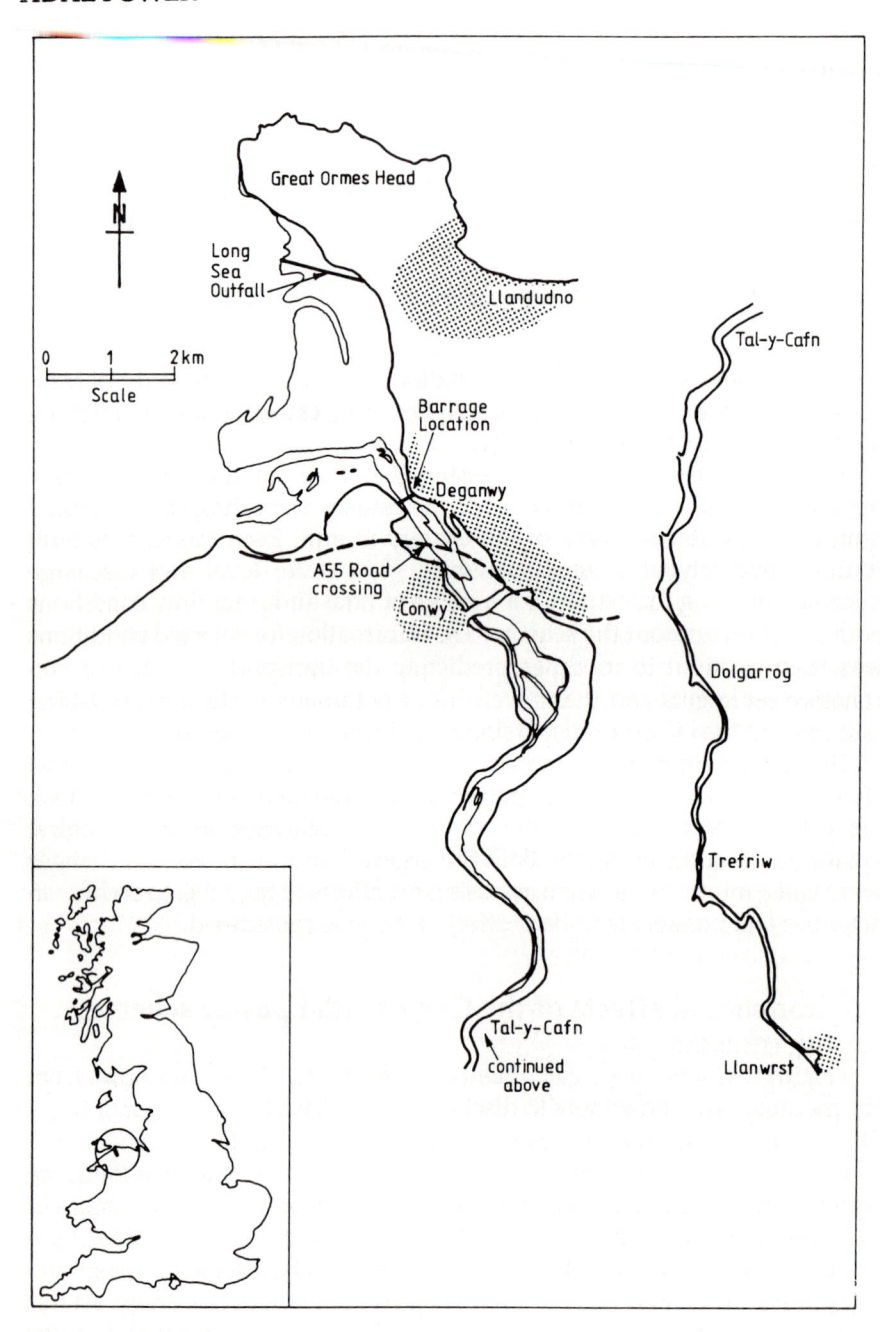

Fig. 1. The Conwy estuary

750 m for some 2 km until the mouth at Deganwy, where it narrows again to some 200 m. Extensive sandbanks lie seaward of the mouth, through which the estuary is drained by a relatively shallow low-water channel about 4 m in length.

The tidal range at Conwy varies from about 6.5 m to about 3.8 m at mean Springs and mean Neaps respectively, the corresponding high water levels being +3.85 m OD and +2.08 m OD. Extreme water levels in the lower estuary are caused by a combination of marine surges and high tidal levels but above Tal-y-cafn, some 7 km from Deganwy, river flows are important. The river is subject to flash floods and parts of the adjacent agricultural land are inundated periodically.

Surveys of bed sediments show that the estuary is dominated by sand and gravel with little fine material, such as there is being generally confined to the margins of the upper estuary. Examination of potential sand and gravel transport rates indicates that there is a tendency for the estuary to accrete, although historic observations suggest that the rate of accretion is relatively low, possibly being constrained by the availability of material from the outer estuary.

In 1985 the Conwy river water quality was classified as Class 1 and the condition of the outer and inner estuary as good and fair respectively. Dissolved oxygen (DO) concentrations were practically always at or near saturation. The quality had improved during the previous five years, probably due to improved treatment of sewage in the mid-catchment area. Further improvement in the inner estuary is anticipated as by 1995 the majority of sewage discharges to this reach will be diverted to a treatment plant and long sea outfall off Great Ormes Head.

Algal growth in the estuary is greatest in the upper tidal reaches and shows a marked reduction further seawards, possibly due to salinity killing freshwater species or to reduced light penetration associated with increasing turbidity. No major blooms have been recorded recently although blooms have occurred in the past in the lower estuary. These appear to have been due almost wholly to algae and nutrients brought into the estuary from Conwy Bay rather than having developed in situ. Extensive areas of saltmarsh were found in the estuary, much of it dominated by *Spartina anglica*. However the majority of this has recently been lost due to the construction of the new A55 road crossing. There is an area of brackish water reed swamp near Dolgarrog (some 15 km from Deganwy) which is of sufficient interest to be classified as an SSSI.

Mussels are found in the outer estuary and have been fished since Roman times. The estuary supports a significant salmonid fishery (both salmon and sea trout) which is exploited commercially and for recreation. There has been a gradual reduction in annual catches over the last 30 years although this is

TIDAL POWER

necessarily thought to indicate a declining fish stock. Other fish found in the estuary include eels, flounder, bass, mullet and the rare cucumber smelt.

Some 17 species of wader and 14 species of wildfowl are reported as wintering regularly in the Conwy with an additional 12 species of river corridor bird. Wader and wildfowl populations can reach about 6,000 birds and the estuary is classified as a site of national importance for redshank as in some years it holds more than 1% of the West European total for this species. There appears to be a considerable interchange of birds between the Conwy and the larger inter-tidal feeding areas on the Lavan Sands and the Anglesey shore to the west, with the Conwy possibly providing sheltered feeding during winter storms. The effect of the loss of salt-marsh due to the new A55 road crossing on bird populations had yet to be determined at the time of the study.

There is a small harbour at Conwy, mostly used by pleasure craft. Recreational sailing in the area could increase significantly if proposals for the development of a marina near Deganwy are realised.

Environmental impacts

The tidal power scheme proposed would contain six 4.0 m diameter turbines, six sluices with a gross water passage area of 240 m^2, two fish passes and a 9 m wide boat lock. The low water channel would be dredged to provide access for the caissons used to build the barrage and to improve flow conditions for the turbine discharge at low water. The scheme would operate as an ebb generation scheme with the turbines being used as pumps at high water to raise water levels with the basin.

Under normal circumstances high tide levels landward of the barrage would be raised slightly (due to pumping at high water) and the duration of the high water stand increased from about half an hour to about three hours. Low water levels would be raised to about mid-tide level at the barrier site but progressively less further up the estuary (reflecting the rise in low water level that naturally occurs. Velocities would be reduced at all states of the tide due to the greater volume of water impounded. Water levels to seaward would be largely unaffected at high tide but would be raised at low water due to head losses along the channel as the turbines discharge.

Although under normal conditions high tide levels would be slightly raised the barrage would be operated to reduce extreme water levels in the lower estuary by excluding marine surges and by providing sufficient storage to absorb fluvial floods. Extreme levels above Tal-y-Cafn are controlled principally by high river flows and would not be reduced significantly by the barrage. The rise in low water levels would impede drainage from the adjacent agricultural land which currently is discharged through gravity outfalls fitted with tidal flaps.

The change in the tidal regime would alter the sediment budget of the estuary. The rate at which sediment is transported landward under normal conditions is likely to be reduced but the effectiveness with which material is flushed seaward during fluvial floods would also be reduced. On balance the barrage is unlikely to increase the present rate of sedimentation in the estuary and could reduce it. There is likely to be a net erosion of material from the low-water channel seaward of the barrage and there may be a need for bank protection to prevent channel migration.

The water quality in the estuary will be affected more by the proposed diversion of discharges to a long sea outfall than by a barrage. Taking the conditions after the outfall is built as a base, salinity would be reduced in the tidal river as seawater would not penetrate quite so far upstream. In the middle and lower estuary average salinities would be raised but the changes would be modest in comparison with the natural range of salinities that occur in response to variations in river flow. DO levels show a slight reduction in mid-estuary due to greater rates of BOD and ammoniacal nitrogen removal after impoundment and the lower rates of re-aeration resulting from the reduction in water velocities. However, the lower velocities would favour increased settlement of suspended solids and thus greater light penetration which could encourage the growth of algae and raise DO levels through increased photosynthesis. There is a possibility that conditions favouring algal blooms would occur more frequently although it is not possible to quantify this without further study. The increased plant production would raise phytoplankton and other invertebrate populations and consequently increase the food supply available for fish stocks and birds.

The development of salt-marsh in front of the areas reclaimed as part of the A55 road crossing would be inhibited by the changed tidal regime associated with a tidal barrier unless sediment accretion occurs to raise bed levels. Further studies would be needed to show whether this would happen. It is thought unlikely that the reed beds at Dolgarrog would be significantly affected but again further studies would be needed to confirm this.

The principal effect of the scheme on fish using the estuary would be the physical obstruction offered by the barrier. This would affect all fish that normally enter or leave the estuary but would be most significant for salmonids. The fish passes could be used in both directions but their efficiency in estuarine conditions is as yet unproven. This would be less important for fish travelling upstream as they could swim through the sluices when open and would be unlikely to be able to swim against the discharge from the turbines. Fish travelling downstream, however, may pass through the turbines and be killed or damaged as a result. A high proportion of the large fish passing through the turbines would be affected but this would be reduced for smaller fish and the majority of very small fish would probably survive. Since it is reported that adult salmonids returning to a river may enter and leave an estuary several times before travelling further upstream this could have a significant effect on fish stock levels.

The reduced inter-tidal area available following construction of a barrage would undoubtedly affect the numbers of waders and waterfowl using the estuary although it is not possible to quantify this at present, not least because the effect of the loss of saltmarsh due to the new A55 road crossing on bird number has not yet been determined. Further studies on this and on the inter-relation between the usage of the Conwy and other nearby feeding or roosting grounds would be needed if a scheme is to proceed.

The visual appearance of the estuary would be affected, both by the barrage itself and by the raised low water levels. The latter would provide improved conditions for water sports activities such as sailing and water skiing. This increased activity may affect bird populations, although the greatest recreational use would be in the summer whereas the estuary is most heavily used by birds in winter.

The human activity that would be most seriously affected by the scheme is undoubtedly fishing, in particular fishing for salmonids, due to the possible reduction in fish stocks. The mussel fishery is also likely to be affected, however, particularly if existing beds in the outer estuary are affected by the proposed dredging of the low-water channel, as seems likely to be the case.

Ameliorative measures

The potential effect on fish stock levels is one of the more serious impacts identified. A number of ameliorative measures have been considered, including the provision of fish screens and fish diversion devices. It is anticipated that a combination of one or more of these devices with appropriate re-stocking would enable fish stocks to be maintained although further study would be required to confirm this. Means of maintaining the mussel fishery would need to be considered but this could be associated with the introduction of higher yielding culture techniques being used elsewhere.

Another major impact is the effect on land drainage. This can be overcome by the improvement of the existing drainage arrangements and the provision of pumping stations. In addition it would be necessary to establish barrage operating rules during times of extreme sea levels and high river flows to ensure that the potential flood defence benefits of the scheme are realised.

A number of other ameliorative measures may be possible. They would be examined in any further studies.

Environmental effects of the Wyre tidal power scheme
Existing conditions

The River Wyre rises in the Forest of Bowland in Lancashire and discharges to the Irish Sea at Fleetwood, draining an area of about 450 km^2 (Fig.

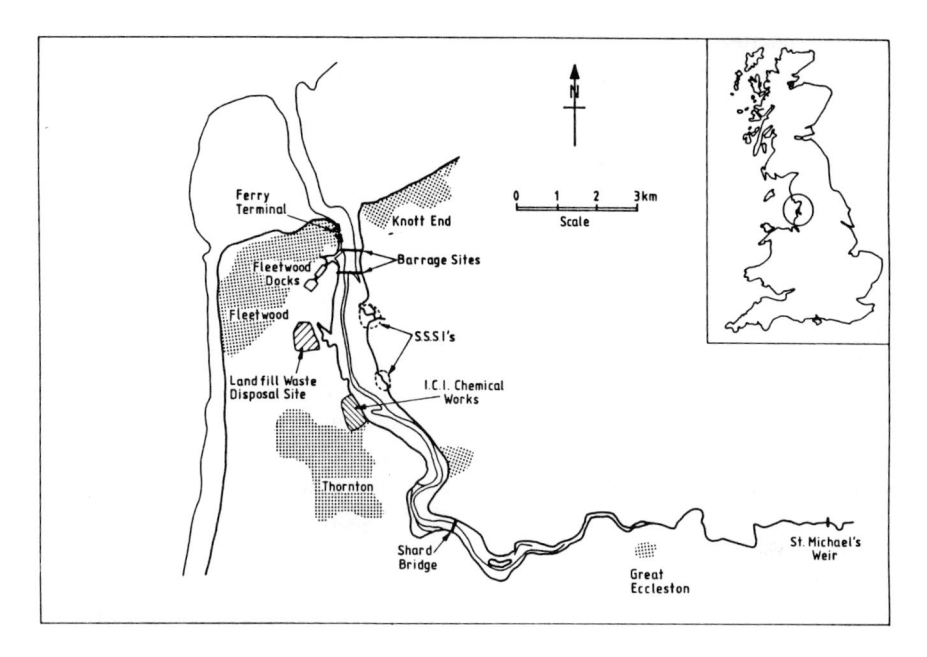

Fig. 2 The Wyre Estuary

2). The tidal limit is at St Michael's, a distance of some 22 km from the mouth. Initially the estuary flows to the west through agricultural land, parts of which have in recent years become affected by drainage problems due to peat shrinkage. At Shard Bridge, about 8 km from Fleetwood, the estuary turns to the north and broadens to a width that varies between about 700 m and 1 km but narrows to about 500 m near the mouth. The coast at this point is fringed by sandbanks extending about 3 km offshore at low tide through which the low-water channel takes a fairly direct route to deep water.

The tidal range at Fleetwood is 8.6 m on mean Spring and 4.5 m on mean Neap tides, the corresponding water levels at high tide being +4.6 m OD and +2.5 m OD. As in the Conwy, extreme high water levels in the lower estuary are caused by a combination of marine surges and high tidal levels. In the upper estuary extreme water levels are caused by fluvial floods, although the completion of the Garstang/Catterall flood alleviation scheme has meant that the maximum flow at St. Michael's Weir during a 1 in 50 year event is now limited to about 180 m^3/s.

The sediment contained in the estuary and offshore consists largely of fine sand with silt and mud being found in the upper estuary and at the margins. The pattern of sediment movement is complex with the low-water channel in the lower estuary migrating regularly from side to side. This appears to be the mechanism whereby accumulation of sediment within the estuary is

TIDAL POWER

avoided as there is no apparent tendency for the curvature to decrease with time. Various reaches of the channel have been trained in the past. Some 15 years ago one of the training walls near Fleetwood was by-passed, which appears to have led to an increase in the rate of accretion in the low-water channel near the ferry terminal.

The estuary currently receives considerable pollution loads from sewage treatment works and industrial effluent discharges, the latter largely from the ICI works at Hillhouse. In the 1960s this led to severe eutrophication but very considerable improvements have been achieved since then, to the extent that in the national survey of 1985 the estuary water quality was classified as fair. Further improvement is anticipated once the proposed Fylde Peninsula Scheme is implemented as this includes the diversion of effluent from the largest sewage treatment plant on the estuary.

Suspended material in the estuary is dominated by non-volatile mineral solids, although the available data suggests that significant algal growth occurs in the mid-reaches. No severe algal blooms have been recorded in recent years, however. Heavy metals are found in the estuary, in particular mercury which has been used in the ICI Hillhouse plant for many years (although this is due to be phased out in the near future). Biological evidence suggests that the estuary biota are unlikely to have been affected by metal contamination, a conclusion that is supported by the high mobility of the sediments in the area which would lead to the rapid re-distribution of any pollutant concentrations. Further study is needed to confirm this however.

There are well-established salt-marsh communities along the banks of the estuary, two areas of which have been designated as SSSIs as, unusually for the area, they are ungrazed and contain plants rare in the North West of England. Surveys of invertebrate animals on the foreshore suggest that the range of species present and the population sizes are not unusual. There is a possibility that populations could decrease following completion of the Fylde Peninsula scheme due to reduced nutrient levels.

The marine fish species and numbers present in the Wyre are as expected in an estuary of this size and character. Salmon and sea trout runs occur but numbers are small compared with nearby rivers, for example the Ribble. Nevertheless the National Rivers Authority are hopeful that runs will increase as the estuary water quality improves.

The Wyre estuary is at the southern end of the Lune-Wyre sandbanks which are part of Morecambe Bay, an extremely important site for waterfowl in winter and for passage migrants. It is considered to be nationally important for black-tailed godwit (holding more than 1% of the British population) while teal, redshank, turnstone, golden plover, sanderling and pink-footed goose have occurred in individual months in numbers which are locally or nationally important. The precise relationship between the use of this estuary and the much greater bird populations found in the nearby Morecambe

Bay, Lune-Wyre and Ribble areas is unclear at present although there is some evidence that it is used as shelter during southwesterly gales. As a result the whole estuary is being considered for classification as an SSSI.

The major industrial activities bordering the estuary are the extraction of brine on the east bank and its processing in the ICI works at Hillhouse on the west bank and a ferry terminal at Fleetwood. The town remains an important fish landing site despite the reduction of the deep-sea fishing fleet to about 80 registered vessels. The ferry terminal is currently operated by Pandoro Ltd and takes vessels with an overall length of 150 m. There is also a land-fill solid waste disposal site, located between the ICI works and Fleetwood Docks, which is operated by Lancashire County Council who plan to continue its use for many years hence. The estuary is not heavily used for recreational purposes but the ongoing redevelopment of the Fleetwood Docks will increase the number of pleasure craft.

Environmental impacts

Two sites were considered for a tidal power scheme, the more northerly being the one preferred. This would contain four 6.2 m diameter turbines, ten sluices, two fish passes and a 10 m wide lock. As with the Conwy scheme would operate as an ebb generation scheme with the turbines being used as pumps at high water to raise water levels within the basin. However, the turbines would also be opened on the rising tide to allow a reverse flow of water to the basin and so increase the effective sluicing capacity of the scheme.

As with the Conwy, the effect of the scheme landward of the barrier would generally be to raise high tide levels slightly, extend the duration of the high water stand and raise low water levels. Water velocities would be reduced throughout the tidal cycle. Seaward of the barrier water levels would be unaffected at high water but raised at low water due to the discharge from the turbines. Water velocities would be reduced on the flood but increased on the ebb.

During marine surges or periods of high river flow the barrage could be operated to maintain water levels within the estuary at or below the levels that would occur if the scheme had not been built, thus providing an improved standard of flood protection. Land drainage would be affected by the rise in low water levels as many of the channels draining land adjacent to the estuary discharges by gravity through flapped outfalls.

Sediment movement patterns would be significantly affected. The transport potential ont he ebb would be greater than that on the flood in the low-water channel seaward of the barrage, leading to a tendency for sediment to move seawards. Nevertheless, some material would be carried through the barrage into the basin, where it would be trapped. As a result the basin capacity would gradually decrease, the length of time needed to

TIDAL POWER

reduce the capacity by half being conservatively estimated as 150 years. The reduced velocities and lower circulation landward of the barrage would result in the bed below the current mid-tide level being reworked only in unusual conditions. The prolonged high water stand would lead to the salt marsh and upper mud-flats being more exposed to wave activity which could result in a general regression of upper shore levels.

The average salinity would be slightly raised in mid-estuary after a barrage is built, due to the presence of a saline impounded pool at low water. Generally, however, salinity variations would remain within those currently experienced. BOD and plant nutrient levels should be slightly reduced compared with conditions after the Fylde Peninsula scheme is completed (when levels should be considerably lower than at present), due to increased dilution. Suspended solids should also be reduced due to the more quiescent conditions. Dissolved oxygen levels would be about the same as before. Present information suggests that heavy metal contamination is unlikely to be of concern although confirmation of this is needed.

The lower nutrient concentrations will mean that algal blooms are less likely to occur in the estuary than at present, although this may be counteracted to a degree by the greater light penetration resulting from lower suspended solids concentrations. Further study would be needed to confirm this. Salt-marsh vegetation would be influenced by the more frequent and longer submergence with more tolerant species, particularly *Spartina* probably being favoured. Mobile invertebrates (shrimps, crabs) should increase in the estuary due to the greater volume of water retained at low tide. The distribution of other invertebrates may be affected locally by changes in submergence and in bed sediment sizes.

The effect of the barrage on fish would generally be as described for the Conwy except that the turbines would be larger (6.2 m compared with 4.0 m) so mortality rates would be lower for fish of a given size. Some damage may be caused to fish entering the estuary through open turbines being used to refill the basin. Overall mortality rates are unlikely to be sufficiently high to prevent the development of healthy populations of marine fish in the estuary although salmonid stocks could be affected unless effective ameliorative action is taken.

Bird populations in the estuary may be affected by the reduction in nutrient levels associated with the proposed Fylde Peninsula Scheme, whether a barrage is built or not. Changes associated with a barrage could also affect bird populations, although preliminary considerations suggest that for none of the species considered of major importance in the Wyre are the effects likely to be particularly adverse. However further studies are needed to confirm this and to establish whether any changes in bird numbers in the Wyre would affect the larger bird populations of the Ribble, Lune/Wyre and Morecambe Bay areas.

The ICI brine extraction and chemical processing works are unlikely to be significantly affected by the scheme unless more detailed studies show that their pattern of discharges to the estuary will need to be altered. Access for the fishing fleet to Fleetwood Dock should, taken overall, be improved and the use of the ferry terminal should be unaffected, although model studies will be needed to confirm this. The dredging requirement for the terminal should not be increased and indeed, may be reduced if the discharge from the turbines can be oriented appropriately. The land fill solid waste disposal site may however be affected by the raised ground water levels caused by the scheme which could cause leaching out of contaminants unless appropriate ameliorative measures are taken.

The presence of the barrage will provide an opportunity to increase the use of the estuary for recreation, particularly for water-based activities. This could lead to conflicts unless an estuary management system, possibly including zoning is established.

Ameliorative measures

The ameliorative measures proposed to maintain fish stocks in the Conwy (screens, diversion devices and restocking) would also be required for the Wyre, as would the improved land drainage arrangements (including the provision of pumps where necessary) and the establishment of operating rules for extreme floods and surges. In addition, however, it would be necessary to undertake work to prevent groundwater contamination from the landfill waste disposal site near Fleetwood. Further studies would be needed to determine the most appropriate method of doing this.

Other ameliorative measures or opportunities for improvement that have been identified include management of salt-marsh and inter-tidal mudflat areas, development of wetlands in poorly drained low-lying areas adjacent to the estuary, screening and landscaping and the establishment of an estuary management organization.

Discussion and conclusions

As may be expected, the direct physical effects of the two schemes, the changes in the tidal regime and in the pattern of sediment movement for example, are similar in character. Many of the other impacts are also similar, however their relative importance differs due to differences in the nature or condition of the two estuaries. The effect on fish is of particular significance in the Conwy due to the importance of the salmonid fishery but is less of a concern on the Wyre, partly due to the larger turbines but also because of the limited variety of species present and the smaller populations. Heavy metal contamination is unlikely to be an issue in the Conwy but may be important in the Wyre, as may the effects of increased submergence of the

TIDAL POWER

salt-marsh and of raised groundwater levels on the landfill waste disposal site.

This pattern is likely to be repeated in other estuaries considered for tidal power schemes; the direct physical effects being similar to those described here but the importance of the various impacts being dependent on the specific features of the estuary itself. Larger estuaries are likely to contain a greater number and variety of features that will be affected by the scheme making an impact assessment more complex although it is unlikely that there will be any fundamental difference from the impacts outlined in the two previous sections.

Discussion on Papers 11 – 13

E. HAWS, Consultant
Concerning the Wyre Estuary and the stated boulder clay foundation, could Mr Matthews please give more information. At the Mersey site two distinctive bounder clay deposits have been identified. The upper is soft, and not very competent for foundations. The lower is, however, stiff and can accommodate the smaller caissons.

Secondly, what reaction has there been from the navigation interests concerning the proposed barrage?

M. E. MATTHEWS, Trafalgar House Technology Ltd
(a)The borehole report described the boulder clay as 'firm to stiff brown clay with a little sand and sub-rounded to sub-angular fine, medium and coarse gravel'.

The minimum effective friction angle $0°$ is 25 of $61KN/m^2$ and the maximum effective friction angle is $32°$ with an effective cohesion of $25KN/m^2$. The saturated density of the clay is $21KN/m^2$.

The extent of site investigation at the proposed location of the Wyre Barrage has been limited to date. Further extensive investigation would be required if a decision is taken for the scheme to proceed.

(b)The reaction from the navigation interest to the proposed barrage has been caution. there has been no appreciation that the construction of a barrage at the north location would greatly improve the access periods to Fleetwood Docks. It is proposed to undertake ship manoeuvring studies, after consultations with the navigation interest in the Wyre, which would utilise the physical and numerical models and be part of the programme of further studies.

Dr E. M. WILSON, Consultant, Renewable Energy Associates Ltd
Why are the sluice caissons made so deep? It appears that a lot of the estuary bed is being dug at to be replaced with expensive concrete, which is not required either structurally or hydraulically.

At the energy costs indicated, which arise from 1700 per KW installed, is there any prospect of any of these schemes being built without substantial government subsidy?

M. E. MATTHEWS, Trafalgar House Technology Ltd

(a)The proposed power-station for the Wyre Barrage would comprise of two identical but handed combined turbine sluice caissons. The turbines have been located beneath the sluices and their depth and hence the depth of the caisson is governed by submergence requirements to prevent cavitation at the minimum water levels.

(b)It is anticipated that government financial support would be required for a tidal energy project to proceed.

S. CHARLES-JONES, Laing Civil Engineering

I would question the time scale to completion which Mr Matthews indicated in his presentations on the Conwy or Wyre barrages. The advantage of desk studies using data already collected is that the opposition has not been able to marshall its cohorts.

I am convinced that consultation is the only way to bring these major projects to fruition by going out to convert the objectors.

I understand that the Mersey Barrage Company went out of its way to involve all of the local interests in their company to ease its progress. It is surprising where opposition can come from.

Could Mr Matthews comment on what action he has taken in this respect or has it been left to the next phase. The consultation process is vital and takes a long time.

M. E. MATTHEWS, Trafalgar House Technology Ltd

The necessity and advantages of full consultations to assist a tidal energy scheme to proceed are appreciated. A programme of consultations would be implemented during the next phase of further studies.

R. F. ALLEN, University of Swansea

What is the depth of submergence, below low water level at Spring tide, of the top of the turbine tube and what is the chance of cavitation occurring at the turbine blade tip at operational speeds?

R. M. YOUNG, Binnie & Partners

The possibility of cavitation is a major factor in the design of any tidal power scheme, as it controls the axis level of the turbines and therefore the overall height of the caissons and the depth of excavation. For the work on the Wyre and the Conwy we set the turbine axis 0.75D below the present low water level on a mean spring tide at the barrage site, where D is the turbine diameter, and then checked for cavitation during the various tidal cycles that we modelled. In both cases we found this criterion to be satisfactory, although it should be noted that head losses in the channel seaward of the barrage meant that the low water level at the barrage site when the turbines were discharging was rather higher than the present low water level.

Professor B. O'CONNOR, University of Liverpool

Could Richard Young comment on a number of issues, namely

(i) Given the important role played by wave action in shaping the banks in the immediate vicinity of the seaward side of the Wyre Barrage, could he indicate if the effects of wave action have been considered in looking at channel changes/situations.

(ii) Given the changes in tidal levels that can occur on the seaward side of barrages due to reflection of tidal energy, where was the seaward boundary of the hydrodynamic models located and did the boundary conditions used in the model allow reflected energy to pass through the seaward boundary?

(iii) Is energy to be generated on Neap tides? If so, the reduction in tidal range behind the barrage in the Wyre (say 38 m to 1.9 m) may well produce highly stratified conditions, with consequent bad affects on pollutant levels.

(iv) If the Conway and Wyre Barrages are to be used as river flood retention basins, what consideration has been given to the possible total retention of river borne sediment in the estuary and its contribution in the loss of intertidal volume over the long term?

R. M. YOUNG, Binnie & Partners

(i) We are very aware of the importance of wave activity on the banks seaward of the Wyre estuary and of the implications this may have on sediment movements and on the need for dredging. The subject was not addressed in detail during the study we described in the Paper (which was essentially a pre-feasibility study) but would need to be considered carefully during any future studies.

(ii) The seaward boundary of the model was located in deep water in Morecombe Bay beyond the Wyre Lighthouse. We assumed that the tide curve at this point would be unaffected by the presence of the barrage and allowed no reflected energy to pass through the boundary.

(iii) We assumed that energy generation would take place on all tides. This will naturally lead to the tidal range behind the barrage being reduced, which will affect water quality in the impounded pool. We recognise that it may be possible for the pool to become stratified under some conditions but again did not address the subject in detail during the study described in the Paper. Again this wold need further consideration in any future studies.

(iv) The problem of sediment deposition behind the barrage and the long-term effect of this on the intertidal volume (and therefore on the energy output of the scheme) is common to all tidal power schemes. Our studies indicated that in both the Conwy and the Wyre estuaries the quantity of sediment brought into the estuary from marine sources was likely to be much more significant than the quantity brought down by the rivers.

14. Prospective tidal power projects in the Kimberley region of Western Australia

E.T. HAWS, MA, FEng, FICE, FIPENZ, Consultant, N. REILLY, BSc(Eng), FICE, Senior Assistant Director, Rendel Palmer & Tritton and P. WOOD, OBE, BSc, DipTP, MRTPI, Consultant

Introduction

The Select Committee on Energy and the Processing of Resources presented its final report to the Legislative Assembly of Western Australia on 12th November 1991 (ref.8). In presenting the report the Hon. Ian Thompson, MLA (Chairman) emphasised the scale and potential of tidal energy in the Kimberley region and noted that two of the generating sites were ranked in the top ten worldwide. The report, which was accepted by all parties recommended that

(a) the Kimberley tidal resources be re-assessed using methods developed in the UK adapted to the unusual tidal profiles found at some Kimberley sites

(b) the tidal height, range and tidal stream velocities and wave heights should be monitored at prospective tidal power station sites in the Kimberley

(c) the possibility of substituting tidal power for part of the electricity supplied by diesel power stations at Broome, Derby and Wyndham should be investigated

(d) the possibility of using free stream turbines mounted under moored pontoons to harness part of the tidal energy in constricted channels should be investigated

(e) long term plans should be made to harness Kimberley tidal power as one of the range of renewable energy resources that Australia will need to employ to achieve sustainable development and to reduce greenhouse gas emissions

(f) any future natural gas pipelines from the North West of Australia to markets in the south or east of the continent, should be designed to allow the later transmission of hydrogen produced by electrolysis from tidal power in the Kimberley, and

(g) Consideration should be given to methods of financing which take

TIDAL POWER

into account the benefits of a 120 year amortisation period for the civil works of a tidal power scheme and 40 years for the machinery and give full credit for the social and environmental benefits of tidal energy relative to fossil fuels.

As part of its deliberations the Committee made a visit to the Mersey and a desk study carried out in preparation forms the basis for this Paper.

Some aspects of eventual study

To help define requirements for the desk study the following list was prepared of aspects for eventual study.

- Topographical mapping
- Geotechnical investigation
- Determination of navigational features
- Hydraulics
- Estuary closure
- Determination of the most suitable configuration and construction method for the barrage (sluices, turbines, navigational features etc)
- Identification of any necessary accommodation works
- Economics, including energy and non-energy benefits and costs on and off-site. Consideration of firming power by complementary gas turbines or pumped storage, and of production of hydrogen
- Local environment – Physical impact on tidal propagation; erosion; sediment transport; current strengths and directions, water levels; salinity; water and sediment quality
- Local environment – Ecological impact: a change within the ecosystem (flora, fauna and marine environment) would arise from potential changes in the above and in turbidity and bacterial decay; primary production and the food chain; fish life and migration; crocodile population; benthic fauna support of large birds and fish; landscape implications
- Collection of baseline data should start early
- Global Environment – a tidal power barrage will avoid atmospheric emissions from thermal generation, and fuel and decommissioning problems of nuclear generation. These advantages have an economic value and perhaps a financial value
- Heritage and aboriginal interests – conservation and any necessary resettlement should be arranged sympathetically after consultation. Resettlement should be to standards above those presently enjoyed

Data now available

The following sections give some available information used for a preliminary assessment of tidal power potential.

Geography and demography

The Kimberley region is extremely remote and sparsely populated. It lies over 2,000 km from Perth (population 1 million) whilst the more populous Eastern States are some 3,000 km distant. The region has few inhabitants and the largest settlements at Broome, Derby and Wyndham have populations below 10,000. Much of the region is described as a wilderness area and there would be pressure from conservation groups to retain this. The area has strong aboriginal associations with several archeological sites protected under the Western Australia Heritage Act. The region has rich mineral reserves, particularly bauxite, and salt is produced by evaporation of seawater. To the south, the Pilbara region has major gas, oil and iron ore extraction. The processing of the mineral resources could create significant energy demands.

Tides

Table 1 gives tidal information for the project area.

Table 1. Tidal levels at selected locations (Fig. 1) predominantly semi-diurnal and mixed tides

LOCATION	HAT	MHWS	SPRINGS RANGE	MHWN	MSL	NEAPS RANGE	MLWN	MEAN RANGE	MLWS	LAT
(Tide Tables)				metres						
o Broome	+10.0	+8.5	8.2	+5.4	+4.5	1.8	+3.6	5.0	+0.3	-1.0
o Cockatoo Is.	+11.3	+9.9	8.5	+6.7	+5.6	3.1	+4.6	5.8	+1.4	+0.1
o Derby	+10.9	+10.0	9.8	+7.0	+4.5	4.3	+2.7	7.05	+0.2	-0.6
o Wyndham	+ 8.5	+7.4	6.6	+5.6	+4.4	2.9	+2.7	4.75	+0.8	-0.6
(Admiralty Charts)										
Adele Island (15°31'S;123°09'E)		+7.0	6.2	+5.4	+3.9	3.0	+2.4	4.6	+0.8	
Degerando 1 (15°20'S;124°11'E)		+8.9	7.9	+5.7	+4.9	1.5	+4.2	4.7	+1.0	

Climate

Rainfall: This is highest (760 mm) in the far north (Table 2). Averages are high in December to March. There is a decline after March to a low in September - October. Rainfall in individual months often varies greatly from the average, usually below.

TIDAL POWER

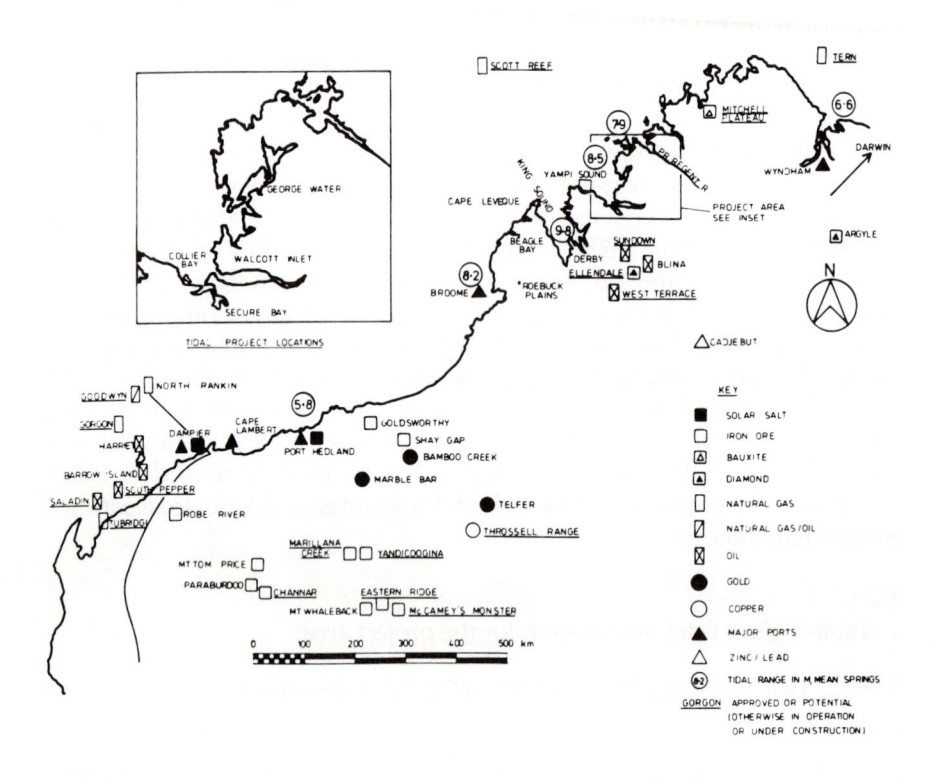

Fig. 1. Kimberley and Pilbara : resources, tides, climate stations and project locations

Table 2. Average monthly and annual rainfall: mm (ref. 1)

Station (Fig1)	Years	Jan	Feb	Mar	Apr	May	June	Jly	Aug	Sept	Oct	Nov	Dec	Total
Beagle Bay	66	182	175	140	49	34	15	9	2	1	3	18	93	721
Broome PO	62	154	140	102	30	19	23	5	3	1	1	13	76	567
Cape Leveque	48	187	193	119	42	47	18	15	2	1	1	6	74	705
Roebuck Plains	55	177	151	101	35	12	17	3	2	1	2	17	87	605

The region is influenced by both the northern tropical rainfall systems, which are responsible for the heavy summer falls and also by the southern systems which bring winter rains to the southern State.

Cyclones and thunderstorms are comparatively common, particularly from October to April, and heavy falls occur at times as shown below.

Date	Station	Amount (mm)	Time (h)
Jan. 1917	Roebuck Plains	924(360+564)	48
May 1931	Cockatoo Island	406	9

Near the west coast the average thunderstorm frequency is 15 to 20 per year. Cyclones affected the area 16 times between 1940 and 1969 and storm winds up to 144 mph have been recorded.

Temperature: Table 4 shows average monthly temperatures at Broome together with extremes. It is hot everywhere in the region in summer. In the hot months the maximum temperatore is very seldom below 90° F (32° C) and in the cooler months it is very seldom above this.

Table 4. Average maximum, minimum, and extreme temperatures : °F

Station	Nr. of Yrs	Jan	Feb	Mar	Apr	May	June	Jly	Aug	Sept	Oct	Nov	Dec	Annual
Broome														
1894-53)	56	92.1	92.0	93.5	93.6	88.1	82.8	82.3	85.4	89.0	91.0	92.9	93.4	89.7
1894-53)	56	79.1	78.8	77.3	71.8	64.7	59.5	57.4	59.4	65.1	71.6	76.4	79.2	70.0
1896-37	41	115.5	108.8	107.0	107.0	101.0	97.2	95.0	100.5	103.5	109.1	111.2	112.7	115.5
1896-37	41	68.2	59.3	55.0	54.0	45.1	43.6	40.2	40.6	49.0	52.8	61.8	63.0	40.2

Humidity: In the north of the region relative humidity is high in the first three months of the year, averaging between 70% and 76% at 9 am on the north Kimberley coast. Subsequently there is a decrease until August, when humidity commences to rise again.

Evaporation: Evaporation is high at some 2250 mm in northern coastal areas, the peak being in December and January.

Geomorphology and geology (Fig. 2)

The north-west coast of Australia from Derby to Wyndham comprises rocks of the Kimberley Block, a stable Precambrian craton. The Kimberley strata represent a basinal succession of sandstones, shales and intermediate

TIDAL POWER

to basic volcanics. The rocks have undergone strong deformation during the precambrian and metamorphism ranges from greenschist to granulite facies.

In the King Sound and Wyndham areas, around the perimeter of the Kimberley Block, rocks of Halls Creek Province crop out. These are part of a Proterozoic mobile belt and deformation is intense, with structure mainly parallel to the trend of the mobile. The coastline (ref. 2) is drowned with submerged valleys. Strong control is exerted by lithology and structure with inlets following faults extending some distance inland. An impressive example of this is the Prince Regent River (Fig.1).

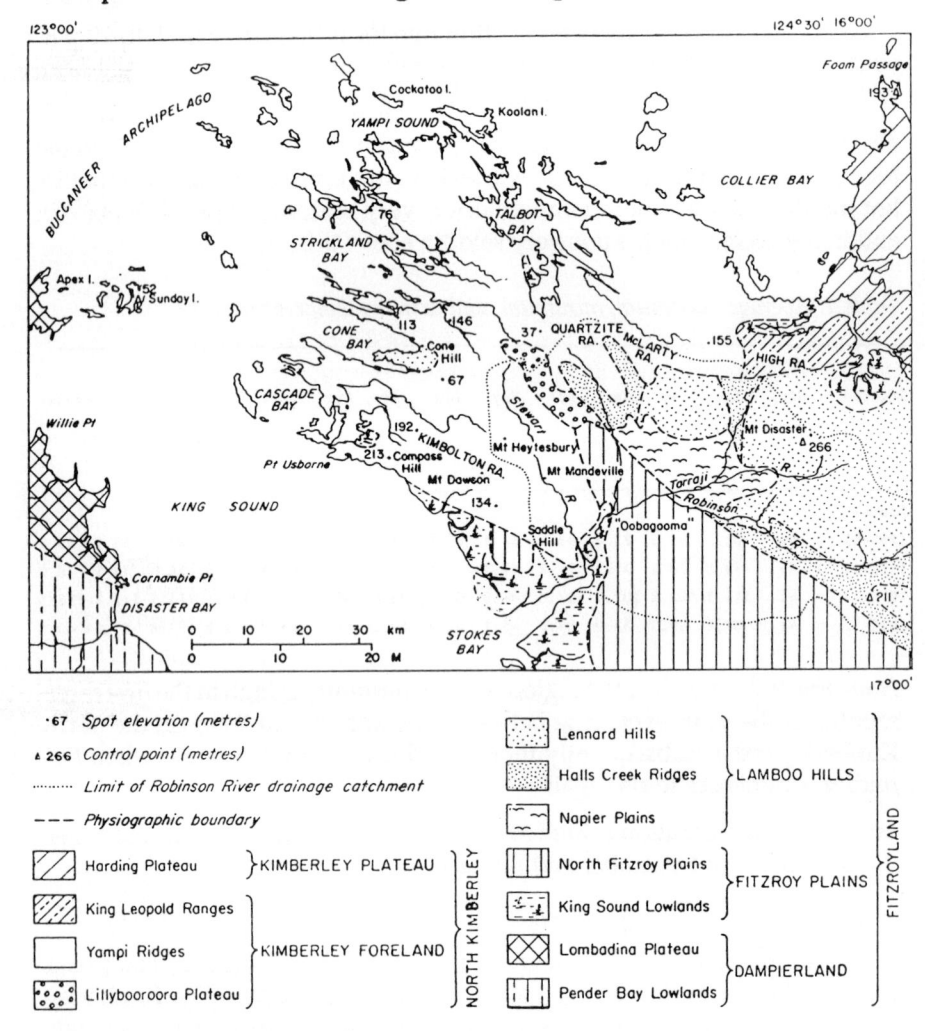

Fig.2. Yampi physiographic sketch map

Coastal features

Ria features are well developed north of Point Usborne where dissected Kimberley Group rocks strike at right angles to the coast. The coast is characterized by sandstone promontories, narrow tidal channels and numerous islands along the strike of the mainland structures. It has been estimated that the depth of submergence in the Buccaneer Archipelago was approximately 60 m.

Limestone reefs fringe parts of the north mainland coast and are extensive about the islands. Beaches are scarce and confined mainly to restricted bays of the outer coast and islands. Mangrove mud-flats occupy sheltered bays, inlets, and estuaries. Other small bays, partly cut off from the sea by sand-bars, have formed intratidal lagoons.

The coast of Collier Bay is controlled by the structural trend of sandstones of the Kimberley Group. Drowned river gorges and strike valleys in the interbedded basic rocks are now narrow tidal channels. Examples are Yule Entrance and The Funnel, giving access to broad mangrove-fringed reaches of Walcott Inlet and Secure Bay (ref. 3).

Infrastructure

The two principal settlements of Collier Bay (Fig.2) are the iron-ore producing centres of Cockatoo and Koolan Islands. The only other settlements are the cattle stations of Oobagooma at the tidal mouth of the Robinson River, and Kimbolton, a recently established station 22 km west-northwest of Oobagooma. Derby, the main town, seaport, and centre of communications, supplies, and administration for the West Kimberley region, is situated south of Stokes Bay near the head of King Sound.

Graded inland and coastal roads from Oobagooma meet north of Meda homestead and connect with the Derby Kimberley Downs road 48 km east of Derby. There are light aircraft strips at all centres of permanent habitation. Communications with Derby are provided by radiotelephone, the Royal Flying Doctor service, and air charter.

Mineral resources & development

Natural gas

The identified recoverable gas reserves and production for Western Australia at 1988 are given in Table 5. Gas from North Rankin is piped ashore to Dampier, whence a long distance pipeline delivers to Perth and the south. Some LNG is exported from Dampier to Japan.

Of relevance to the tidal power prospect are the potential gas fields of Scott Reef and Tern (Fig. 1). Scott Reef has special environmental problems and Tern may be more readily developed although its proved reserves are much less.

TIDAL POWER

Table 5. Identified recoverable gas reserves at 31 December 1988

Natural Gas

Probability of Recovery	Billion Cubic Metres (BCM) P90%	P50%
PROVED Developed		
North Rankin	184.719	211.719
Barrow Island, Dongara, Mondarra, Yardarrino, Harriet, Rosette, Woodada Total	4.924	7.177
Total	**189.643**	**218.896**
PROVED Undeveloped		
Goodwyn Main	70.60	82.70
Goodwyn North	36.50	45.80
Campbell, North Rankin West, Saladin, Tubridgi Total	5.74	11.77
TOTAL	**112.84**	**140.27**
PROBABLE		
Angel	14.90	35.90
Brecknock	92.00	141.00
North Gorgon	11.13	130.89
Scarborough	70.00	350.00
Scott Reef	306.99	499.00
Tern	15.08	17.81
Tidepole	12.80	14.80
West Tryal Rocks	11.30	80.77
Bambra, Brewster, Central Gorgon, Gorgon, Rankin, Spar, Wilco Total	21.37	149.18
TOTAL	**555.57**	**1,419.35**
TOTAL RESERVES	**858.06**	**1,778.52**

P90% : 90% probability of recovery
P50% : 50% probability of recovery

Mineral Resources

General: Mineral resource projects are shown on Fig. 1. As potential markets for tidal energy the iron ore site at Yampi and the Cadjebut zinc/lead site are closest, with future developments for diamonds at Ellendale and for bauxite at Mitchell Plateau. However, it is understood that Yampi Sound is due to close by 1993. Cadjebut already produces concentrate and so increased energy demand can relate only to scaling up. The diamond mine at Ellendale is unlikely to be developed before 2000, but that could suit a tidal energy programme.

Mitchell Plateau bauxite may be developed in some 5-10 years with an initial capacity of 700,000 tonnes /a, perhaps moving to 2 million. An alumina plant and particularly a smelter would have significant energy demands.

Energy demand from mineral developments: Mt. Newman mining has a power demand of 30 MW, and the Robe River Iron facility at Cape Lambert supplies sufficient gas fuelled power for Robe and into the domestic NW grid.

Table 6. Gas production in Western Australia for 1988

Field	Operator	Natural gas (m^3)
Barrow Island[3]	WAPET	101,575,000
Dongara	WAPET	144,334,000
Harriet[1] and Rosette	Bond Corp	97,242,000
North Rankin[2]	Woodside	3,780,581,000
Woodada	Consolidated Gas	21,470,000
North Herald[1] and South Pepper[1]	Western Mining Co.	106,161,000
Total		4,251,363,000

[1] Flared
[2] Excludes gas re-injected for enhanced condensate production.
[3] Re-injected for oil recovery

The energy requirements for ore beneficiation are quite variable but the ranges are: iron, 10 - 40 kWh/t; and copper, 15 - 60 kWh/t. A typical figure for NW Australian operations over both ores for budgeting would be 30 kWh/t.

There is an energy demand in converting bauxite to alumina but that for smelting alumina to metallic aluminium is much greater at some 15,000 kWh/t.

Power systems

Regional: Electrical generation from tidal energy in the Kimberley Region could permit extension of the Pilbara grid system, to meet future demand arising from increased development of resources. This would release non-renewable fuels (natural gas), for which a transmission system exists to serve the more populated and industrialised south west region of the State. The transmission of tidal-generated electrical energy from Kimberley to the south west of Western Australia or interstate would not seem warranted at present, given the ratio of demand to distance.

Existing power installations: Power stations already serving the region are given in Table 7.

Hydrogen

The Select Committee referred to the possibility of using tidal power to generate hydrogen by large scale electrolysis of water. Such clean burning

TIDAL POWER

Table 7. Power installations in the North West

Location	Fuel	Capacity kW	Acquired or Commissioned From	Energy Generated GWh 1988/9
NON-INTERCONNECTED SYSTEMS				
- Broome	Light Fuel Oil	11,020	1976	37.38
- Derby	Light Fuel Oil	11,100	1973	23.24
- Fitzroy Crossing	Light Fuel Oil	2,564	1976	5.05
- Halls Creek	Light Fuel Oil	2,668	1970	4.53
- Kununurra	Light Fuel Oil	12,400	1970	27.55
- Lake Argyle	Distillate	570	1985	0.42
- Marble Bar	Distillate	1,386	1973	1.84
- Nullagine	Distillate	370	1973	0.75
- Port Hedland	Heavy Fuel Oil	71,360	1967	36.33
- Wittenoom	Distillate	800	1975	1.97

Source: SECWA report of 1989

gas could be blended into natural gas and the Committee suggested that such a possibility should be taken into account in the design of new transcontinental pipelines.

Tidal power prospects
General

A desk study has been made of potential tidal power projects in the Kimberley Region. Topography and coastal physiography has been adopted from enlarged naval charts.

Parametric methods (ref. 4) have been adopted to establish very preliminary power and energy capacities of four of the most promising sites. Comparative costs have been used to achieve an outline ranking for the sites. This analysis for the leading three is presented in Table 8 along with transmission distances to possible load centres.

For a comparison with a project at a more advanced stage, data for the Mersey is also included in Table 8. Mersey has two entries, one on the same parametric basis as used for Kimberley, and one for the latest project parameters. The actual Mersey Project has a greater installed capacity and

Table 8. Ranking and comparative costings of Kimberley potential tidal power sites: $A 2.3 = £1

Site	Barrage Length m	Area of basin HWS, km²	Mean tidal range m	Maximum water depth at barrage	Installed capacity MW	Annual production GWh	Unit Energy cost x/kWh (Comparative Only)	Inferred capital cost (Jan 1991) Million $A	Transmission distances Km Mitchell Plateau	Derby
Secure Bay	900	94	5.2	22	590(740)	1070(1400)	2.7	(1800)	300	130
Walcott Inlet	2000	264	5.5	38	1750	3310	3.4	-	290	140
Mersey (Parametric)	1700	60	6.45	23	560	1070	3.4	-		
Mersey (Actual with pumping)	1700	60	6.45	23	700	1400	-	2220		
George Water	1500	112	3.35	27	230	480	6.0	-	200	210

St. George Basin has attractive features but an apparently excessive water depth at all potential barrage sites.

Figures in brackets are indicative for Secure Bay with high installed capacity and flood pumping cost based on Mersey detailed estimates at January 1991.

St.George Basin has attractive features but excessive water depth at all potential barrage sites; 1 Figures in brackets are for high installed capacity and flood pumping cost based on Mersey detailed estimates at Jan. 1991.

energy output than the parametric version. A factor is the adoption of flood pumping at Mersey, which would probably also be advantageous at Kimberley.

Of the parameters used, that of maximum water depth is most uncertain although it is known that the ria entrances are often deep and rocky. In the case of St.George Basin apparently excessive water depth has led to its rejection in the present study.

Earlier assessments

Tidal power schemes were examined previously by the Public Works Department (ref. 5). The most promising in terms of power output per kilometre of dam was Walcott Inlet, with an area of 250 km and a mean effective tide of 11 m. Secure Bay was considered a further possibility. Geological aspects are discussed in a preliminary report (ref. 6) which concludes that handling tidal conditions is going to create more problems than local geology.

In earlier State assessments for a West Kimberley deepwater port, Collier Bay and particularly Secure Bay, were considered possibilities.

Construction materials
Sand and gravel: The beds of most major rivers draining western Yampi Peninsula contain medium-grained clean quartz sand suitable for concrete. Those draining the east and south contain feldspar and are less suitable.

TIDAL POWER

Small deposits of quartzite cobble and boulder gravels are present close to outcrops of conglomerate in the central Yampi Peninsula.

Rock-fill: Beneath the zone of surface leaching most of the sandstone in the area is compact and silicified, and suitable for rock-fill material, as would be the local igneous rocks.

The projects
Secure Bay Project (Fig. 3)

Parametric work suggests 1070 GWh p.a. of energy from 590 MW of installed capacity and cost comparisons with Mersey appear favourable. Water depth appears good for construction purposes and the setting of turbines to avoid cavitation. An installation similar to that proposed for the Mersey may be appropriate.

Secure Bay, at the south end of Collier Bay, is large, with a narrow entrance named The Funnel. The bay is entered between two islets 560 m apart, and the tide runs at 2 to 3 m/s creating violent eddies. There are depths of 18.3 m within the north shore of the bay, the south being shallower. The surrounding hills are mainly steep and rugged. In the south east corner of the

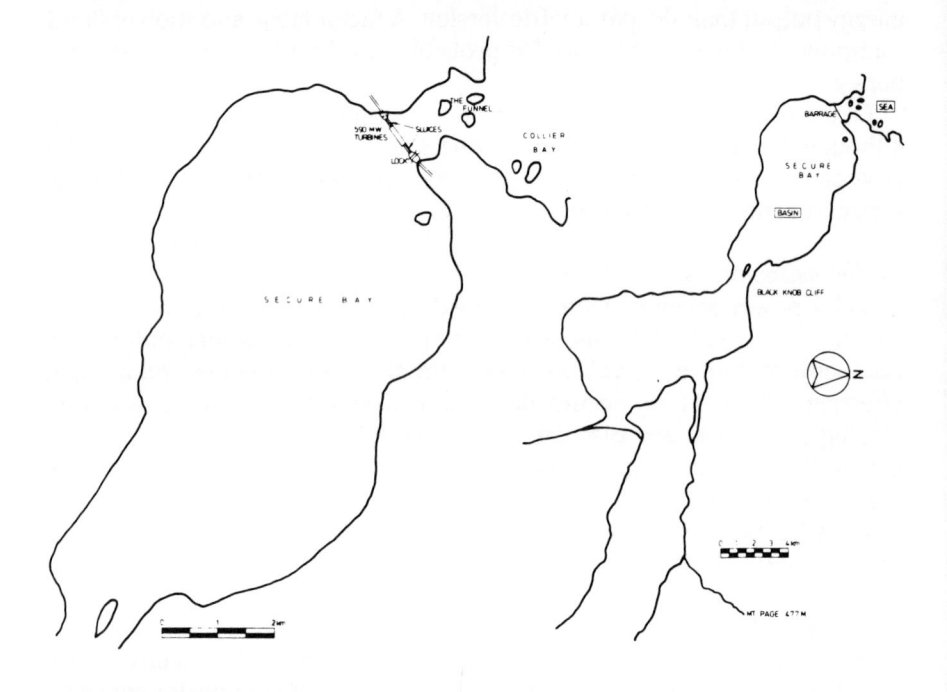

Fig. 3. Secure Bay Barrage

Bay there is a narrow channel with very strong tidal streams leading to a basin of similar size, bounded by mangrove banks and creeks.

Any placing of caissons in the Funnel during construction may require particular care. Transmission distances to Derby and Mitchell Plateau of 130 km and 300 km respectively are not too daunting. Scott Reef might provide gas for turbines to firm up the power installation.

Walcott Inlet Project (Fig. 4)

This would be a significantly larger project than either Secure Bay or Mersey with potential installed capacity and energy generation of 1750 MW and 3310 GWh/a respectively

Yule Entrance, which leads into Walcott Inlet, is obstructed intermittently by a sand bar which dries. The channel through the entrance has been recorded as wide and deep with a dividing island about 3.2 km long. The north channel is 3.2 km long and 1500 m wide, with depths of from 27 m to 55 m, and velocities up to 3 m/s. The channel south of the island had similar depths. The narrowest part of Yule Entrance is just beyond the channel junction with depths of over 91m, velocities of about 4m/s, and whirlpools.

Walcott inlet extends 51 km east and then 10 km north and is bordered by hills. For the size of this project and the apparent water depths available, larger size turbines, say up to 9m diameter, might well be appropriate. Tidal

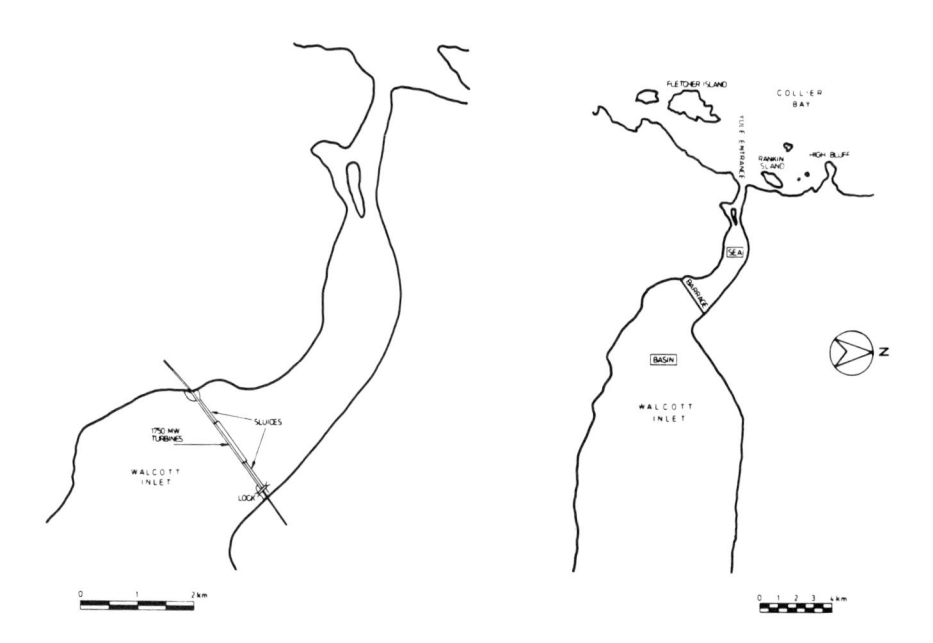

Fig. 4. Walcott Inlet Barrage

TIDAL POWER

streams are again such as to cause construction difficulties. Studies including mathematical modelling would be needed concerning the permanence of the entrance channel.

Parametric costings assess this project as about the same as those for the Mersey. Transmission distances of 290 km to Mitchell Plateau and 140 to Derby are manageable. Fuel for gas turbines to firm up the power could be from Scott Reef.

George Water Project (Fig. 5)

This project of 230 MW installed capacity and 480 GWh /a is somewhat expensive in parametric comparisons. It could comprise two separate barrages each side of an island, the larger installation being in Success Strait.

Success Strait, is the western and wider of two passages leading from Doubtful Bay to George Water. It is approximately 10 km long, 370 to 560 m wide with depths of from 18.3 m to 31 m and considerable tidal velocities. The northern approach is through a large shoal area which dries in places.

The Eastern Passage is 3.2 km long, and 1,300 to 2,800 m wide. The tidal streams are stronger than in Success Strait. Water depths could be so great as to cause construction difficulties.

George Water, a landlocked bay north of Doubtful Bay, has general depths of 27.5 m. A mud bar extends across the northern end with a depth of 11 m to 22 m. It could be economic to dredge this bar to use the northern storage.

Transmission distances to both Derby and Mitchell Plateau are 200 km. Scott Reef gas could firm up power.

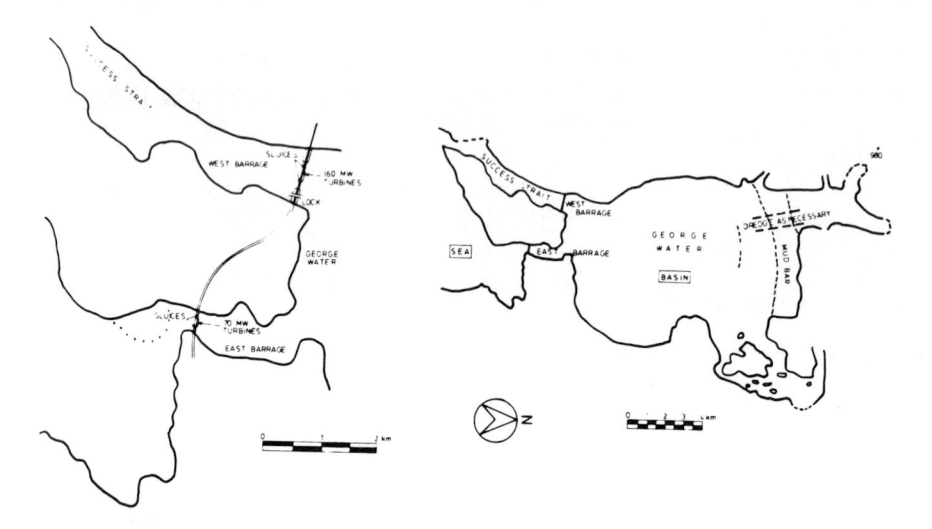

Fig. 5. George Water Barrage

Kimberley tidal power prospects

Tidal Power Projects at Secure Bay and Walcott Inlet may be attractive in a world scale. Climatic and hydrodynamic conditions are severe and infrastructure is negligible. However parametric assessments indicate that further studies should certainly be worthwhile.

The strongest doubt for such developments exists in unkown future power demand in the region. For this, macroeconomic planning will have a vital role to play. The long distance transmission of electrical energy from Kimberley to the south west of Western Australia or interstate does not seem warranted. However, the Select Committee has recommended study of production by electroysis of hydrogen for blending with natural gas to be transmitted by future pipelines.

Proven gas resources in the broad region would meet current demand for some 200 years. Other mineral resources are substantial and their extraction and processing could create a local demand for energy. The favourable long term economics of tidal power projects could have a significant part to play in the regional development package and in husbanding the finite hydrocarbon resource. The hydrocarbons themselves permit firming up of tidal energy with gas turbine plant of low initial capital cost.

References

1. AUSTRALIAN BUREAU OF METEOROLOGY. *Climatic Survey, Northwest Region 6-WA*. Australian Bureau of Meteorology.
2. GELLATLY, D. C. and SOFOULIS, J. *Geological Memoir Yampi, Sheet SE/51-3*. 1973.
3. McWHAE, *et al*. The stratigraphy of Western Australia. 1958.
4. BAKER, A. C. The development of functions relating cost and performance of tidal power schemes and their application to small scale sites. *Tidal Power*, Thomas Telford, London, 1986.
5. LEWIS, T. G. The Tidal Power Resources of the Kimberleys. *J. Inst. Eng. Australia*. 1963.
6. GORDON, F. R. Secure Bay - Walcott Inlet tidal power scheme. *Geol. Survey of Western Australia Rec.*, 1964/6 (unpublished).
7. VARIOUS AUTHORS. *Papers on the Mersey Barrage. Proc.4th Tidal Power Conference, Institution of Civil Engineers, London, 1992*, Thomas Telford, London.
8. Hon. IAN THOMPSON. *Report of the Legislative Assembly Select Committee on energy and the processing of resources*. Western Australia Legislative Assembly, 1991.

15. Update on Severn Barrage project studies

H. J. MOORHEAD, Project Manager, Severn Tidal Power Group Studies, R. POTTS, Director, International Research & Development Ltd, T. L. SHAW, Shawater Ltd and C. J. TEAL, Principle Environmental Scientist, Wimpey Environmental Ltd

This Paper summarises the various activities and studies which have been in progress on the Severn Barrage since the last Tidal Power Conference. These have included: a public consultation exercise following the publication of the major reports in 1989/90, further studies on the sediment dynamics of Bridgwater Bay and on primary productivity and bird populations in the estuary, development of energy capture algorithms for use in a two-dimensional mathematical model, and examination of the effects of the construction and long term operation of the barrage on the region.

Introduction

At the time of the 3rd Tidal Power Conference (November 1989), a major study had just been carried out by the Severn Tidal Power Group (STPG) into the technical, environmental, regional and economic feasibility of the Severn Barrage Project. The General Report on the studies had been published by HMSO as *Energy Paper No 57* in October 1989 (ref. 1) and the Detailed Report volumes were shortly to be published by the Energy Technology Support Unit (ETSU) in the Renewable Energy series of Contractor's reports (ref. 2). Subsequent to publication of this major series of reports the studies have been progressed on a number of fronts.

(a) In the foreword to EP57, the Secretary of State for Energy announced a period of consultation during which views and comments on the project could be expressed in writing. The views and comments have been collated by STPG and published by ETSU in early 1991, also in the Contractor's report series (ref. 3).

(b) In accordance with recommendations in the report, a number of environmental issues which lie on the "critical path" to project implementation have been studied further.

(c) The energy output has been checked by developing the computations for incorporation into a 2-dimensional mathematical model.

(d) Following approaches by the Standing Conference of Severnside Local

TIDAL POWER

Authorities (SCOSLA) to the Department of Energy (DEn) further studies have been initiated into the effects of the construction and long term operation of the barrage on the region.

(e) Due to the privatisation of the Electricity Supply Industry (ESI), the organisation and financing studies were deferred from the EP57 work. Discussions are now in hand between the STPG, DEn and ETSU on processing the financing aspects of the project.

This Paper outlines the results of the work concluded to date and indicates the progress achieved on the other aspects noted above.

Report on responses and consultation

Following the announcement in EP57 of the consultation period and direct approaches to leading organisations, over 300 responses had been received by the end of 1990. For the purposes of analysis these were grouped as follows

Subject	Number of References
Technical, scientific, regional and general comments	[a]110
Letters to the Press	14
Requests for information about the project	[b]60
Requests for speaker or presentation on the project	31
Requests for information for educational purposes	100
Promotion of services or materials supply	21
Attendance at seminars or exhibitions	13
Participation in technical conferences	6

[a] Plus SCOSLA.
[b] Including one reference which lists copies of Detailed Report volumes issued to over 80 organisations.

The consultation period proved an extremely valuable exercise for estuarine studies in general and the Severn Barrage Project in particular. Technical, scientific, regional and general comments comprised over one third of all the responses. All the main subject aspects of the report were well covered by the responses and many recognised the very considerable amount of work already carried out. A representative selection of contributions together with an outline of their content is summarized in the Table 1.

The responses received form a valuable source of background information for ongoing and future work. With few exceptions, the most favourable responses came from industry, commerce, regional and leisure groups, followed by local and public authorities who tended towards a neutral

PAPER 15: MOORHEAD, POTTS, SHAW AND TEAL

Table 1. A representative sample of consultation contributions

Source of Response	Favourable	Neutral	Critical	Remarks
Local Authorities				
Avon County Council	✓			Many well-balanced comments.
Bristol City Council		✓		Comprehensive response.
Cardiff City Council	✓			Comments/recommendations.
Standing Conference of Severnside Local Auths	✓			Very comprehensive and constructive.
Stroud District Council			✓	Critical, undesirable changes caused.
Sully Community Council		✓		A limited response.
Statutory National Public Authorities				
Countryside Commission (SW Region)		✓		Very constructive response.
National Rivers Authority	✓	✓		Very comprehensive and constructive.
Nature Conservancy Council		✓	✓	Comprehensive, mainly critical.
Environmental and Ecological Groups				
Bristol Ornithological Club			✓	Views as for SECG.
Council for the Protection of Rural England (with CPR Wales)	✓			A constructive response/suggestions.
Friends of the Earth (Bristol & Severnside)	✓	✓		Opposes, but makes suggestions.
Royal Society for the Protection of Birds			✓	Comprehensive, lack of energy strategy.
Royal Society for Nature Conservation			✓	Critical, refers-lack of energy strategy.
Severn Estuary Conservation Group		✓	✓	Comprehensive partially critical.
Severnside Green Party	✓			Comprehensive favourable response.
Somerset Trust for Nature Conservation		✓		Comprehensive with suggestions.
Wildfowl & Wetlands Trust		✓	✓	Supports and supplements SECG.
Regional and Leisure Groups				
Country Landowners Association		✓		Constructive comments/suggestions.
Royal National Lifeboat Institution	✓			Lifeboat and safety issues.
Royal Yachting Association	✓			Yachting and boating issues.
Salmon & Trout Association			✓	Concerned about fish passage.
South Western Council for Sport & Recreation	✓			Recommends management plan.
Sports Council for Wales	✓			Recommends management plan.
Wales Tourist Board	✓			Supports further studies.
Welsh Federation of Coarse Anglers	✓			Comprehensive constructive response.
Industry and Commerce				
Associated British Ports			✓	Concerns re Cardiff, Newport, Barry.
ASW Holdings plc	✓			Supports concept - gives suggestions.
Brain & Co Ltd	✓			Supports concept - raises questions.
Bristol Initiative	✓			Supportive.
Burley House Group of Companies	✓			Supportive.
Cardiff Chamber of Commerce & Industry	✓			Supportive with further research.
Chemical Bank	✓			Supportive comments.
Confederation of British Industry - Wales	✓			Supportive but queries power market.
Dale Owen	✓			Supportive with suggestions.
Newport & Gwent Chamber of Commerce	✓	✓		Supportive with comments.
Severn Estuary Ports		✓		Requests "no detriment" commitment.
South Wales Institute of Engineers	✓			Offers general support.
South Western Electricity Board	✓			Supportive comments/questions.
TUC South West Regional Council	✓			Supportive comments/suggestions.
Wales TUC	✓			Supportive comments/suggestions.
Welsh C'ttee for Econ. & Indust. Affairs	✓			Supportive comments re further work.
Welsh Water	✓			Supportive - lists issues for action.
Wessex Water	✓			Supportive - lists issues for action.

Note: Where more than one ✓ is shown it indicates that the comments do not fall wholly within one category.

position, while the more critical response were generally from environmental and ecological groups. On balance, comments were in favour of the project which many urged the government and STPG to progress.

The open consultation procedures used in the studies involving Advisory Panels and Working Parties incorporating persons from outside the STPG were welcomed and it was recognised that the Severn was now probably one of the best studied estuaries in Britain. Nevertheless, responses noted that much work remained to be tried out, particularly on sedimentation and ecological matters. The strongest reservations came from those organisations representing conservation interests and the ports: some conservationists wished to retain the hyper-tidal estuary because it was unique in Britain, and the ports were concerned at the potential damage to their trade due to the uncertainty about whether the project would proceed and any reduction in the high water levels.

Various organisational matters were mentioned in many responses and there were suggestions for an energy policy embracing renewables, local planning requirements, coastal zone management and a Management Authority for the Estuary.

The following recommendations were made.

(a) To continue with further ecological studies on the critical path to an environmental assessment.
(b) To carry out further regional studies on the effects on the local authorities and ports; this would require continued liaison with SCOSLA and the port authorities.
(c) To consider carrying out organisational studies at an early date.

These recommendations are being implemented as shown in the following sections.

Further environmental and energy capture work

In accordance with EP57 and while the Responses and Consultation work was in progress, with joint DEn/STPG funding, the STPG commenced work in August 1990 on three items which could usefully be taken forward on an interim basis

- *Sediments:* Studies to advance understanding of the sediment and turbidity regime of the Severn Estuary and Bristol Channel including their relevance to the ecological features of this zone, focusing particularly on the Bridgwater Bay area.
- *Energy capture:* A recheck of the optimisation of the energy capture and value of the barrage output using algorithms specially developed for use in 2-D mathematical modelling.

- *Primary productivity and bird populations:* A study of key aspects of primary productivity including phytoplankton in the water column and algae in the intertidal area; also further work on waterfowl feeding behaviour including the specific requirement of dunlin, the species which makes up more than 75% of the Severn Estuary's bird population.

Sediments

In the General and Detailed Reports published in October 1989 and March 1990, it was noted *inter alia* that tidal flows over Bridgwater Bay would be modified by the Barrage, although the Bay itself lay outside the proposed alignment for the Barrage. There was concern that sediments mobilised particularly during westerly storms could be carried through the sluices nearest the coast and instead of being carried back on succeeding ebb tides as at present could tend to remain in the basin (ref. 4). The overall objective of this latest programme was to advance the understanding of the sediment and turbidity regime of the Severn Estuary and Bristol Channel focusing particularly on the Bridgwater Bay area and including their relevance to the ecological features of this zone.

The field investigations included three main tasks under the initial stage.

(*a*) Setting up and survey of three east-west transects from the high water mark out to the limit of the sub-tidal mud area in order to accurately measure the profile.

(*b*) Collection of a series of core samples in the inter-tidal zone. These cores to be analysed to examine stratigraphy and physical properties of the material.

(*c*) Surveys of a series of north-south transects by dual frequency echo sounder to establish the variability of the superficial sediment layer during typical Spring/Neap cycles.

Subsequently a series of repeat surveys of the defined trensects were conducted to check the seasonal variability of the profiles and the effects of episodic events.

Correlation has also been sought where possible between information coming from the present studies and information and data from earlier work by STPG and others; in particular

(*a*) site investigations carried out by STPG in 1988 as part of barrage foundations and alignment studies (ref. 5)

(*b*) submarine cable-laying operations carried out in 1988 in the areas of Berrow Sands

(*c*) coring and x-ray examinations of deposits in Bridgwater Bay by IOS (ref. 6)

TIDAL POWER

(d) cross-sections surveyed by the Somerset River Authority in 1954.
(e) bathymetric surveys carried out over the past 100 years for the preparation of Admiralty charts.

Arrangements were also made for a continuous record to be kept of wind speed and direction on a daily basis throughout the period of the investigations. Where possible, repeat surveys were arranged to be carried out as soon as practical after major storm events.

The field work was divided into three main sub-tasks

- Inter-tidal bed levels
- Sub-tidal echo sounding
- Coring and bed densities.

Inter-tidal bed levels: Three transects were established at Brean, Berrow and Stert at the locations shown in Fig. 1. A permanent 3 m long horizontal datum bar was installed approximately 500 mm above the bed on each transect. These datum bars were levelled to OD Newlyn by conventional land levelling techniques. The transects were surveyed over high water on a spring tide using a shallow draught vessels fitted with a high frequency echo sounder. This data was processed using the datum bar height above the local bed level (measured during the survey) and the known rate of change of water level. The latter obtained from a temporary tide gauge installed at Burnham for the purpose. These individual readings were averaged over 50 m to produce a profile free of wave motion and localised irregularities.

Sub-tidal echo sounding: A series of echo sounding surveys using a dual frequency transducer were carried out along predetermined lines in the sub-tidal zone of Bridgwater Bay during a Neap to Spring tidal cycle. The purpose was to check for the existence or build-up of significant areas of fluid mud during neap tide periods and monitor possible re-entrainment as tidal ranges increased towards spring tide conditions. Two particular areas which had previously been identified by IOS as "accretional" and "erosional" were selected for repeat surveys. These areas are referred to as Western Box and Eastern Box respectively on Fig. 1.

An indication of the superficial soft mud thickness was obtained by registering the differences between the recorded depths from the two frequencies (33 kHz and 210 kHz) of a Deso 20 echo sounder. Although the difference between the two frequencies does not necessarily indicate that the layer is fluid, it does however, indicate a lower density deposit which is less consolidated and probably ephemeral.

*Coring and bed densities:*The coring and bed density work was sub-divided into the following

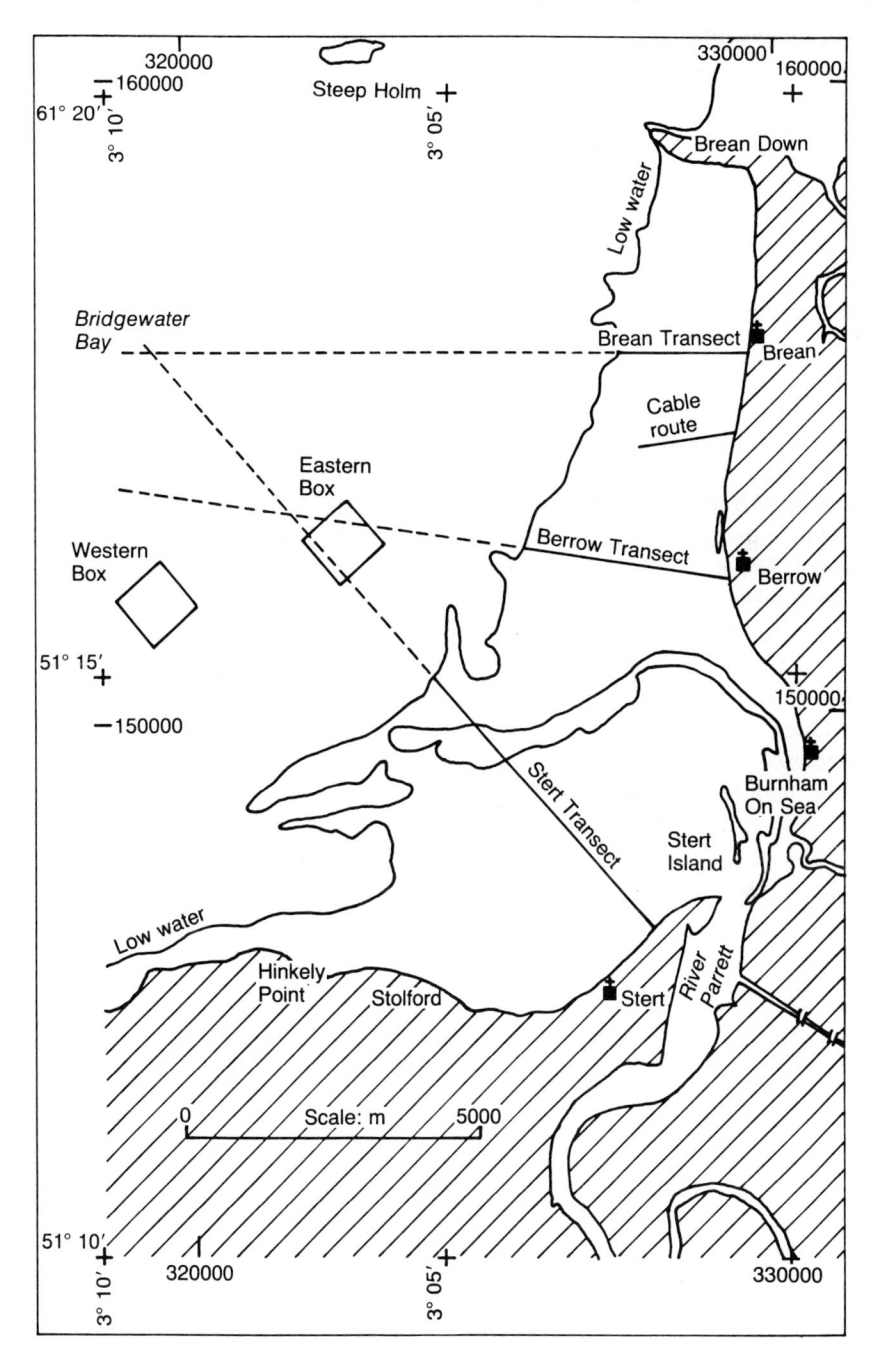

Fig.1. Location of bed level transects and echo sounding areas

TIDAL POWER

- core collection
- X-ray radiography
- gamma spectrophotometry
- in-situ bed density profiles.

A number of core samples were collected in Bridgwater Bay to provide data on the history of accretion/erosion in the area. By subjecting the cores to x-ray radiography, core descriptions, in terms of the layers making up the top 0-2 m of the sediment were detailed. The recent history of these bed sediments was also examined by the use of Caesium-137 (^{137}Cs) and Caesium-134 (^{134}Cs) gamma spectrometry on sub-samples taken at different depths through each core. These isotopes of Caesium have been introduced into the environment over the last 40 years or so from the nuclear industry and from nuclear weapons testing. Thus the presence or absence of these isotopes in the sediment sub-samples could help to determine whether the sites are accretional, erosional or stable, at least during the last 40 years.

The coring was carried out using a Kasten Corer from a shallow draft vessel. The Kasten Corer is driven into the sea bed by gravity like a conventional cylindrical barrel corer, but, having a much lower area ratio, yields substantially better cores: entry ratios better than 95% are achieved compared to 60-70% for cylindrical barrel corers. The corer also yields more material for testing.

Once the core had been taken, the barrel was split on the diagonal and, by means of thin formica sheets and a thin wire, a slice 20 mm thick and 90 mm wide was taken, with one face on the diagonal. The core slices were then taken to the laboratory for x-ray radiography. Sub-samples were taken at approximately 50 mm intervals down the core length for possible 137 TCs gamma spectrophotometry analysis. These sub-samples consisted of about 200 g of sediment from each level. Each sub-sample was stored in a plastic bag and deep frozen as soon as possible until analysis.

Results: The results of this further programme of work on the sediments of the Bridgwater Bay area will be presented at the Conference.

Energy capture

From energy capture studies carried out on the Severn Barrage Development Project (ref. 7) it was apparent that whilst the results, obtained from the use of computer based 0-D and 1-D mathematical models, were in reasonable agreement with regard to the energy output from ebb generation only, the corresponding results for ebb generation plus flood pumping showed significant disparity. In particular there was a factor of three difference between the energy benefit attributable to the addition of flood pumping; with the results from the 1-D model being the smaller. Both the 0-D and 1-D models were capable of optimising the operation of the barrage to maximise the energy sent out. In terms of computing time this proved to be very time consuming with regard to the use of the 1-D model. Within the timescale of the work it was not possible to resolve the disparity in the results but it was noted that the 1-D model results showed apparent exaggerations of the water surface movement with time, local to the barrage. It was considered that the comparative reduction in energy benefit from flood pumping was directly attributable to these exaggerations. It was also noted that the results from the 2-D modelling work (ref. 8) carried out as part of the Estuary hydrodynamics studies showed only relatively small effects on the water surface movement as a result of significant perturbations to the operational status of the barrage. However it was not considered practicable to investigate energy capture using a 2-D model capable of optimising the operation of the barrage to maximise the energy sent out. Arising from the above energy capture studies it was proposed at the 3rd Tidal Power Conference (ref. 9) that to a close approximation the optimum operation of the turbines could be represented using a simplified algorithm. With this it was practicable to investigate energy capture using a 2-D model which would much more closely model the hydrodynamics of the estuary.

At the beginning of the present studies algorithms were developed using 0-D models to cover the turbine and pumping modes of operation. Using this approach a number of investigations were carried out to establish how well the 2-D model predictions for water levels and energy output per tide compared with those coming from the 0-D model. Ebb generation only was considered in the first instance. The 0-D model was used initially to determine the "optimized" timings for the start of generation, start of sluicing, etc. for a given tidal range at the barrage. Thereafter the 2-D model was run, driven by a tidal range at Port Isaac, suitably adjusted to give the required tidal range at the barrage. The results were compared for a 5.5, 8.5 and 10.5 m tidal range at the barrage in order to establish how well the results from the simple algorithmic 0-D model, with a sinusoidal tide form, compared with those from the 2-D hydrodynamic model.

When the pump algorithm had been established, using the 0-D model and the previous results from the 1-D modelling, which optimized the pump operation within the total cycle a further comparison was made to establish how close the 2-D results compared with the previous 1-D and 0-D results.

Results: It is intended to present the results of this further programme of energy capture work at the Conference.

Primary productivity and bird populations
Primary productivity: The results of the ecological studies published in the General and Detailed Reports identified areas requiring further attention (ref. 10). A principal deficiency was concluded to be a shortage of key field

data in the areas of primary productivity and on long run biological variability.

During the present programmme of studies two key aspects of primary productivity were made the subject of further data collection and study.

The first refers to the species composition, standing crop and productivity of phytoplankton communities in the water column according to season, tide range and depth below the water surface. The principal issue here is how turbidity constrains productivity, hence it was proposed that the necessary field monitoring work should be closely co-ordinated with the work on sediments.

The second refers to intertidal algae. The species and the extent and timing of their development about the estuary through an annual cycle are poorly recorded. A basic survey linking observations to substrate type and how changes in sediment regime are reflected in the algal species present was therefore proposed.

Bird populations: The above report also drew attention to the need for further monitoring of bird species numbers and location. The waterfowl populations, primarily waders and shelduck, had been previously studied by several different organisations, generating substantial data sets. However, each organisation and project had used different methodology. Consequently, it had not been possible to make direct comparisons of distribution patterns within the Severn. The only studies which yielded comparable results for a run of years was the Birds of Estuaries Enquiry (BoEE) of the British Trust for Ornithology which monitors total bird populations.

The above studies carried out by the British Trust for Ornithology during the winters of 1987/88 and 1988/89 gave the first detailed comparative data for distributions of waterfowl throughout the Severn Estuary. The low tide counts which were the main thrust of the distribution data for the Severn and which were done in a rigorously standardised way, can be used for detailed comparisons of distribution patterns and habitat preferences for each species wintering on the Severn.

There were three main objectives to the present studies. Firstly, to obtain another winter's intensive survey of the low tide distribution of waders and shelduck on the Severn. Secondly, to look in detail at the sediment on one site, Clevedon, and examine the relationship between sediment mobility and bird distribution. And thirdly, to assess the variability in habitat use through the tidal cycle at selected sites around the estuary.

The 1989 General and Detailed Reports also draw attention to uncertainties which existed regarding dunlin, in a with-barrage situation, as it appeared that in south western estuaries there was little spare capacity for dunlin.

Preliminary findings from earlier studies done in the Severn Estuary had suggested that dunlin occur on only part of the intertidal area available to them and tend to concentrate on the muddier substrates. Wider evidence confirmed that the numbers of dunlin wintering on estuaries generally are as variable as the substrate types themselves. However, these factors had not been correlated in earlier work. In view of the changes to the sediment regime of the intertidal area which may follow barrage construction it was decided to assess the main sediment types used by dunlin on UK estuaries, and to establish from this information the extent to which those estuaries are at carrying capacity in respect of this species.

The further studies had therefore three main objectives. First, to assess the densities of dunlin throughout British estuaries in relation to their likelihood of being at carrying capacity. Second, to assess the main sediment types used by dunlin and relate this to their dispersion patterns on British estuaries. Third, to predict the likely effect of changes in substrate type on the numbers of dunlin that the Severn could accommodate post-barrage.

Additional regional studies

As a result of the Responses and Consultation work and of the representations made by SCOSLA, there arose a perceived need to develop further the earlier regional studies carried out by STPG and reported in EP57, with particular regard to the impact of the construction and long term operation of the barrage on the local Authorities and their areas, and the Ports, including consideration of legal and administrative issues.

Work on these items commenced in May 1991 under joint DEn/STPG funding and is sub-divided into the following tasks.

- 1. Impacts of the construction of the barrage

    (*a*) Labour and staff availability
    (*b*) Effects of indigenous and in-migrant labour use
    (*c*) Effects on infrastructure of labour and materials transportation and the effects of any potential local sourcing of materials
    (*d*) Local development pressures.

- 2. Impacts arising from the existence and operation of the barrage

    (*a*) Effects of potential increased tourism and leisure activities
    (*b*) Development pressures and land use demands
    (*c*) Possible port and shipping development
    (*d*) Effects of the barrage road crossing and the additional traffic generated
    (*e*) Drainage consequences.

TIDAL POWER

- 3. Initial consideration of possible arrangements for estuary management and conservation
- 4. Initial consideration of studies on legal and administrative matters
- 5. Further work on estuary management and conservation
- 6. Further work on legal and administrative matters.

By December 1991 Tasks 1(a), 3 and 4 had been completed. All sub-sections of Task 1 are expected to be concluded about March 1992 by which time most parts of Tasks 2, 5 and 6 will be in hand. Discussions were held with Government Departments, SCOSLA and other interested parties as part of the work in Tasks 3 and 4 which was to define Tasks 5 and 6. This work concluded that there was general support, although with reservations, that Task 5 should consist of the development, and subsequent appraisal, of a number of models for a Severn Estuary Authority having, essentially, the following responsibilities

(a) To act as the focal point in discussions with the developer, before commitment to the project, to determine the safeguards necessary for maintenance or improvement of the environment both during construction activities and subsequent operation of the barrage for electricity generation.
(b) To exercise control over subsequent, non-energy, developments expected to be promoted as a consequence of construction of the Barrage.

The reservations related to the possibility of transfer of statutory powers from existing authorities to the new authority and, in the case of the ports, for the new authority to be accountable for the financial consequences of decisions reached on conservancy and related port operations.

Various recent conservancy acts might form the basis for the development of models but account would need to be taken of the larger scale of the Severn Barrage and the recently increased public concern for the environment.

Task 6 would be limited to the review of the known legal position regarding the ports, Local Authority powers and boundaries, legal or "ancient" rights and rules on compensation, so far as the needs of Task 5 require it at the present stage.

The Additional Regional Studies are expected to be concluded towards the end of 1992 and the corresponding report should be available in Spring 1993.

Financial studies

The financial and organizational elements originally intended to be covered in the Severn Barrage Development Project study were omitted by mutual agreement of all three funding parties due to the decision by the

Government to privatise the ESI. As the privatization has now been concluded, DEn and the STPG are considering whether a jointly funded definition study should be put in hand to review the extent and nature of possible financial studies.

Acknowledgements

The Authors are pleased to acknowledge the substantial contribution made by the Department of Energy towards the cost of these studies.

References

1. ETSU. *The Severn Barrage Project, General Report.* Energy Paper Number 57.
2. ETSU. *The Severn Barrage Project, Detailed Report, Volumes 1-V, Contractor Reports.* ETSU TID 4060, p1-5.
3. ETSU. *The Severn Barrage Project, Report on Responses and Consultation, Contractor Report.* ETSU TID 4090, p1 -2.
4. ETSU. *The Severn Barrage Project, General Report.* Energy Paper Number 57, Section 3.1.5.
5. THE SEVERN BARRAGE DEVELOPMENT PROJECT. *Site Investigation Report.* Supporting Document Number SBP81.
6. KIRBY, R. AND PARKER, W. R.. *Settled Mud Deposits in Bridgwater Bay, Bristol Channel.* IOS Report No 107.
7. ETSU. *The Severn Barrage Project, Detailed Report, Volume 11, Chapter 1, Contractor Report.* ETSU TID 4060, p2.
8. ETSU. *The Severn Barrage Project, Detailed Report, Volume I, Chapter 2, Section 2.11 Contractor Report.* ETSU TID 4060, p1.
9. E. Goldwag and R. Potts. Energy Production. *Proc. of Third Conf. on Tidal Power, Institution of Civil Engineers, London.* 1989, Thomas Telford, London.
10. ETSU. *The Severn Barrage Project, Detailed Report Volume IV, Chapter 10, Contractor Report.* ETSU TID 4060, p4.

16. The environmental effects of tidal energy

S. J. MUIRHEAD, ETSU

Introduction

A tidal energy barrage placed across an estuary will inevitably lead to changes in the character of that estuary. However, the degree to which such changes may be regarded as detrimental to the environment, or environmentally acceptable will vary with both viewpoint and the estuary concerned. Some may regard any changes to the estuarine ecology as detrimental, while others may see the changes to the system having some benefits in broad environmental terms.

The environmental assessment of a tidal energy barrage is a complex and difficult task, because we are dealing with the dynamic estuarine environment on a site specific basis, and a renewable energy technology which provides environmental benefits in terms of savings of CO_2, SO_2, NO_x and waste disposal, arising from the displacement of fossil fuels which would otherwise be burned. A difficult balance has to be achieved between possibly detrimental changes to an estuary and a range of benefits which, in addition to the gaseous emissions already mentioned, include various opportunities for regional development and other socioeconomic factors. The assessment of many of these factors involves a degree of subjectivity in the perception of benefits and disbenefits to the environment, and in particular there is, as yet, no readily agreed-method for placing a monetary value on all the factors involved.

The aim of this Paper is to outline the environmental effects of tidal energy in relation to the estuarine ecosystem, and therefore will not deal with the regional environmental and socioeconomic implications of tidal power. Before starting a detailed environmental assessment of tidal energy some appreciation is needed of the nature of the estuarine environment and how, as a system, it works.

The Estuary

There have been various definitions of estuaries given, one of the most comprehensive being as follows

TIDAL POWER

An estuary is an inlet of the sea reaching into a river valley as far as the upper limit of tidal rise, usually being divisible into three sectors
- *a marine or lower estuary, in free connection with the open sea*
- *a middle estuary subject to strong salt and freshwater mixing and*
- *an upper or fluvial estuary, characterized by freshwater but subject to daily tidal action.*

The limits between these sectors are variable and subject to constant changes in river discharges. (Day et al. 1989)

Many definitions of estuaries are based for the most part on their physical and geological characteristics. This is probably because these are the most obvious features of an estuary, and they include the rise and fall of the tide, complex water movements, high turbidity levels, and different salt concentrations. The nature of landforms such as beaches, mudflats and the geometry of the basin are also very noticeable.

An understanding of the physical regime of an estuary will assist in leading to an understanding of the biotic (biological) processes, as the physical and biotic processes are closely linked within an estuary. From the seaward end of the estuary through to the inflowing river changes in, for instance, salinity, currents, water clarity, chemical concentrations and oxidizing and reducing conditions in the sediments influence the nature of the biota present along that estuarine gradient.

The alternate flooding and wetting of the intertidal area with each tidal cycle affects the assemblages of intertidal plants and animals, with the frequency, depth and duration of submergence of any given part of the shore being of particular importance to algal and salt-marsh plants.

Another important feature in estuaries is light and its availability. The euphotic zone, which is that part of the water column which is penetrated by sufficient light for plant growth, varies in depth with turbidity. Since turbidity varies spatially along the length of the estuary, and temporally, with the state of tide, plant productivity will be affected accordingly.

For both chemical (e.g. metal exchange) and biological processes within the estuary the gradient from oxidizing conditions (where oxygen is present) to reducing conditions (where oxygen is absent) has a significant influence on both water quality and animals within the estuary. The estuarine water column is usually oxygenated, but estuarine sediments are usually anaerobic a short distance below the sediment surface.

The amount of oxygen in the sediments is related both to the rate at which oxygen moves into the sediments and the rate at which it is consumed. The activities of the biota, including construction of burrows, facilitate the movement of oxygen and water through the sediments. Many plants that grow in oxygen-depleted conditions, including *Spartina*, the cord grass, actively pump oxygen into the soil and create a thin oxidized zone around their roots.

A longitudinal section of an estuary demonstrates the attributes that result from a mixing of fresh and salt water. Salinity gradually increases from that of fresh water to that of the sea. The isohalines often show that salinity increases with depth as a salt wedge moves under the less dense fresh water. The depth of the euphotic zone usually decreases towards the freshwater and the most turbid waters (the turbidity maximum) usually occur at salinities from 1 to 5 ppt.

The estuarine food dynamics are characterized by a variety of primary producers, grazing and detrital food chains, a high degree of interaction between the water column and bottom, numerous secondary and tertiary consumers forming a complex food web.

This brief review of the main estuarine processes illustrated the complexity of the estuarine ecosystem. It is necessary for tidal energy purposes to be able to understand the basic functioning of the system so that this information can be used to predict the post-barrage conditions. As different parts of the system are currently understood to different levels, predictions made now rely on a mixture of conceptual and mathematical modelling with gaps in our knowledge leading to the use of empirical formulae that may not take into account all the complexities of reality. There is a need to improve our predictive capability in mathematical modelling, so that an ecosystem model can be produced which can be used with some degree of confidence for portraying conditions post-barrage, or will at least allow assessment of the probability of particular post-barrage conditions occurring.

Superimposed on the natural cycle of estuaries has been the use and manipulation by man of estuaries through the centuries. Man has developed towns, industry and agriculture on the estuarine shore and along the water catchment, all of which have changed the nature of the estuarine environment. Two especially significant influences have been the reclamation of land for agriculture and the development of sea defences which have constrained the estuary within artificial boundaries.

Despite man's activities, many of the UK estuaries are of significant conservation interest, holding internationally and nationally important numbers of wintering waders and wildfowl. In recent years, however, continued pressure on the estuarine environment by development for leisure, industry or agriculture has led to a real concern for their continuing value as wildlife habitats. Protection for estuaries has been initiated through various conservation designations, notably two international designations for the protection of wetlands, the RAMSAR convention and the European Directive on the Conservation of Wild Birds (Special Protection Areas).

Environmental assessment of tidal energy barrages

The environmental assessment process, eventually culminating in the production of an Environmental Statement for a tidal energy barrage, has to be initiated at the very start of the project, firstly to identify any potential 'show stoppers' and secondly, due to the complicated nature of the estuarine system, there is a considerable amount of ecological data to collect and analyse. It may also be necessary to develop further the predictive tools necessary to assess the environmental impact. It is also appropriate to initiate consultations with the relevant authorities and interest groups at an early stage in the project to obtain feedback and input to the environmental assessment of the development. As the hydrodynamics are central to many of the other processes this is an appropriate point at which to start discussing the effects of a tidal energy barrage.

Hydrodynamics

The suitability of sites for tidal power is limited to those estuaries with a large tidal range and appropriate topography. These large tidal ranges lead to strong currents, and together with wave action, result in significant sediment movement.

The Severn Estuary, is often considered to be a "stressful" environment with respect to hydrodynamics (currents and waves), to the sedimentary regime and to salinity variations (gradients between freshwater and open sea and fluctuations during tidal cycles). Central to much of the work on the physical and biological environment is the need for an adequate understanding of the present complex tidal and sedimentary regime in the estuary.

A tidal energy barrage would cause a reduction in the tidal amplitude upstream of the barrage, although the tidal range would be affected less further upstream, and would tend towards that currently found. A change in the tidal curve would be seen with an extension of high water slack and of ebb flow.

In an estuary, especially one with water quality problems, it is important to determine the effect of a barrage on the flushing time of the estuary, as this time will increase post-barrage leading to longer retention times of potential pollutants.

Tidal current velocities would be also be reduced except in the immediate vicinity of the barrage. Reduced velocities may increase stratification of the water column (separation into layers of different salinity or temperature), and although these effects may be small within the Severn, they may be more significant in other estuaries. Ecologically they may be significant in relation to phytoplankton productivity, for example.

Mean distribution of salinity in the estuary would be changed but this would occur against a background of larger natural fluctuations. For the

Severn Barrage, for example, changes might be of the order of an increase of one part per thousand (ref. 1).

Suspended sediment loads and resulting turbidity of the estuarine waters would decrease and this is likely to have ecological effects.

The effect on the wave climate may also be of environmental importance particularly at high water slack when any potential erosive force may be concentrated at one point for a longer period of time.

Sediments

The sediments of an estuary have a major influence on its characteristics whether aesthetic, biological, or commercial. The combined action of tidal currents and waves is largely responsible for determining sediment features. River flows and density-induced currents are also important when and where they have sufficient strength to influence the regime established by tides and waves. The hydrodynamic changes caused by a barrage will alter the current velocities and wave action, with consequential effects on the sediment movement and the biological processes which these influence. Therefore in many estuaries where the environmental regime centres on and in the sediments it is important to identify the likely post-barrage conditions.

Sediment transport mechanisms mediated through wave and current action can only move material if it is available but how and to where this sediment moves is of both engineering and ecological importance. Coarse sediment transport has been relatively well defined as transport is primarily by bedload. However, the exact movement of fine cohesive sediments is not easily defined, due to the range of particle size, their reactivity and their transport as suspended particulate matter. Features of fine cohesive sediments such as erodibility are also not well elucidated. Laboratory measurements made in the past to determine critical erosion thresholds have been shown not to represent totally the processes going on in the field where an interaction of organic, inorganic, physical and biological factors affect sediment stability.

Although results to date give indications of the order of magnitude of possible changes, there is considerable scope for improvement. To make accurate predictions of ecological change a much more precise knowledge is required. Therefore generic studies under the Department of Energy's tidal enery programme are underway to improve our understanding of the processes involved.

Turbidity

Due to the high tidal range, and hence vigorous current activity, many of the estuaries currently being considered in the UK for tidal energy are highly turbid. Post-barrage, with a reduction in current velocities, some of the suspended material may settle out resulting in a reduction in turbidity.

A reduction in suspended solid concentrations in the water column would increase the depth of the euphotic zone and consequently the standing crop of phytoplankton.

The estuarine food chain in these estuaries is presently dominated by detritus, which is often referred to as particular organic matter, comprising of either natural or man-derived materials. This material is generally available to organisms such as zooplankton throughout the year. Post-barrage, a reduction in the availability of this detritus, but an increase in phytoplankton production due to increased light penetration, may shift the estuarine system to a more phytoplankton-based, and hence seasonally orientated one.

The key question for post-barrage predictions and water quality issues is whether the concentrations of suspended matter will decrease sufficiently to allow autotrophic primary production to dominate.

Such a development could be beneficial by providing improved resources for filter feeding invertebrates, e.g. mussels, which in turn could increase the food sources available for birds and fish.

Since phytoplankton growth in the estuaries under consideration appears only to be limited by low light availability, in clearer water adverse effects could occur if the algal growth became excessive, resulting in "nuisance blooms" or "red tides". However, to predict accurately the type of bloom that may appear post-barrage, knowledge is required of the species present in the estuary and how these species respond to temperature, salinity, nutrient concentrations and turbidity. Laboratory investigations such as the current Laboratory Mesocosm study at Plymouth Marine Laboratory are attempting to answer some of these questions.

Water quality

Considerable attention is given, both by the public and by regulatory bodies, to the water quality found in the UK's rivers, estuaries and coastal waters. The concept of a tidal energy barrage produces, in the minds of some, the image of a "dam" retaining polluted water with high nutrient loadings, leading to the formation of algal blooms of a "nuisance" or toxic nature. However, as yet there is no evidence to support this image.

The significance of potential water quality problems that may arise will depend on the estuary. For any barrage built in the future, however, water quality is likely to have been improved on the current status as pollutant inputs to estuaries decrease with the implementation of the Urban Waste Water Directive. However, nutrient inputs may be difficult to control as not all sewage improvement schemes will require nutrient stripping, and there will still be nutrient inputs from agricultural runoff. The issue of algal blooms will, therefore, still need to be considered. Increased retention of faecal bacteria may result from a barrage scheme due to the increased estuarine flushing time, but this could be offset to some degree in certain estuaries by enhanced light-induced mortality of bacteria as reduced turbidity allows greater penetration of ultraviolet radiation.

Another impact that may result from the reduced currents, and hence reduced sediment mobility, i.e. the alteration of sediment/water interactions. This would affect metals and organohalogens both in terms of binding to sediment particles and their availability to biota.

Post-barrage, increased low water volumes would give greater dilution of pollutants, and larger surface/volume ratios resulting in improved dissolved oxygen concentrations. Severn Barrage studies have illustrated the greater dispersion of effluents in the vicinity of the barrage, due to the "mixing action" caused by the operation of the scheme (ref.1).

In terms of preparing an environmental assessment there are a number of successful water quality models available to assess bacterial dispersion from an outfall. Potentially more difficult to assess is the long term fate of both present and historic metal and organic pollutant inputs which bind to the sediments. Good forecasts will be needed for the movement and future pattern of fine sediment and sediment water partition coefficients will need to be determined for each pollutant.

Fish

There is uncertainty about the effects of a tidal energy scheme on fish behaviour in the estuary and passage across the barrage line.

Commercial fish have been highlighted in this context, for example salmon and sea trout. In estuaries such as the Severn there are also small cod and flounder fisheries, primarily for sport, and there may be species of conservation importance. An estuary may also be an important nursery area for commercial stocks and certain species, such as shrimp, which form a significant link in the estuarine food chain, migrate seasonally within the estuary and may therefore have to pass the barrage. Studies on fish and shrimp are needed to determine seasonal, temporal and spatial patterns of movement in the estuary as this could influence their passage route through the barrage. On the ebb tide the route downstream is through the turbines unless there is success in installing fish passes and diversion techniques. There is the potential of multiple passage of salmon through the turbine, as the salmon on returning to its natal river may "look" into several estuaries or wait within the estuary until riverine conditions are suitable to move upstream to spawn.

Observations of fish damage in turbines has primarily concentrated on river hydroelectric schemes and published mortality rates have varied from

0 to 60%. Few studies have been carried out on tidal power barrages and work here has mainly consisted of initial exploratory studies (ref. 2).

The main types of damage attributed to hydroelectric schemes and the Annapolis Royal tidal energy scheme include gashes, severed heads, removal of scales, various forms of eye damage, fractured backbone and external and internal bruising (ref. 3). These main sources of damage are believed to be due to physical impact, pressure effects and shear or turbulence effects within the draft tube. Many of these types of damage are believed to be due to low plant operating efficiencies. However, the La Rance tidal energy scheme during its operation over the last twenty five years has reported no evidence of fish mortality although mortalities have been reported for shad and herring at Annapolis.

There is currently insufficient information available to provide adequate forecasts of fish mortality at tidal energy barrages, as past experiments have often been ill-designed and carried out on hydroelectric schemes with vertical axis turbines. An experiment-based, rather than an analogous approach is needed to predict conditions and the associated mortality in tidal schemes.

Recent laboratory experiments are beginning to show how certain types of damage are caused, and how fish vary in their vulnerability. Field experiments are now needed to evaluate potential mortality rates for different species and size of fish. Suitable projects are being considered under the Department of Energy's generic environmental programme on tidal energy. Other projects will examine aspects of fish behaviour and potential deterrence methods for diverting fish from the turbines to fish passes.

Birds

Birds are the most obvious and publicised component of the animal life in British estuaries. UK estuaries hold wintering populations of waterfowl of national (1% of the British population) and international (1% of the Northwest Europe population) importance. Although many estuaries hold large numbers of birds, they may only use a certain proportion of the intertidal feeding area. Studies on the Mersey and the Severn have shown bird populations to be highly concentrated on certain areas within the estuary (ref. 3). Such use is primarily related to food availability, but determining how that availability of food may change post-barrage is difficult to assess without detailed knowledge of sedimentary changes.

Other factors may also be critical in the overall survival of bird populations, for example use of estuaries as refuelling sites on migration, influence of organic inputs or severe weather. Changes in productivity post-barrage will influence the resources available for birds. A study carried out on the Severn estuary suggested that there would an increase in productivity of invertebrates, and that despite a reduction in the intertidal area, the populations of most species would be maintained (ref. 6). Densities of invertebrates on the Severn are generally low and a doubling of invertebrate density on existing feeding areas, compensating for a halving of the intertidal area available for feeding, would be within the range of prey densities that already occur on other estuaries. So in principle similar bird numbers could be achieved post-barrage. An increase in productivity of areas that are currently little used, for example through changes in sedimentation, or the creation of new areas could further increase the capacity of the estuary to support the necessary bird numbers. The main species thought to be impacted by the barrage was the dunlin, due to a reduction in numbers of prey items of a suitable size for this species, despite an increase in productivity. A more recent study characterizing the types of sediment preferred by dunlin for feeding purposes suggests that this decline in numbers might not occur.

The natural variability found in the density and distribution of wintering waterfowl populations illustrates both the difficulty of prediction and the influence of summer breeding success. A study was undertaken to attempt to assess the potential monitoring period needed before predictions of numbers could be made. The emphasis in this study was the detection of decreases rather than increases since this is of most interest to the barrage situation (ref. 7). The results showed that the time required to detect a 30% change was species dependent and ranged from 4 years for redshank to 20 years for widgeon. Four years would also be required to detect a similar change in the total population of wildfowl and waders.

In a study on links between waterfowl and organic enrichment, community composition of waders was found to be affected by salinity, ammonia and oxygen concentrations and biochemical oxygen demand (ref. 8). It is unclear whether there is a direct causal link between organic enrichment and bird populations or whether the effects are exerted indirectly through their impacts on invertebrate prey species.

Conservation

In the UK, where the land surface is intensively used, there is a shortage of areas upon which wildlife conservation can be practised for its own sake. It is therefore necessary to consider how conservation interests can be incorporated into other activities or in an extreme case how land or water of low present value can be modified to create or recreate ecological interest.

Whilst the ecological value of an estuary can be based upon the observable characteristics of that biological system, the evaluation of conservation worth may also need to include information about economic and social factors. There are considerable difficulties in achieving a comprehensive, quantitative statement of ecological value for the estuarine system, and so it follows that there are even greater problems in assigning conservation value. Conservation will be one of the factors considered in the planning decision that will eventually determine the building or otherwise of a tidal energy barrage. It is important, therefore, that consideration of possible changes in the conservation value of the estuary is based upon sound predictions of the ecological effects of the barrage, both negative and positive.

Summary

The effects described here illustrate the complexity of the estuarine ecosystem and show that any barrage scheme will require extensive and site-specific studies. It is impossible to describe in a short space all of the environmental considerations. This Paper has concentrated on the ecological effects within the estuary but any environmental assessment also needs to take into account various possible effects of a barrage on man's activities such as drainage, groundwater effects, public health effects, archeology, visual impact and noise during construction. In addition it is recognised that to take account of all interests involved in an estuary some degree of estuarine management will be required to co-ordinate the many uses.

It is worth recognizing that all forms of energy have some environmental effects, and the positive benefits of renewable energy should not be forgotten, especially in relation to the greenhouse effect, diversity of supply and resistance to fuel price increases. Tidal energy barrages would prevent flooding of low-lying land by storm-surges and sea-level rises, and do not add to the greenhouse effect. The Severn Barrage could save 8 million tonnes of coal equivalent per annum and up to 17 million tonnes of CO_2 per annum, has no SO_2 and NO_x emissions, and no waste to dispose of. However, although the financial benefits of a tidal energy scheme as an electricity generator are readily evaluated, there is as yet no simple mechanism for taking environmental factors into account when considering the real cost of electricity.

References

1. THE SEVERN TIDAL POWER GROUP. *Severn Barrage Project, Detailed Report, Volume 1. Tidal Hydrodynamics, Sediments, Water Quality, Land Drainage and Sea Defences*. 1989, ETSU TID 4060, p1
2. EPRI. *Turbine-Related Fish Mortality: Review and Evaluation of Studies*. 1987, AP-5480 Research Project 2694-4
3. STOKESBURY, K.D.E. and DADSWELL, M.J. Mortality of Juvenile Clupeids during Passage through a Tidal Low-Head Turbine at Annapolis Royal, Nova Scotia, *North American Journal of Fisheries Management*, 1991, 11, 149-154

4. CLARK, N.A *et al. Waterfowl Migtration and Distribution in North West Estuaries.* 1990, ETSU TID 4074
5. THE SEVERN TIDAL POWER GROUP . *Severn Barrage Project, Detailed Report Volume IV. Ecological studies, Landscape and Conservation.* 1989, ETSU TID 4060, p4
6. NATURAL ENVIRONMENT RESEARCH COUNCIL AND RAVEN-SRODD CONSULTANTS LTD. *The Predicition of Post-Barrage Densities of Shorebirds.* ETSU TID 4062, p1
7. DAVENPORT, T. *et al. Monitoring Requirements for Detecting Tidal Barrage Induced Changes to Estuary Bird Populations.* 1990, ETSU TID 4087
8. GREEN, P.T., HILL, D.A. and CLARK, N.A. *The Effects of Organic Inputs to Esutaries on Overwintering Bird Populations and Communities.* 1992, ETSU TID 4086

Discussion on Papers 14 – 16

S. CHARLES-JONES, Laing Civil Engineering
I speak as a layman in environmental matters. Has any work been done on the adaptability of birds to changes in their environment?
In a different context, whilst working on a motorway or a certain firm some time ago, we encountered, in a quarry which we wished to develop, a certain type of rare blue orchid which allegedly would not grow anywhere else. One evening, as a trial, we dug up a bucketful of the earth with a large machine and transported it elsewhere, where the orchid grew quite happily. Nature has been going for a long time and has adapted to many of the horrible things that humans have done to it.

S. MUIRHEAD, ETSU
Many plants and animals have successfully adapted to the changes that man has imposed upon their environment. However, the changes caused by a tidal energy barrage in terms of reducing the intertidal area and hence causing a potential reduction in the quantity of food available would cause problems for waterfowl. The difficulty is that these effects may be overcome if the productivity of the estuary as a whole is increased and more food becomes available albeit on a smaller area. Wading birds themselves are often site-faithful and could not adapt to obtaining food in deep water!

E. A. WILSON, Mersey Barrage Company
A layperson in environmental matters such as myself is tempted to equate the predicted loss in bird numbers on an estuary with a tidal power barrage resulting from the time integrated loss of feeding area (allowing for any productivity increases) as meaning a reduction in bird population. Presumably this may be true if availability of inter-tidal area is a limiting factor upon population size. What are the major influences on populations of birds using UK estuaries and to what extent is the inter-tidal area important? Is sufficient information available in this respect or are detailed studies required?

S. MUIRHEAD, ETSU
Many wildfowl and waders spend the summer in their breeding grounds in the Arctic. Breeding success is mixed between years and may even correlate with the lemming cycle. However, waders are long-lived and can

sustain a mix of good and bad breeding seasons without detriment to the population as a whole. The winter feeding grounds are therefore a potential control on numbers, especially in relation to juveniles coming to estuaries for the first time. Most adult waders tend to be site-faithful and the juveniles have to find their own niche. Each estuary may therefore reach capacity at a certain point, but this point will also depend on winter weather conditions with cold winters in the east driving birds westwards. Therefore, under such circumstances the amount of intertidal area available may be critical, but the use of that intertidal area will depend on its nature, e.g. mud and sand. Ongoing studies are attempting to define more closely both the controls on juvenile mortality and the influence of cold weather on the population of wading birds as a whole.

M. JEFFERSON, World Energy Council

The World Energy Council's interests cover all forms of energy and the total energy/environment interface. Dr Muirhead departed a little from her prepared Paper to conclude that no insurmountable barriers to any of the tidal barrage projects under consideration had as yet been identified. There were no 'show stoppers'. This is a view she expressed in an earlier Paper (ETSU L.31).

What would be required to produce a 'show stopper'? To the ornithologist, the international importance of the main sites, RAMSAR and other international and national designations, the relationship of the various estuaries to each other in the context of the main migratory flyway, etc. might suggest that the environmental work going on is 'shadow boxing', and that there is no intention of stopping any show for environmental reasons. In the light of the various interesting studies done for ETSU in this field, what are Dr Muirhead's comments on this point?

S. MUIRHEAD, ETSU

In relation to potential 'show stoppers' for birds, the most detailed work that the Department of Trade and Industry and Severn Tidal Power Group have carried out on the Severn has suggested that birds may actually benefit in most cases to a barrage being present. In the Mersey, detailed work on this aspect is still ongoing so I am unable to comment on that particular scheme. I certainly do not think that the work that has been carried out is shadow boxing and if there was an environmental impediment, such as severe loss of bird populations, it would be recognized, especially in relation to an internationally designated site, and without some overriding national need intervening, in that particular case a project would be extremely unlikely to go ahead. Clearly, those developer's studies that are proceeding are not felt to have identified any 'show stoppers', or they would have been stopped.

There is opportunity for this view to be countered when any proposal finally comes under scrutiny, at a private member's bill or public inquiry stage.

C. J. A. BINNIE, W. S. Atkins Consultants

We have heard about the loss of inter-tidal mudflats and that this could affect the wading bird feeding grounds and hence bird population. It would be technically possible to dredge mud from below the low water mark and raise part of the area to above the new low water mark thus providing replacement feeding areas. Two questions; could this be made suitable as a replacement bird feeding area and would it be economic to do so?

S. MUIRHEAD, ETSU

There are now many suggestions that a mudflat could be created, although it would need careful design and thought given to such issues as drainage and colonization by fauna and flora. In certain circumstances it may well be economic, especially if it forms a disposal route for dredgings, and dredgings may be increasingly difficult to dispose of easily at sea with tightening legislation. I would not of course advocate it as an easy route for the disposal of polluted dredgings, but it may provide an alternative and potentially useful use for dredgings required during the construction process. If this was the case and there were difficulties due to legislation of disposing at sea then such schemes could easily be economic. Also, depending on the scheme, it may be that recreation of habitat is written in as a condition to consent and this would therefore have to be fulfilled in some manner.

Professor B. A. O'CONNOR, University of Liverpool

Dr Shaw has shown correlations between bird density and sediment type in intertidal areas. Clearly, high density being associated with high concentrations of mud may also indicate that the muds are associated with high levels of nutrients from pollutant discharges, such as sewage. Given that Dr Muirhead has referred to the 'clean-up' measures being taken on estuaries, such as the Mersey, would Drs Shaw/ Muirhead like to comment on the possible changes in bird populations due to 'clean-up' operations, which are independent of barrage construction.

S. MUIRHEAD, ETSU

It is difficult to assess the effects of a 'clean up' on bird populations, much of the data collected to date has been sketchy and often anecdotal in relation to observed effects of sewage or industrial discharges on birds. It has been suggested that in highly polluted estuaries invertebrate and hence bird populations are low. Also, diversity of invertebrates and even birds may be low. As the area concerned becomes cleaner, invertebrate numbers both increase and the species diversity also show an increase, as do the birds. However, this increase may be sustained by a small amount of pollution. The question is that as the 'clean up' progresses further there may be a decrease in the productivity of the invertebrates, as there is less food available and therefore such an area would probably be unable to sustain high bird populations. It may therefore be the case that the current levels of bird populations on the Mersey cannot be sustained as the estuary is cleaned up, regardless of the construction of a barrage.

P. ROTHWELL, RSPB

There are clearly still a large number of uncertainties in forecasting the environmental impact of tidal power barrages. How long will it take to remove most of these uncertainties?

Is there not a danger that the concrete pouring industry will push the Government to give economic comfort to barrages before the environmental assessments are complete?

S. MUIRHEAD, ETSU

It is well recognized that due to the highly dynamic and variable nature of the estuarine ecosystem an environmental assessment of a tidal energy barrage is a complex task. The time-scales for an environmental assessment will vary depending on the knowledge already available on that estuary and the intensity of work undertaken during the environmental assessment process. Therefore the time taken will be site specific.

There is no danger that pressure from the construction industry will allow a barrage to be built without an environmental assessment, since an Environmental Statement, based on such an assessment will be a prerequisite for the granting of consent to build. As with any other development, the onus will be on the developer to present a believable Environmental Statement or run the risk of consent being withheld.

D. COWIE, *Binnie & Partners*

I would like to make a short presentation as a supplement to Ted Haw's excellent Paper. In 1976, Maunsell, Merz and McLellan, and Binnies were asked by the State Energy Commission of Western Australia to consider the 1965 report of SOGREAH on the potential for tidal power in the Wilcott Bay area of the Kimberley Coast and in particular to examine the tidal closure technique assumed and to develop alternatives to a point where their feasibility and cost could be established. It was considered that the then recent experience with North Sea Oil structures could have significant impact on costs previously put forward.

From aerial photographs of the proposed barrage sites at Secure Bay and Wilcott Inlet taken from a light aircraft by Professor Peter Ackers, the whirlpools and overfills caused by the strong tidal streams can clearly be seen. It was calculated that towards the completion of the rockfill side closure dams at the Secure Bay site velocities rose progressively to about 6 m/s and the final closure would require rock of about 1 m size. It was also found that a double basin scheme including Wilcott Inlet could generate twice as much power over a year by operating the same installed capacity at a higher load factor for perhaps a 50% increase in capital costs. The cost of firm power for Secure Bay was then admitted to be about 3.1 pence/kWh.

There is undoubtedly great potential for tidal power development in NW Australia. The tyranny of distance, always a factor in Australia's economy, holds back this development.

17. Composite construction of caissons

C.J. BILLINGTON, MSc(Eng), PhD, DIC, FCGI, FICE, FIStructE, MASCE, Director, Billington Osborne-Moss Engineering Ltd and
H.M. BOLT, BSc(Eng), PhD, DIC, ACGI, MiCE, Principal Engineer, Billington Osborne-Moss Engineering Ltd

Ten years ago the Bondi report confirmed the feasibility of caissons as the primary construction element for tidal power barrages. At the time, attention focused on concrete caissons, but following the development of generic steel designs by the Steel Construction Institute, recent studies such as those for the Mersey and Conwy rivers have considered both steel and concrete as alternative options.

In fact the generic steel study utilised not only stiffened steel panels but also steel-concrete-steel sandwich construction. Depending on panel loading and draft constraints this option, despite conservative design assumptions, was often found to offer the most cost-effective solution. The conflicting aims, to achieve a structure with shallow draft yet sufficient mass to be self-founding, favour first steel then concrete. Nevertheless, judicious combination and sequencing of steel and concrete elements is being recognised as offering the greatest potential.

The composite sandwich construction concept is being developed further in a major research and technology development programme by Billington Osborne Moss Engineering Ltd (BOMEL) in which component testing and analysis is being undertaken to develop rational design methods. The design, construction and installation aspects are all being considered in the subsequent design comparisons with steel and concrete caissons.

Such approaches are not without precedent and the combined use of materials and the integration of trades in constructing the Sydney A. Murray caisson on the Mississippi in the USA are studied.

The debate is not one of steel versus concrete, rather it centres on the most efficient and cost-effective use of the primary construction material is available.

Introduction

This Paper concerns the use of composite steel/concrete/steel sandwich construction for the construction of caissons forming barrages for tidal power generation. Steel and concrete have both been used separately, or have been proposed, for the construction of tidal or low head hydrostatic power generation stations and in the case of the Sydney A. Murray Jnr power plant on the Mississippi in the USA the caisson housing the main turbines makes structural use of concrete ballast to resist in-situ loading conditions.

Composite construction is, of course, widely used in building and civil engineering but not normally in double skin form. However, Composite double skin panel construction has been proposed for a wide range of structural uses particularly in the marine environment where structures tend to be large and the construction sequence is easily adapted to benefit from this form of construction. The benefits arise from both the design and construction viewpoints and can be summarised as

- containment of concrete or grout by steel leading to higher triaxially contained strengths
- support of steel plating by concrete largely eliminating the need for expensive stiffening
- use of steel skins as permanent shutters for the concrete, thus saving on construction operations
- provisions of external abrasion and impact resistant surface
- use of concrete primarily to resist compression loads and steel to resist tension loads
- ability to resist reversal of out of plane forces (e.g. hydrostatic pressure)
- panel boundary condition restraints reduce or eliminate the need for shear connection and shear reinforcement
- ability to absorb fabrication imperfections within concrete infill
- Insensitivity of design strength to geometric imperfections.

These benefits were explored extensively in relation to arctic drilling and production facilities in the early 1980s with both testing and analytical development work. The literature survey later in the Paper describes the scope of this work. BOMEL continues to be involved in design development studies on arctic caissons and two other applications, namely composite floating systems and submerged tube tunnels. In each case earlier schemes for these applications have either been in stiffened steel or reinforced/prestressed concrete and, when compared to both methods, composite construction can offer technical and/or cost benefits. On investigation it quickly becomes apparent that the structural component dimensions and hydrostatic loading are very similar for all four applications (see Fig. 1). The basic cross-sectional dimensions are built up from approximately 10m x 10m units and hydrostatic heads are in the range 0-50 m.

The interest in this area was rekindled when, in 1987, a major study commenced on the use of steel for large caisson construction for tidal power generation barrages (refs 1 and 2). The caissons are constructed from assemblies of flat and curved panels and three forms of construction were studied

(a) Floating Production System

(b) Ice Resistant Walls of Arctic Offshore Structures

(c) Tidal Power Generation Barrage Caisson

(d) Submerged Tube Tunnel

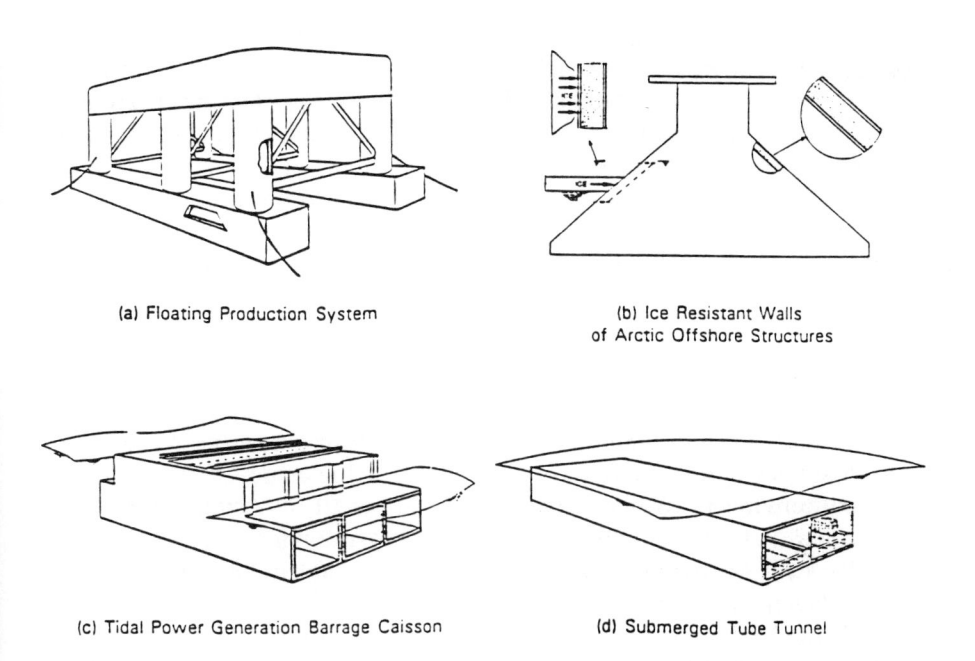

Fig. 1. Applications of composite steel/concrete sandwich construction

- single skin stiffened steel, double skin stiffened steel and double skin composite steel/concrete (Fig. 2). In order to assist in the large number of panel calculations necessary to generate and optimum design a computer programme was written to carry out routine plate panel design calculations on the basis of known panel loading and overall dimensions. This program which optimises on cost, and therefore has to be capable of allowing for variations in material, manpower and equipment costs as well as productivity is known as BAGPUS (BArrage Generic Programme for the Use of Steel) and is available from the Steel Construction Institute.

Following interest in deep water exploration and production, particularly in the Gulf of Mexico, and the announcement of a concrete TLP scheme for Conoco's Heidrun field in Norway, it was recognised that conventional orthogonally stiffened steel shell construction was complicated and expensive and alternatives were sought.

Another strand in the development of the technology concerns submerged tube tunnels where composite sandwich construction was investigated for the Conway tunnel in North Wales (ref. 3). Preliminary work showed significant potential economy. In the event, construction proceeded with a steel lining to provide water tightness and impact resistance but otherwise the structural benefit was not accounted for in the design. In this case there was insufficient time to investigate and develop design methodologies for the composite approach.

With four major applications of the technology being investigated in the Authors' organisation, it was decided to develop an integrated programme to cover testing, development of analytical tools and preparation of design guidance, which would then feed into applications studies. This Paper describes the integrated research programme and provides preliminary weight and cost comparisons for single skin and double skin stiffened steel construction and double skin composite construction.

Literature review

Double skin composite construction benefits from the confinement of the cementitious material which enables its mechanical properties to be fully developed with greater strength and ductility being achieved. The steel at its maximum lever arm eliminates the weight penalty associated with concrete cover. Furthermore the fixing of complex reinforcing steel is obviated and the steel skins not only provide permanent formwork but give an impervious membrane.

The problems associated with predicting the failure of complex ring stiffened cylinders led Montague (ref. 4) to propose the use of a double skinned composite shell for sub-sea structures as early as 1975. The concept led to a comprehensive programme of work which is continuing into the 1990s. Double skinned cylinders, with external diameters between 140 mm and 500 mm, have been tested, firstly to demonstrate the inherent strength of the structural form and thereafter to study the effect of imperfections, steel/filler bond, creep, end restraint, penetration/damage, slenderness, mechanical bonding, sustained pressure and different core materials including light-weight concrete. Tested within hyperbaric chambers at Manchester University in the UK, hoop stresses from the hydrostatic pressure loading were of the order of twice the axial stresses. The response of the composite cylinders generally remained elastic until the inside skin yielded precipitating a reduction in stiffness until yield of the outer skin occurred. Beyond that

(a) Single Skin Orthogonal   (b) Double Skin Stiffened Steel   (c) Double Skin Composite

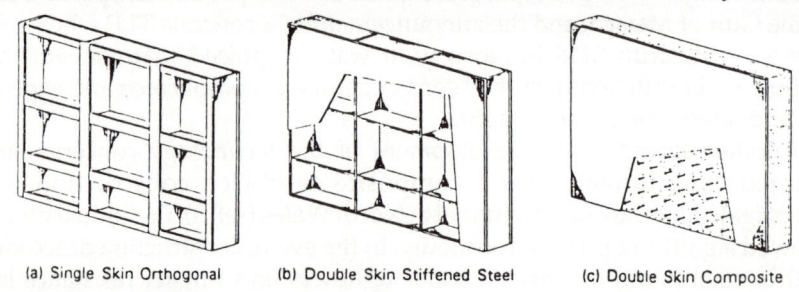

Fig. 2. Alternative forms of construction

point it was the highly-stressed filler which supported the circumferential loads induced by the increasing pressure. Once the concrete crushed, failure was complete.

Two theoretical approaches have also been pursued with success. One assumes an orthotropic filler in conjunction with the tangent modulus approach and accounts for non-linear material characteristics. The second method for predicting failure-pressure is a limit state membrane analysis which assumes the steel is at yield with a uniform distribution of circumferential stress across the filler, failure being governed by concrete failure criteria. The two methods have been shown to give comparable results.

The application of composite construction in grouted pile/sleeve connection and repair clamps for fixed offshore structures is well established (refs. 5 to 9). The ability of the cementitious material to fill the annulus between imperfect steel tubulars and thereafter to transmit loads makes it an efficient and economic method of connection and repair which, following extensive testing, is now a fully accepted technique in the offshore industry.

Other areas of offshore structural design have generally not explored the benefits to be gained from application of composite construction. The potential exception is in the Arctic where local ice loading, possible 10 MPa over an area of a square metre, demands heavy and thus costly stiffening of steel structures. The substitution of composite ice-resistant walls is shown schematically in Fig. 1.

In 1977, Matsuishi (ref. 10) presented the results of composite beam tests which served to demonstrate the excellent strength, ductility and energy absorption characteristics of the new material with a view to its application in ice-resistant structures. Promising results of finite element analyses were also presented.

The structural advantages were readily accepted by the industry (ref. 11) and numerous proprietary studies have sought to extend the early work (ref. 12). Attention has been paid to the provision of shear keys between concrete and steel and Fig. 3 gives a selection of the schemes tested. The heads of the studs have been shown to be important for the development of bond with the concrete whereas simple bars precipitate local cracking, reducing the effective depth of the concrete section (ref.10). Given an adequate key, Najiri (ref.13) has shown that the specific arrangement of steel is of lesser importance than the percentage in determining ultimate strength. However, Matsuishi (ref. 14), again testing beams, has demonstrated a considerable strength enhancement in the presence of longitudinal stiffeners which serve to spread the applied load.

From the earliest tests, axial stresses were identified to be carried by the steel plates with the shear stress being carried from the point of loading directly to the supports along the principal stress trajectories forming a tied arch as shown in Fig. 4. A linear response was obtained until yielding of the

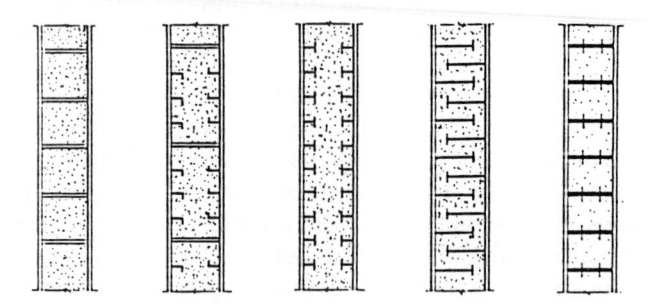

Fig. 3. Alternative shear connection arrangements

bottom plate commenced. A ductile response followed as the plate elongated and diagonal shear cracks propagated through the concrete. Final failure coincided with crushing of the concrete and buckling of the compressive steel face.

The application of this composite technology to Arctic structures has been the driving force behind most of the test data generated to date. The Centre for Frontier Engineering Research has run a major programme covering almost 30 beam-type tests (ref. 15 and 16) on simply supported specimens 375 mm wide, with span to depth ratios of 4 to 6 and a depth of 250 mm, which were loaded at 16 points (8 longitudinally by two transversely). Of the two tests reported in ref. 15, the only difference was in support conditions, whereby the horizontal restraining forces in the presence of friction gave a higher flexural capacity than the steel plate alone. The specimens utilised the simple diaphragm web as shown in Fig.3 and the vulnerability of the heat affected zone to tensile failure at the outer skin interface was revealed. The corresponding programme (ref. 16) covered a range of parameters to investigate the effects on concrete capacity. A significant database has resulted from this work at C-FER leading to an empirical prediction of shear capacity

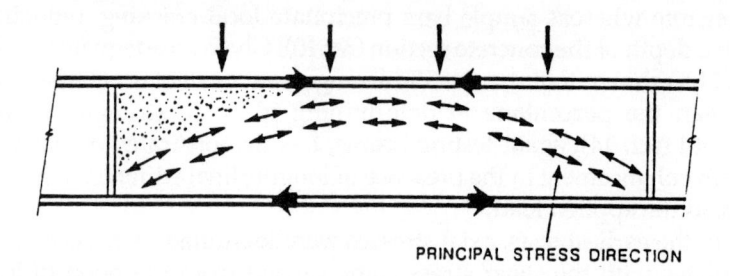

Fig. 4. Idealisation of arch action within concrete core of composite steel/concrete section

(ref. 15). This is complemented by an energy-method approach which simply gives an upper bound prediction of capacity. Alternatively, a lower bound plasticity approach based on a plastic truss model is put forward and compared with the same database in ref. 16.

The static tests described above are complemented by cyclic loading of beams 2 m long with a 200 mm by 300 mm cross-section, conducted to simulate freeze-thaw cycling (ref. 14). Despite the deterioration in the concrete due to cracking, the confinement afforded by the steel plates ensured little change in capacity. Failure was still caused by yielding of the steel plates.

From these beam tests the double skin shell and slab specimens tested by Taywood Engineering (ref. 17) mark a notable advance. The single spanning slabs, some 2.3 m by 2.8 m in plan demonstrated a dependency on the web orientation for load distribution. The webs were welded to alternate faces in the test but the structural advantage of a through web was realised. The shells, as the slabs, were loaded centrally out-of-plane. This ultimate capacity was some 90%, higher than that of earlier reinforced concrete shell specimens which had three times the steel content by volume.

The demand for ice-resisting structures has successfully driven much of the test work in the field of composite structures. The studies have largely focused on out-of-plane beam loading and the effect of three-dimensional continuity has yet to be fully explored. In-plane stresses appear potentially significant in light of the C-FER tests and furthermore direct comparison between beams and slabs is required as the applications to offshore structures are developed.

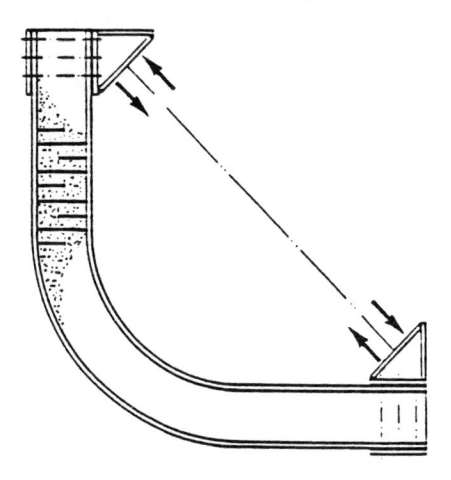

Fig. 5. Pull-in and push-out tests of full-scale models of tunnel corner section, simulating normal loading conditions and an internal explosion

TIDAL POWER

The Authors have already identified the economic potential associated with tidal power barrage caissons. Early work (ref.1) has necessarily adopted conservative assumptions demonstrating the need for a design methodology for panels loaded with in-plane direct end shear forces and out of plane bending.

Submerged tunnels require an outer steel skin to ensure water tightness and the new approach utilising this membrane in composite action with the concrete lining is now being developed in the UK (ref. 18). A series of tests on half-scale tunnel walls has been conducted with specimens some 4 m long by 250 mm deep and 800 mm wide; one full-scale beam test has also been conducted. The design concept was verified by these preliminary tests and the two dimensional plane-stress finite element programme developed in parallel was shown to give satisfactory results. Particularly unusual were the corner section tests (Fig. 5). A pull-in load was applied up to the maximum design value and an elastic response recorded. On reversing the load to simulate an explosion, failure occurred as the stud/concrete interlock at the corner was broken locally although the rest of the structure remained intact. Further testing work in this area is to be undertaken under the direction of the Authors.

Integrated composite development programme

The review of available data and the application of composite technology to offshore floating production facilities, immersed tube tunnels and tidal barrages demonstrated that the structures for each application consist of an assembly of flat panels, curved shell panels or tubular members which are subjected to loading of similar magnitude and which have similar overall panel dimensions.

For each component there is a long list of variables which affect performance. The principal parameters are

For flat rate panels:

Geometry
- aspect ratio
- steel and concrete thickness
- boundary conditions
- penetrations
- through thickness stiffeners
- shear connection geometry

Materials
- steel strength
- type of cementitious filler - e.g. concrete, cement grout, grouted pre-packed aggregate

Loading	- hydrostatic
	- in-plane direct and shear
	- out-of-plane bending and shear dynamic/fatigue

For curved shell panels: (as for flat plate panels plus)

Geometry	- curvature
	- local supports
Loading	- in-plane/bending interaction

For composite tubulars:

Geometry	- diameter
	- length
	- steel and concrete thickness
	- transitions (e.g. draught tubes)
	- boundary conditions
	- diaphragms/bulkheads
	- longitudinal/circumferential
	- stiffening, penetrations
	- shear connection geometry
Loading	- hydrostatic (external or internal)
	- axial/bending/shear
	- dynamic/fatigue
Materials	- as above

The overall integrated project covering the four applications is divided into 27 separate work packages. The objectives and deliverables of each work pack are described briefly below.

Data collection and collation. (Work Packages 02-04)

The objective of this group of activities is to collect and collate data on the subject applications, global behaviour, base geometries and loadings to establish the applicable range of the parameters listed above. This work is now complete.

Composite construction fabrication and construction review (WP05)

Construction methodologies are being formulated, to be agreed with construction yards for each application. Detailed consideration is being given to panel construction, assembly, intersections (panel:panel, panel:shell, tube:panel, tube:tube etc) concrete emplacement, launch/loa-

TIDAL POWER

dout towing and ballasting, in order to provide the base line for preparation of detailed costing exercises.

Test programme (WP06-09)

A total of some 48 tests are to be carried out. Approximately half are static tests on flat panels with the remainder equally split between static tests on curved shells, static tests on composite tubulars and fatigue tests. The specimens will be instrumented to measure load/deformation relationships, concrete fracturing, slip and separation at the steel/concrete interfaces and distributions of stress within the steel plating.

Analysis (WR10-11)

Existing non-linear finite element analysis incorporating concrete failure models and cracking is being enhanced to include the effects of triaxial containment and slip and separation at the steel/concrete interface. This analytical package will be calibrated against the test results and then used to carry out parametric studies to develop design relationships between the governing parameters.

Development of design tools (WP12-14)

Costing norms are being developed for each fabrication/construction operation. Simple design rules will be prepared based on the results of the test programme and the analytical studies. The existing panel optimisation program is to be updated to include the new design rules and cost data.

Case study scenario definition (WP15-18)

For each application a representative case study scenario is being established for which comparative design in steel and/or concrete are available. This includes overall geometry, pay load, environmental load, mooring/foundation conditions, construction constraints and installation conditions. The Wyre Barrage is to be used as the case study for tidal barrages.

Generic design and cost studies (WP19-22)

For each application case study, component design and cost optimisation are being performed for input into overall cost studies. The costing includes all aspects of material supply, component fabrication, assembly, concreting, load out, towing and installation. Costs cover structural components and associated equipment for installation, but do not include topsides equipment, M & E requirements etc.

Cost comparisons (WP23-25)

The final stage of the work is to prepare cost comparisons with equivalent design in steel and/or concrete as defined in WP15-17.

Comparison of stiffened plate and composite panel weights and costs

Panel weights and costs have been developed for three modes of construction. These are single skin stiffened plate construction, twin skin stiffened plate construction and composite steel/concrete/steel sandwich construction. Single skin construction comprises steel plates welded together to form a skin of the desired panel dimensions. The skin is stiffened in two directions by T-section, angle or bulb flat stiffeners. Twin skin construction consists of two steel plates separated by a series of flat plate stiffeners positioned in the orthogonal directions. The composite mode of construction comprises two steel plates forming the outer skins of a steel/concrete/steel sandwich. Stud connectors may be welded to the inner faces of the steel plates to provide transfer of shear between steel and concrete and prevent separation. For each mode of construction, optimum cost solutions have been derived by scanning several parameters such as steel plate thickness, stiffener size/spacing, construction depth etc.

The designs of single skin and twin skin stiffened plate constructions used recognised codes and standards. There was, however, little published information available for the design of composite panels and, therefore a conservative design philosophy was adopted. The design procedures are outlined below.

For stiffened steel plate modes of construction, forces due to out-of-plane loading (hydrostatic or live loading) were initially resisted by the plate in bending. Simple elastic plate bending formulae were used for design. The ability of a plate to carry these loads depended on the plate dimensions, the plate thickness, the boundary support conditions, the type and intensity of loading and the permissible bending stress. Thus, for specified loads, and assumed plate thickness and an appropriate allowable bending stress, the maximum plate dimensions and hence the panel stiffener spacing requirement could be determined. Several plate thicknesses ranging from 6 mm to 80 mm were considered along with numerous stiffener spacing solutions. By automation, a multitude of technically acceptable solutions was rapidly scanned and by using unit cost rates, the most economical solutions were identified.

For the design of large composite panels subjected primarily to hydrostatic pressure, little guidance was available and therefore the following conservative procedures were adopted. Maximum bending moments and shears were determined for beam sections of unit width. The beam elements were assumed to span in one direction only and assumed unrestrained at their boundaries. Where it can be confidently assumed that loading will be maintained on adjacent panels, then the benefit of a continuous beam may be taken. The design procedure assumes that bending moments were resisted solely by the steel plates and concrete alone resisted shear. The designs were based on Grade 50 steel (F_y = 354 N/mm^2) and 30 grade concrete (f_{cu} = 30N/mm^2). Optimum cost solutions were identified by considering several designer specified plate thicknesses and a range of construction depths. For a particular steel plate thickness, the minimum specified construction depth was initially assumed. The ability of the steel plates to carry the design moment and of the concrete to carry the design shear was determined. The depth of construction was increased in steps of 25 mm until a technically acceptable solution was achieved. This process was then repeated for other specified steel plate thicknesses.

The transfer of longitudinal shear forces at the steel/concrete interfaces was achieved by the provision of shear connectors. The optimum sizes and spacings of stud connectors were determined from the static strengths of various sizes of headed stud shear connectors.

For each technically acceptable option for each mode of construction, the total panel weight was calculated. For stiffened panel construction the total weight comprised the weight of steel skins and all stiffeners. For composite construction the total weight included the weight of steel skins and the concrete infill. The weights were based on steel and concrete densities of 7.85 t/m^3 and 2.34 t/m^3, respectively.

To derive cost-optimised solutions, unit rates were assembled for material procurement (steel, concrete and shear studs) and fabrication and construction (plate butt welds, stiffener fillet welds, stud welding and concrete placement). Figs 6 to 10 show the comparisons of stiffened plate and composite panel weights and costs. Panel sizes and loading regimes selected in these examples are representative of panel sizes and loading intensities associated with legs and pontoons for offshore floating structures. The figures demonstrate that for optimum cost, composite solutions are heavier than the stiffened plate designs. This effect is predominant at lower plate thicknesses where, to achieve adequate strength, the depth of construction for composite panels becomes large. However, composite panels are often shown to be more economical (even with current conservative design assumptions) than stiffened plate construction.

Figure 11 shows that using current conservative design philosophy for composite construction, panels with spans less than 14 metres are more economical when constructed using composite construction, avoiding the high cost associated with fabrication of stiffeners.

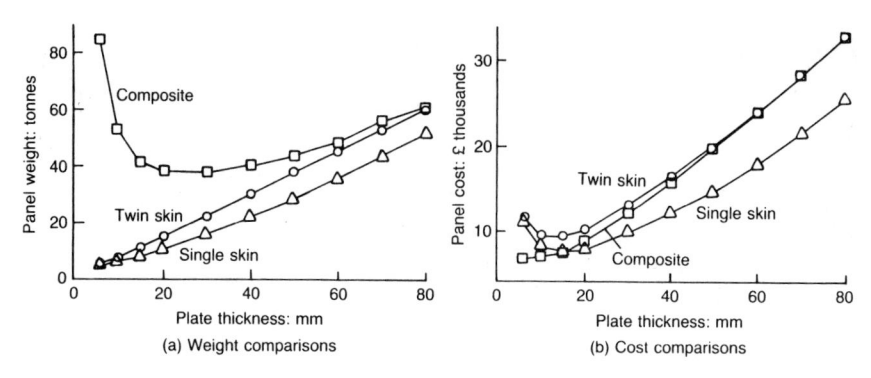

Fig. 6. 4 m x 12 m panel (5 m hydrostatic loading)

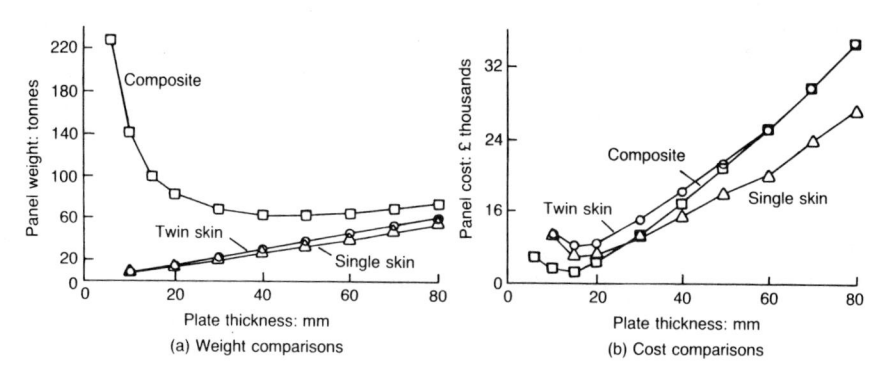

Fig. 7. 4 m x 12 m panel (15 m hydrostatic loading)

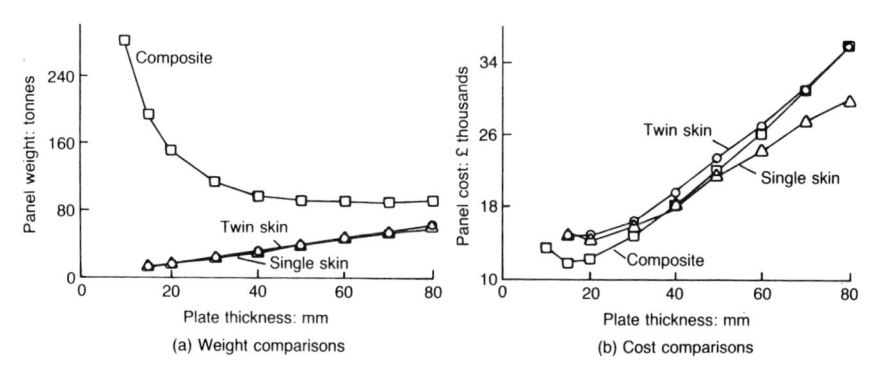

Fig. 8. 4 m x 12 m panel (30 m hydrostatic loading)

TIDAL POWER

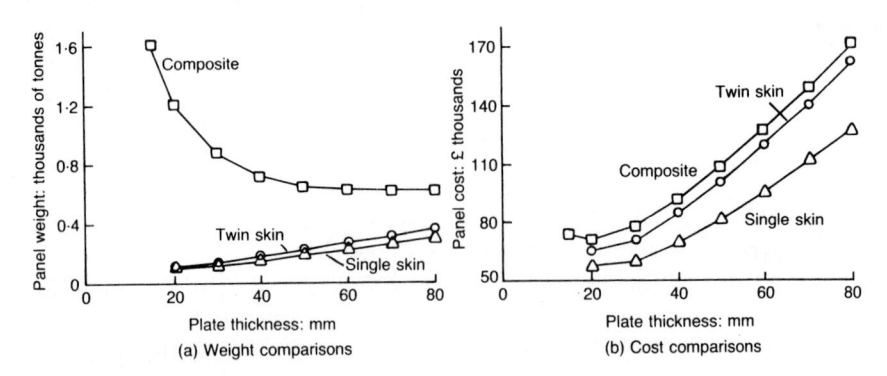

Fig. 9. 14 m x 20 m panel (15 m hydrostatic loading)

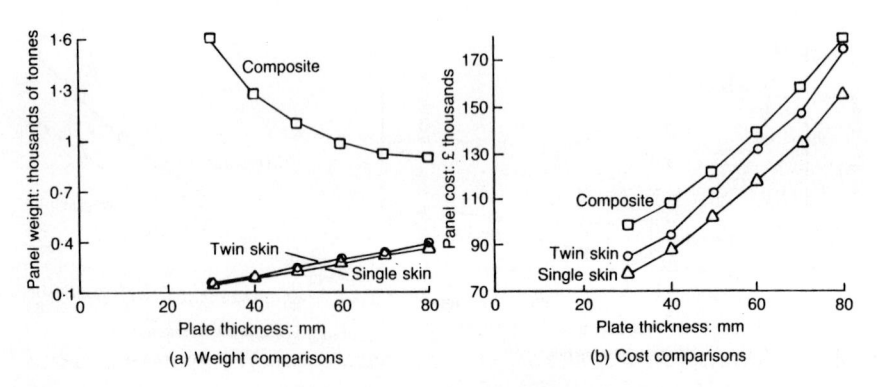

Fig. 10. 14 m x 20 m panel (30 m hydrostatic loading)

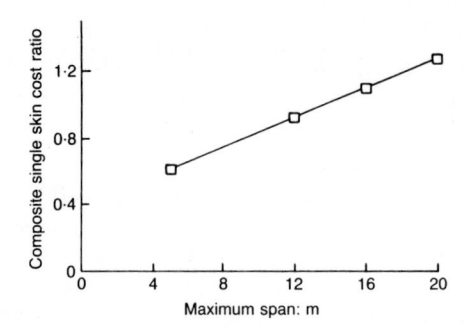

Fig. 11. Composite/single skin cost ratios

Conclusions
This Paper has described the background to a major testing and design development programme in progress at BOMEL. A survey of the literature and collation of available test and analytical data has shown that

- compositely constructed components have much higher strength than that calculated from consideration of steel yield strength and concrete uniaxial compression strength, due to the effect of containment
- composite construction offers significant technical and construction advantages over both stiffened steel or reinforced/prestressed concrete
- available data does not cover the necessary range of loading particularly in respect of combined in-plane and out-of-plane loading due to hydrostatic pressure and global effects
- using a conservative approach to design, composite panels offer significant cost savings for spans up to 14 m.

References
1. THE STEEL CONSTRUCTION INSTITUTE. *Generic steel designs for tidal power barrages.* Department of Energy Report, ETSU TID 4066, 1989.
2. GUY, R.G., SONDHI, N. and BILLINGTON, C.J. The use of steel in tidal barrages. *Proc. Third Conference on Tidal Power, Institution of Civil Engineers, London.* Thomas Telford, London, 1989.
3. NARAYANAN, R. et al. Double skin composite construction for submerged tube tunnels. *Steel Construction Today,* December 1987.
4. MONTAGUE, P. et al. *Tests on composite cylinders under external pressure at Manchester University, 1976-1985.* Simon Engineering Laboratory Report, 1985.
5. BILLINGTON, C.J. Composite structures. *Steel Construction Offshore-Onshore Technology Transfer Conference, The Steel Construction Institute, London, April 1987.*
6. BILLINGTON, C.J. et al. Recent developments in the design of grouted connections. *OTC 4890 Offshore Technology Conference, Texas, May 1985.*
7. BILLINGTON, C.J. and LALANI, M. Underwater strengthening and repair of steel offshore structures. *Subsea Challenge Conference, Amsterdam, 22-24 June 1983.*
8. BILLINGTON, C.J. The use of composite construction in piled offshore jacket structures. *IABSE Colloquium on the effective use of materials in structures, Imperial College, London, September 1981.*
9. BILLINGTON, C.J. Research into composite tubular construction for offshore jacket structures. *J. Constructional Steel Research,* 1, No. 1, September 1980.

TIDAL POWER

10. MATSUISHI, M. et al. On strength of new composite steel-concrete-material for offshore structures. *OTC 2804, 9th Annual Offshore Technology Conference, Houston, 1977.*
11. GERWICK, B. C. et al. Resistance of concrete walls to high concentrated ice loads.*OTC 4111, 13th Annual Offshore Technology Conference Houston,1981.*
12. GERWICK, B.C. and BERNER, D. Utilisation of composite design in the Arctic and sub-Arctic. *9th International Conference on Port and Ocean Engineering Under Arctic Conditions, Alaska, 1887.*
13. NAJIRI et al. Structural behaviour and design method of steel/concrete ice walls for Arctic offshore structures.*OTC 5292, 8th Annual Offshore Technology Conference, Houston, 1986.*
14. MATSUISHI, M and IWATA, S. Strength of composite system, ice-resisting structures. *9th International Conference on Port and Ocean Engineering Under Arctic Conditions, Alaska, 1987.*
15. ADAMS, P.F. et al. Design and behaviour of composite ice-resisting walls. *9th International Conference on Port and Ocean Engineering Under Arctic Conditions; Alaska, 1987.*
16. O'FLYNN, B. and MACGREGOR, J.G. *Tests on composite ice-resisting walls.* Centre for Frontier Engineering Research, C-FER Special Publication No. 1, 1987.
17. SMITH, J. R., and MCLEISH, A. The resistance of composite steel/concrete structures to localised ice loading. *9th International Conference on Port and Ocean Engineering Under Arctic Conditions, Alaska, 1987.*
18. NARAYANAN, R et al. Load tests on double skin composite girders. *Proceedings of the Hennikar Conferenceon Composite Construction, American Society of Civil Engineers, New York, 1988.*

18. Field, laboratory and mode feasibility of a tidal barrage in a muddy hypertidal estuary

K. W. OLESEN, Danish Hydraulics Institute, W. R. PARKER, Blackdown Consultants Ltd, A. J. PARFITT, Rendel, Palmer and Tritton and H. ENGGROB

Field observations in a turbid hypertidal estuary show that modelling requires a 3 dimensional representation of the transport of suspended solids. Laboratory and field measurement allow parameterisation of the transfer of material from suspension to be and evaluation of the erosion of fluid muds. A mathematical model has been developed which represents the observed processes. It incorporates a 1-D dynamic description of the flow, a pseudo 2-D description of erosion and deposition and a pseudo 3-D description of the transport and bed processes.

Introduction

As part of an urban renewal and development programme, Newport Borough Council have proposed the construction of a tidal control structure across the river Usk at Newport, Gwent (Fig. 1). The purpose of the structure is to

(a) maintain water levels through the town at +3.5m O.D. to improve local amenity and encourage redevelopment of waterside areas
(b) exclude high spring tides which, when coincident with high river flow, result in flooding of lowlying areas
(c) provide a crossing for a southern ring road
(d) by associated sewering development, enhance the local water quality.

The creation of such a structure across this hypertidal, turbid, estuarial reach of the river Usk would fundamentally modify the hydrodynamics of the system and create a wide spectrum of environmental impacts. The principle ones are

(a) impacts on the hydrodynamics, sedimentation and water quality seaward of the structure
(b) the development of an impoundment landward of the structure and the water quality, sedimentation and other aspects of this impoundment

TIDAL POWER

(c) the effects of the changes in environment on local fishing, particularly migratory species.

This Paper considers the field, laboratory and numerical modelling studies undertaken to appraise the first of these issues; the effects of the proposed development on sedimentation seaward of the barrage. Because of the turbid nature of this estuary and the link between suspended sediment levels and dissolved oxygen levels, the link between sedimentation and water quality studies has been strong. These issues will be reported elsewhere.

The Usk Estuary

The river Usk (Fig. 1) rises in the Brecon Beacons in South Wales and flows out into the Bristol Channel at Newport. It drains a catchment of approximately, 1,320 sq. km. It is a characteristically spate river with flood events rising above a base flow of 20 - 30 Cumec. which drops to as little as 2 - 3 Cumec in droughts. The tidal influence of the Severn extends upstream to Newbridge on Usk, some 25 km from the entrance to the river at No. 1 Buoy. It is an hypertidal estuary (Table 1).

Table 1. Newport Tidal Data 1990

Mean Spring Range	11.9
Mean Neap Range	6.1
Mean High Water Springs	+ 6.29 m OD
Mean High Water Neaps	+ 3.19 m OD
Mean Low Water Neaps	- 2.95 m OD
Mean Low Water Springs	- 5.61 m OD

The Spring tide intertidal volume is approximately 24×10^6 m and the neap tide intertidal volume is 16×10^6 m. Construction of the barrage to exclude the tide would reduce these by around 38%. On Spring tides water level variations occur beyond Newbridge on Usk. Salt water penetrates some 22 km upstream on Spring tides but only 17 km on Neap tides (ref. 1). The tidal reaches from Spittles Point to Newbridge may be regarded as the upper estuary and from Spittles Point down to low water Springs as the lower estuary. The proposed barrage is sited at approximately low water Neaps.

In the upper estuary the channel is characterised by muddy banks and a gravelly invert which occupies about 1/3 of the channel perimeter. In the lower estuary the gravelly floor of the channel is covered with permanent mud deposits up to 1 metre thick which have grooves and other flow parallel topography.

The large spring tidal range generates strong currents which suspend considerable quantities of fine sediment. The large Spring-Neap changes in

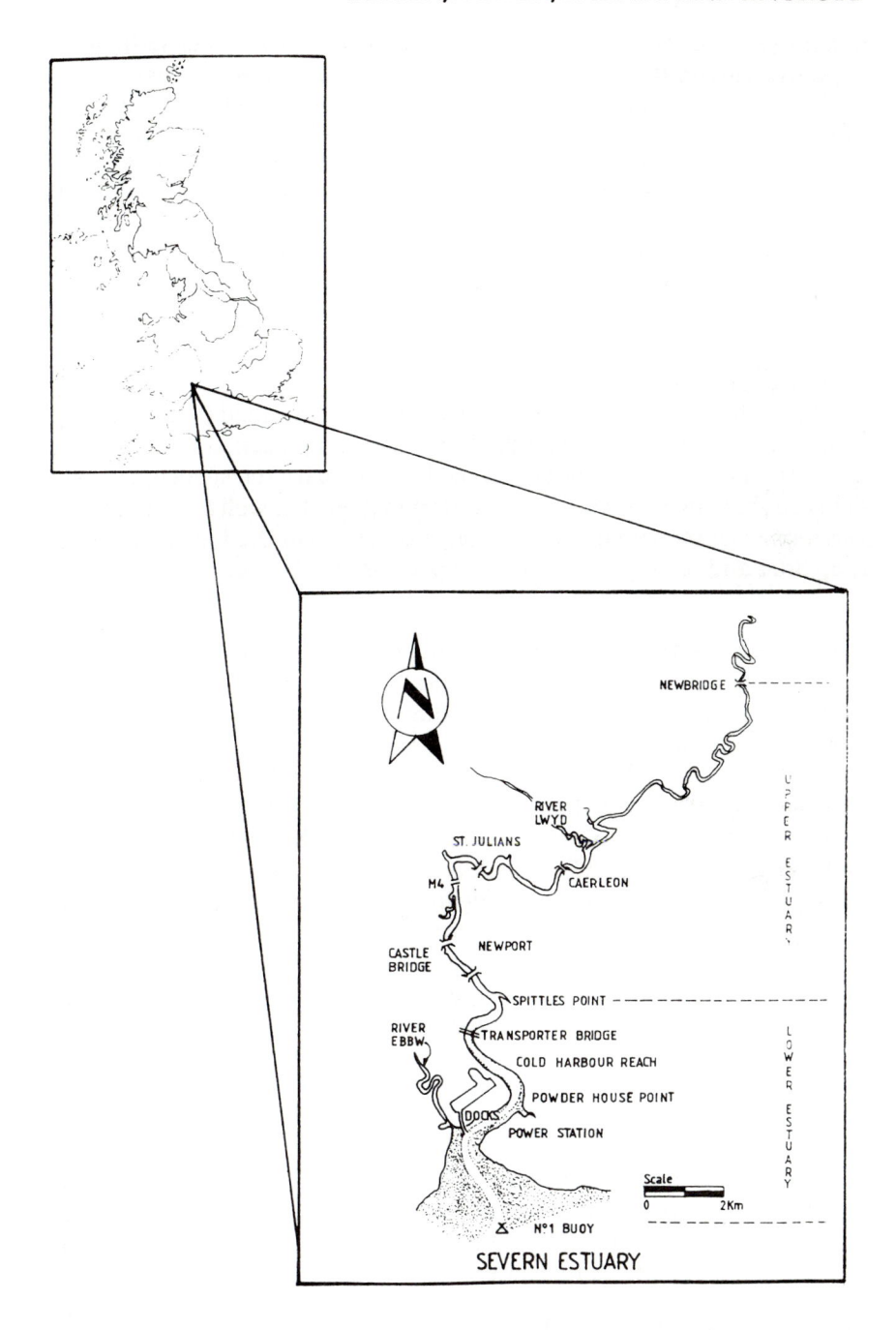

Fig. 1. Location of river Usk and places named in text

tidal range result in the temporary deposition of much of this sediment in the lower estuary during Neap tides as pools of fluid mud - a high concentration slurry of fine sediment (refs 2 and 3). The suspended solids signatures during Spring tides (Fig.2) indicate that the turbidity maximum of the Usk is not directly linked to the Severn turbidity maximum (refs 4 and 5)

During Spring tides the Usk turbidity maximum is advected between the upper estuary at high water and the lower estuary at low water (Fig. 3). During Neap tides mud which settles out at high water is re-suspended during the late ebb and carried down to settle at low water below low water Neaps in the lower estuary. Consolidation of parts of these pools of liquid mud leads to the formation of the permanent mud deposits of the lower estuary.

Although previous studies of this estuary have assumed the system to be well mixed and during Spring tides little vertical statification of temperature or salinity is observed, typically in the lower estuary the suspended solids field can show marked vertical stratification (Fig. 4) as well as cross channel variation. Thermohaline stratification is observed in the lower estuary on neap tides and on higher tides during periods of enhanced river flow.

Field studies and parameterisation relevant to modelling

Adequate representation of the physical processes in a numerical schemetisation of the estuary require specification of a number of sediment related quantities or relationships.

The solids boundary conditions

Existing data was evaluated to derive relationships which might relate tidal range to solids input. Further data collection proved necessary to define adequately the solids field for modelling use. Although several tidal data sets were used no adequate long-term series of velocity and suspended solids was available. By monitoring a network of stations over successive days of a Spring-Neap cycle an empirical relationship was derived for the flood tide input. A seasonal modulation of the total sediment population involved was identified.

Suspended solids transport

Experience and field observation, indicated that the suspended solids field had a strongly 3 dimensional structure and thus a 3 dimensional treatment of transport and dispersion would be necessary. Cross sectional time series of data were collected to provide assessments of relationships between single point depth average and cross sectionally averaged depth average values. These data were collected by free fall profiling systems and synchronous sampling for calibration.

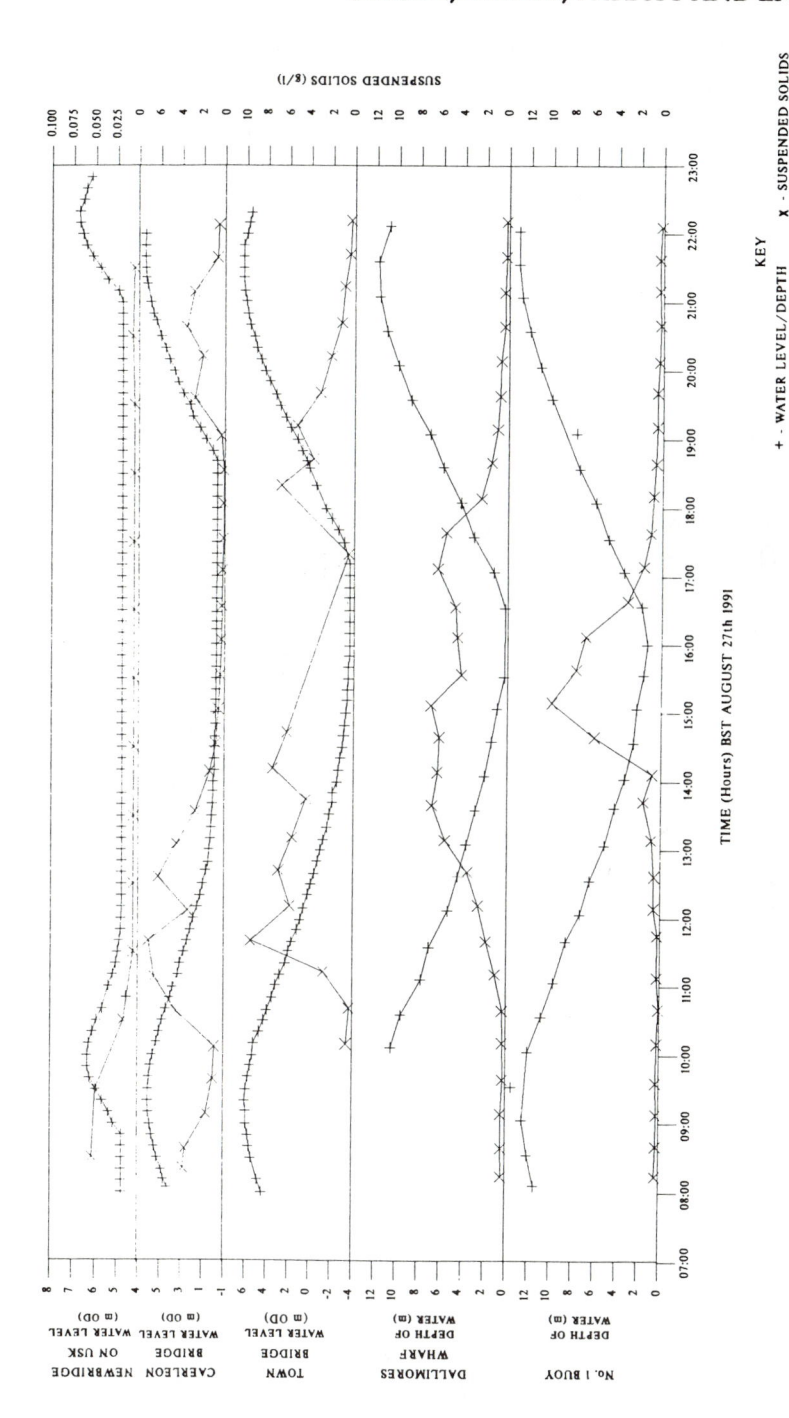

Fig.2. Water surface elevation and suspended solids relationships

TIDAL POWER

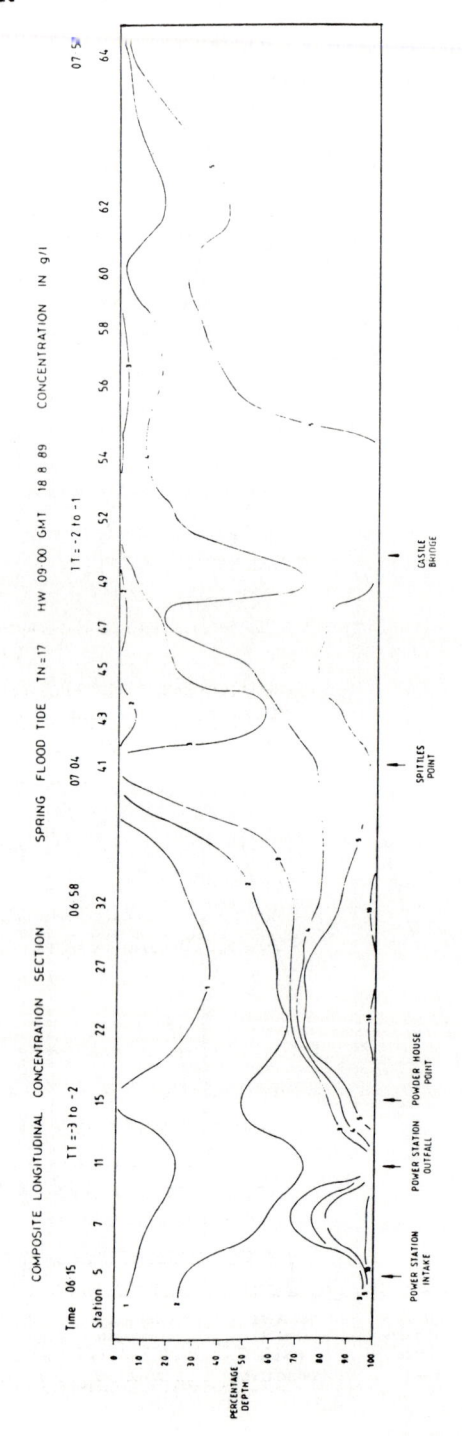

Fig. 3. *Longitudinal distribution of solids at 1 and 2 hours before High Water Springs*

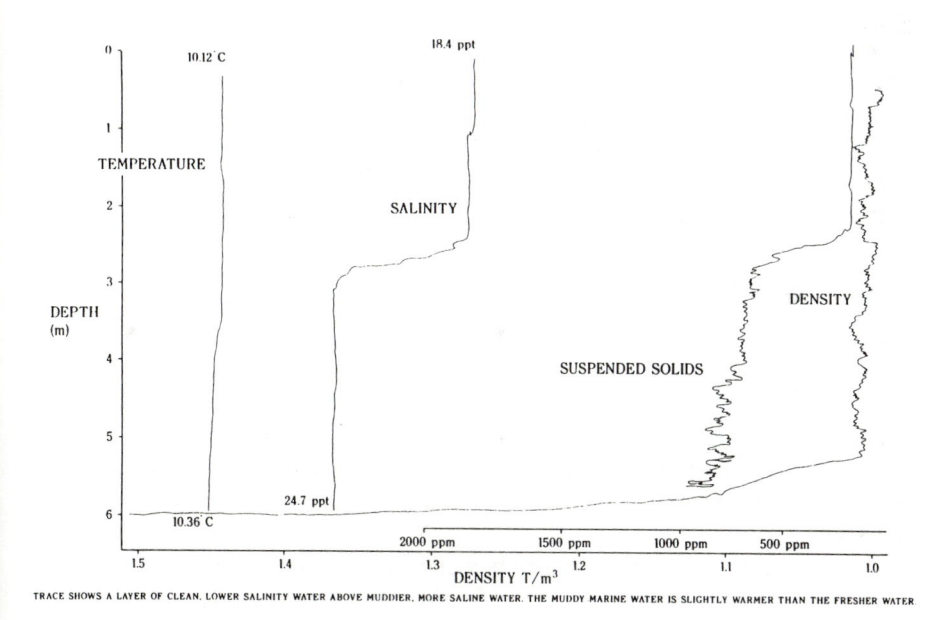

Fig. 4(a). Vertical stratification in salinity and suspended solids

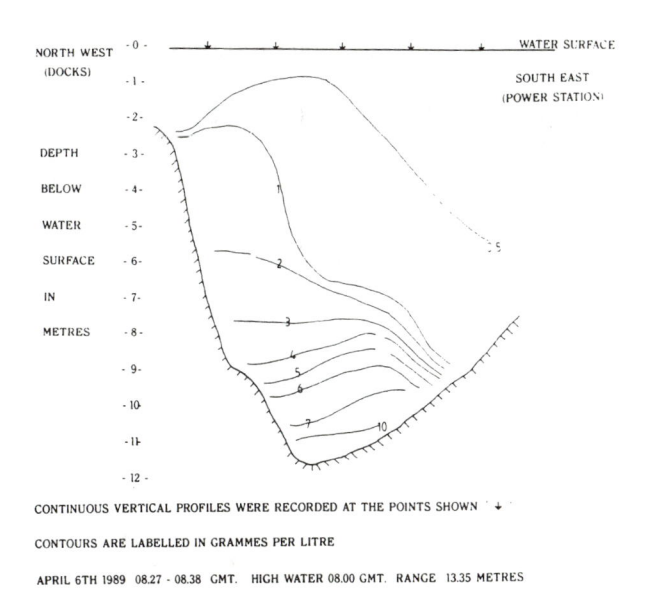

Fig. 4(b). Vertical stratification and lateral variation in ebb tide suspended solids field

Settling velocity

The settling behaviour of aggregated systems requires careful treatment. The settling velocity of flocs depends upon their relative density which in turn is dependant upon local turbulence and both the number and volume concentration of other aggregates.

No single concentration related function could be applied to the Usk because of the strong spatial and temporal anistropy of conditions affecting settling. An effective settling parameter was derived by matching observed and simulated vertical concentration profiles during sensitivity analysis.

Deposition

In turbid environments, classical concepts such as "deposition" have a much more problematical character. Deposition of a sand grain is relatively unambiguous but in circumstances where the water column may grade imperceptibly from very muddy water, through watery mud to mud and where no clear interfaces or boundaries exist, then a different approach is necessary.

It was proposed that the estuary bed should be considered to consist of a number of zones or phases of definable material properties. These were

- *Suspension*: where the solids content was insufficient to produce continuous structure.
- *Weak fluid mud*: material which has definable structure but a very low resistance to erosion.
- *Fluid mud:* structured material with resistance to erosion on neap and some spring tides.
- *Underconsolidated bed*: Material having definable structure and excess pore pressures.
- *Bed:* Material generally resistant to spring tide currents.

The critical solids content of these various zones were evaluated by rheological analyses, laboratory consolidation tests and in situ observation of density. The values used are shown in Table 2.

In addition to these "bed" material classifications it was necessary to assess the rate of transfer of one phase to another. This was evaluated from low concentration settling and consolidation tests. By these means the transfer of material from "suspension" to "bed", as occurs in nature was parametrised.

Table 2. Solids content of material phases

Sediment	Bulk Density (KN/M^3)	Concentration /Dry density (Kg/M^3)
Suspension	<10.00	<30
Weak fluid mud	10.0-10.2	30-60
Fluid mud	10.2-10.8	63-160
Underconsolidated bed	10.8-12.75	160-475
Bed	>12.75	>475

Erosion

Three erosion phenomena were observed in the field.

(a) Tidal erosion of material deposited temporarily over high slack water.
(b) The erosion of fluid mud during the Neap to Spring tidal sequence.
(c) Seasonal of erosion of sediment stored in the intertidal areas and the floor of the channel in the lower estuary.

The most crucial of these related to the erosion of fluid mud (Fig. 5). Laboratory settling and consolidation tests revealed that the time taken for

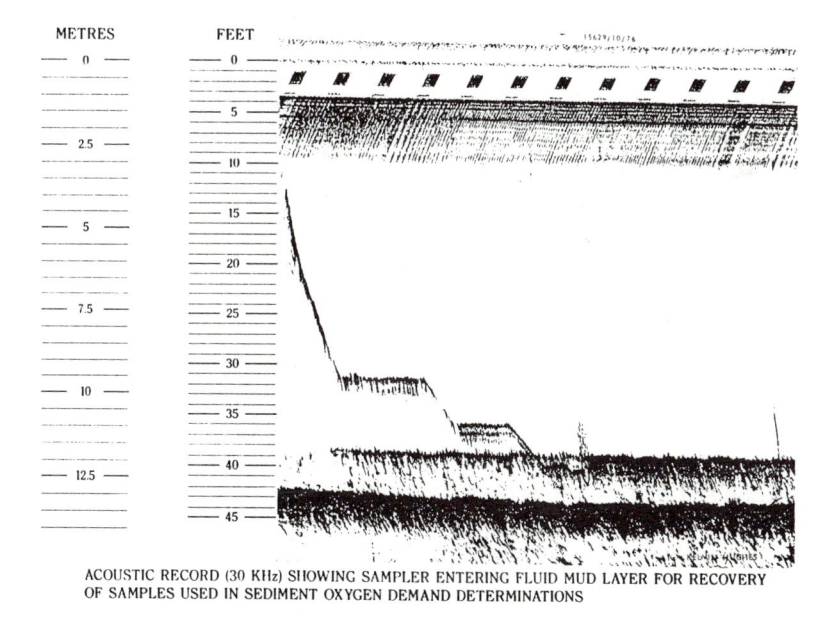

ACOUSTIC RECORD (30 KHz) SHOWING SAMPLER ENTERING FLUID MUD LAYER FOR RECOVERY OF SAMPLES USED IN SEDIMENT OXYGEN DEMAND DETERMINATIONS

Fig. 5. Typical acoustic record of fluid mud

TIDAL POWER

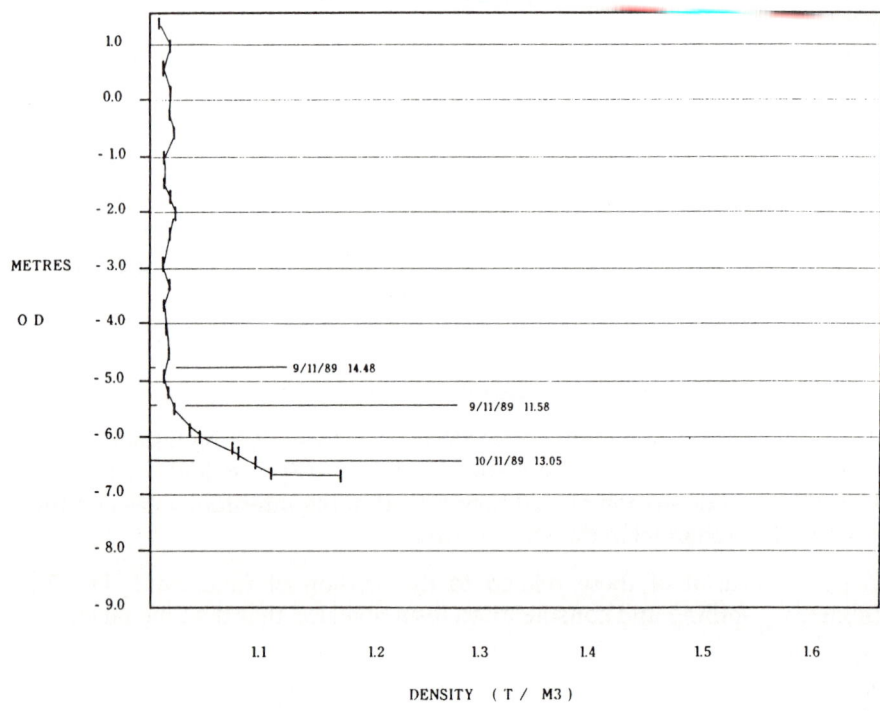

Fig. 6. In situ density profile

a layer of fluid mud to consolidate to densities characteristic of the permanent bed (Fig. 6) exceeded the time between successive periods of Spring tides. Field observation further revealed that large quantities of fluid could be eroded in very short periods of time (minutes to hours).

An initial evaluation of the entrainment of Usk mud was conducted in the Carousel flume at Hydraulics Research Ltd. However, the limiting stress generated in the flume (0.9 N/m^2) is substantially below that generated in the natural situation. Although the HR experiments were presented to give a linear erosion function the field studies suggested that much higher rates occured. Field studies of changes in bed level, determined using UHF sonar systems, combined with velocity measurements and correlated with model derived bed stresses indicated an erosion function which was linear up to approximately 1.0 N/m^2 but was strongly non-linear thereafter.

Laboratory determinations of the erosion thresholds and erosion rates of fluid or very soft muds, suffer from the inadequate representation of significant large scale turbulence effects. Since the response of these materials is dependant on both stress and strain, amplitude field observations was thought necessary. Field evaluation is faced with the problem of relating observed loss of material to a fluid dynamic stress in circumstances where boundary layer models may not work effectively. The essential role of the field observation was to evaluate the general form of the erosion function in the natural stress range.

Material quantities

Field and laboratory measurements provide either a description of relevant processes to be represented in the model, parametrisation of important constants or coefficients or data against which to check the model simulation. A key aspect of this in the Usk study has been the model prediction of fluid mud development over neap tides and its erosion on spring tides. A combination of acoustic surveys and density measurement (Fig. 5 & 6) was used to estimate the quantity of material deposited as fluid mud during Neap tides and to follow its erosion during the Neap to Spring cycle.

Over 200 nuclear transmissance density profiles were used to develop characteristic but conservative density profiles for the mud pools in different reaches of the estuary.

These studies also revealed the importance of fluvial discharge in modulating the quantities of fluid mud generated and determining its survival during the Neap to Spring cycle. The coincidence of large amplitude river discharge with low water Spring tides was revealed as being particularly important and pointed to the importance of the upstream boundary conditions of river flow in accurate sedimentation modelling. River flow data was provided as 15 minute discharge time series by the Wales N.R.A.

Model description

The mathematical model presented here was developed to suit the requirements of the feasibility study for the proposed Usk Barrage. The key requirements of the model were to be able to describe the cyclic behaviour of cohesive sediment (including fluid mud development) during both a tidal cycle and during a Neap-Spring-Neap cycle and to describe the interaction between sediment resuspension and oxygen content in the water.

The large periods to be simulated in combination with the relative large modelling area and the rather extreme tidal variation excluded fully three-dimensional dynamic modelling because of excessive computation time. Even a dynamic two-dimensional simulation model of the entire estuary would have resulted in a model which would be very impractical.

In view of these constraints it was decided to base the model development on a one-dimensional model with some essential properties described in either two or three dimensions.

MIKE 11 - a one-dimensional general river modelling system developed jointly by the Danish Hydraulic Institute and the Water Quality Institute - was used as a basis for the model development. The core of this system is a fully one-dimensional dynamic hydrodynamic model, with add-on modules for water quality and sediment transport modelling. The MIKE 11 modelling system is described in more detail in refs 6 and 7.

The transport of cohesive suspended sediment is described mathematically by an advection-dispersion transport model with erosion and deposition of the sediment represented as source and sink terms in the transport model. It was deemed necessary to utilize a three-dimensional advection-dispersion transport model in combination with a two-dimensional description of erosion and deposition due to large vertical gradients in observed suspended solid concentrations and large lateral variation of flow velocities across the river.

The three-dimensional flow pattern necessary for the three-dimensional advection-dispersion transport model is based on the one-dimensional representation of the flow obtained from MIKE 11 in combination with an assumed lateral and vertical distribution of the flow. The following procedure is applied: The cross-sections are divided into a number of vertical sections each with different depths. In each section the flow velocity is assumed proportional to the square root of the depth of that section. This implies that the lateral variation of flow is assumed to be frictional dominated. In the vertical direction the velocity distribution is assumed to be logarithmic.

The distribution of sediment at the river bed is highly non-uniform in the Usk. The river bed can for instance consist of consolidated mud at the side slopes while there is soft fluid mud in the central deeper parts of river cross-sections. Also vertically, there will be variation in the composition of the bed sediment. A full three-dimensional description of the river bed was therefore required.

A multi-layer bed model consisting of three layers was implemented into the model. The three layers are

- weak fluid mud (top layer)
- fluid mud
- underconsolidated bed.

Each of these layers are associated with different time scales. The weak fluid mud is the sediment which deposits around flood slack tide and is re-entrained as the ebb flow starts to pick up. Fluid mud is the sediment which develops during Neap tides and is re-eroded again during the spring tides. The underconsolidated bed layer increases in period with low energy tides and decreases in periods with large Spring tides. In the Usk the period for this variation is 6 months.

Transfer of sediment from one layer to the next is modelled either as constant rates or as first order processes, i.e. where the rate depends on the weight of the top layer(s).

In natural rivers the weak fluid mud and fluid mud will move due to hydrodynamic forces (shear and pressure gradient) and due to gravity. The latter is included in the model in a rather schematic way via a friction type formulation similar to Chezy's law. This formulation yields a continuous flow of sediment down side slopes. It represents two different processes in the estuary

(a) the deposited material which glides gradually down the side slopes
(b) liquefaction of mud deposits on the side slopes at low water.

Each of the layers in the multi-layer bed model are characterized by having its own sediement properties such as

- critical shear stresss for erosion
- erosion function
- density
- transition rate.

The river bed will erode when the bed shear exceeds the critical shear stress for erosion of the layer which is exposed to the flow. Two types of erosion are implemented in the model, viz.

- *Instantaneous erosion*: In this case all the sediment available in a layer will re-entrain instantaneously when the shear stress exceeds the threshold value. This type of erosion is associated with the instability of the sediment deposit and normally applies to the weak top layer.
- *Gradual erosion*: When the applied bed shear stress exceeds the critical shear stress for erosion the sediment is assumed to erode gradually as a non-linear function of the excess shear stress. Traditionally, the erosion function has been assumed to be linear (ref. 8). However, field observations have indicated that the erosion function can be highly non-liner, as discussed below.

  Due to the non-linearity of the erosion function the gross part of the erosion takes place at the highest stresses, thus an accurate calculation of the flow and bed shear stresses is very important. It also implies that no great accuracy is required for the critical shear stress and hence for the sediment density. Consequently, it is sufficiently accurate to lump a certain range of densities into one type of deposit and use the mean erosion characteristics for the whole range.

In the model, deposition takes place when the bed shear stress is less than a threshold value. In that case the deposition rate is a function of the near bed concentration and the settling velocity. The deposition function was parameterised from low concentration settling and consolidation tests.

TIDAL POWER

Also, deposition is modelled two-dimensionally, so that it is possible to describe deposition in shallow areas while there is erosion at the same time at the deeper parts of a cross-section. The deposited material is introduced into the top layer of the multi-layer bed model.

The settling velocity varies with the suspension concentration due to flocculation, i.e. the fall velocity will increase with the floc size. In suspension, particles will collide due to turbulence motion and differential settling. The frequency of collisions will increase with the concentration, and hence cause an increase in particle aggregation.

The model distinguishes between two settling regimes. In flocculating suspensions with concentrations below 10 kg/m^3 the fall velocity will increase with the concentration. An increase of suspended sediment concentrations above 10 kg/m^3 causes the settling velocity to decrease due to the so-called hindered settling effect.

For high concentration the settling will generate an upward flow of water, which reduces the effective fall velocity. The model proposed in ref. 9 has been implemented.

Model application

A traditional model application programme was adopted in the study, i.e. the model was calibrated on the existing conditions and subsequently used to project the post-barrage conditions.

The calibration of the model was divided into three phases. First the hydroynamic model was calibrated against measured water levels for a Neap and a Spring tide. Subsequently, the sediment transport model was calibrated. This is described in more detail below. Finally the water quality

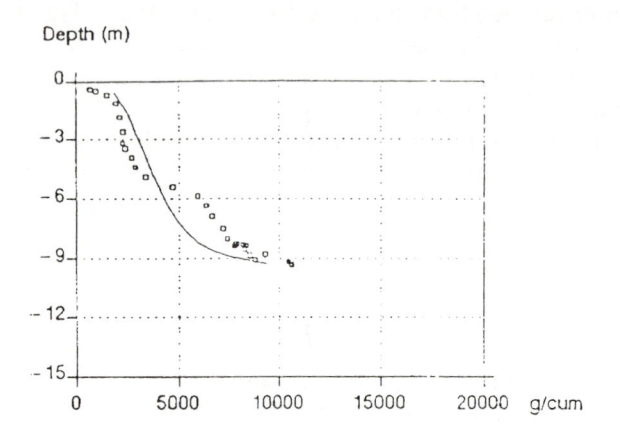

Fig. 7. Related and observed vertical distribution of suspended solids

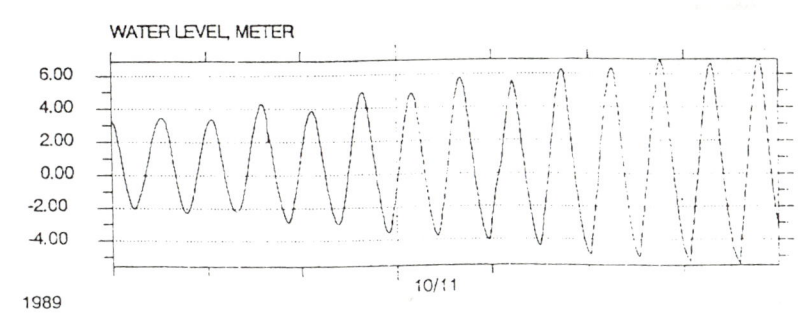

Fig. 8. Tidal variation at downstream model boundary - Neap to Spring

model was calibrated against data from a dry period with critical water quality conditions.

The calibration of the sediment model was facilitated by field and labaratory parameterisation of many of the observed processes described earlier. In fact, only the settling velocity and erosion function was not parameterised directly from observations. The settling velocity was determined via sensitivity runs with the model and observed time series of vertical suspended solids distributions. In Fig. 7 a sample output of the simulated and observed vertical distribution of suspended solids is shown.

Two fluid mud surveys, with one day between them, showed dramatic changes in the amount of fluid mud. The first day (9/11 1989) a total of approximately 20,000 t of fluid mud was observed in the river. This amount of fluid mud should be compared to a total transport in and out of the river during spring tide conditions of approximately 60,000 t. The next day (10/11 1989) nearly all the fluid mud had disappeared, only about 3,000 t of mud were left. The tidal variation at the seaward boundary and the river flow

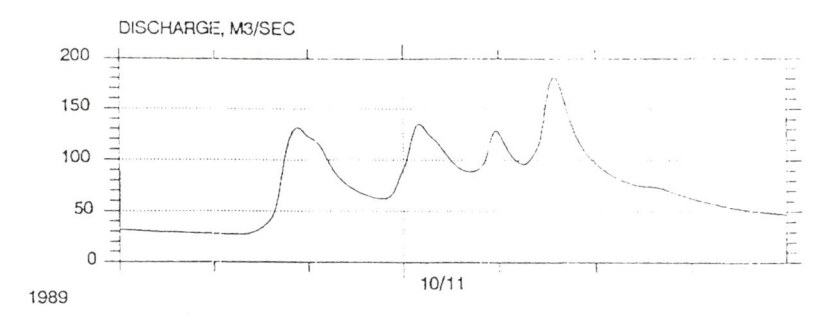

Fig. 9. Observed river discharge at Newbridge

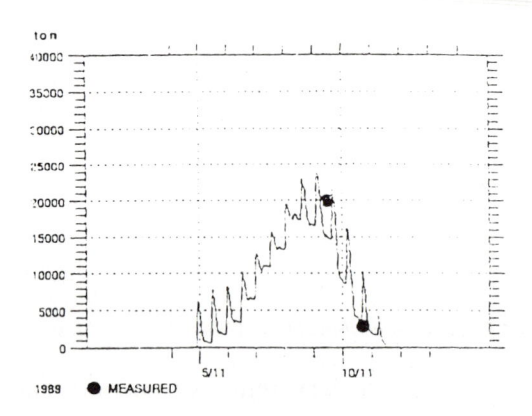

Fig. 10. Simulated and observed amount of fluid mud and weak fluid mud in the

measured beyond the tidal limit for this period are shown in Figs. 8 and 9, respectively.

The figures indicate that the removal of the fluid mud coincided with a peak in the river flow. However, about 28 hours earlier, with only moderately smaller tidal range, a flood with approximately the same peak discharge did not cause any significant erosion of the fluid mud layer. The explanation for this is that the duration of the floods was approximately one tidal period but they occurred at different stages of the tide. In the second flood the arrival of the peak river discharge in the area with fluid mud coincide with low water, thus giving rise to large velocities. This implies that the phasing of the flood river discharge is important, thus emphasizing the need for an accurate hydodynamic description.

The relatively sudden erosion of the fluid mud can only be explained by a high degree of non-linearity of the erosion function. This is also supported by the direct observations of changes in bed levels discussed earlier.

The simulated development of weak fluid mud and fluid mud is shown in Fig. 10. The frequent peaks in this figure represent the deposition and resuspension of weak fluid mud at each flood slack tide.

The largest concentrations of suspended solids in the estuary occurs in connection with the erosion of the fluid mud pool, which generally takes place a few days before maximum spring tide. These large concentrations normally coincide with the smallest oxygen concentration in the estuary due to the relative high sediment oxygen demand (SOD). These features are well represented in the sediment transport and water quality models of the estuary and have also been verified with field observations.

References

1. KNIGHT, D.W. AND WEST, J.R. *A study of the Hydroynamic and Water Quality Characteristics of the Usk Estuary. Vol 1.* Department of Civic Engineering, University of Birmingham, October 1976.
2. PARKER, W.R. AND KIRBY, R. Studies of Fluid Mud in the Severn Estuary and Bristol Channel. *Proc. Challener Soc.*, 1974, IV(6), Abstract.
3. PARKER, W.R. AND KIRBY, R. Fine Sediment Studies relevant to Dredging Practice and Control. *2nd Int. Symp. Dredging Technology.* B.H.R.A. B2.15-B2.26.1977.
4. PARKER, W.R., SMITH, T.J. AND KIRBY, R. Observation of Density Stratification due to Suspended Fine Sediment. *2nd Int. Symp. on Stratified Flow, Trondheim, 1980*, p 955-966
5. KIRBY, R. AND PARKER., W.R. A Suspended Sediment Front in the Severn Estuary. *Nature*, 295, 396-399.
6. BACH, H.K., BRINK, H., OLESEN, K.W. AND HAVNOE, K. *Application of PC-based models in river water quality modelling*, Hydraulic and Environmental Modelling of Coastal Estuaries and River Water, Univ. of Bradford, UK, 1989.
7. OLESEN K.W. AND HAVNOE, K. A Water Quality Modelling Package for Fourth Generation Modelling, IAHR Conf. Ottawa, Canada, 1989.
8. Partheniades, E. 1965, Erosion and Deposition of Cohesive Coils. *J. of the Hydraulic Division*, ASCE, Vol. 91, No. HY 1.
9. VAN RIJN, L.C. *Sediment Transport by Current and Waves*. Delft Hydraulics, Report H 46, 1989.

Discussion on Papers 17 and 18

E. HAWS, Consultant
Could Dr Billington please describe his proposals for making joints between pre-fabricated composite panels?

C. J. BILLINGTON, Billington Osborne-Moss Engineering Ltd
Joints between pre-fabricated panels can be made in a number of ways. The pre-fabricated steel components can be welded such that there is continuity between steel plates in perpendicular planes and then the concrete can be poured to provide continuity through the joint. Alternatively, the individual panels can be completed with steel end plates which can then be connected by various methods to perpendicular base plates, for example.

Within the testing programme we are studying junctions and junction connections in order to come up with sufficient solutions for this particular problem.

R. WALKER, Babtie Shaw & Morton
Will the forthcoming case studies on composite steel caissons for the Wyre Barrage assume the new fabrication techniques cited by the Author (Japanese techniques) and what would be the likely order of magnitude of cost reduction to the overall Wyre Barrage Scheme?

C. J. BILLINGTON, Billington Osborne-Moss Engineering Ltd
The productivity achieved by Japanese shipbuilders results in the steel structure construction man-hours reducing by a factor of 3. This would probably reduce the construction costs of the bare steel caisson by some 45% and of the fitted-out caisson by approximately 20%. The relationship between the caisson cost and the total barrage cost varies from site to site. Unfortunately, I do not have details on the Wyre barrage at this stage that could give global cost reduction. Obviously the cost of the major equipment items and the large site civil engineering costs are not included in the above.

Professor B. O'CONNOR, University of Liverpool
Given that the Authors are driving their quasi-3D model by a 1D hydrodynamic model, could the Authors comment on the ability of the 1D

hydrodynamic model to cope with reflected tidal energy from the barrage when it is fully closed.

A. J. PARFITT, Rendel, Palmer & Tritton

In reply to Professor O'Connor, we are pleased to comment on his points regarding post-barrage hydrodynamic conditions. The hydrodynamic model solves the full non-linear St Venant equations which automatically describe reflection at the boundaries. The wave length of the tide is very large (order of magnitude g x depth x tidal period, i.e. about 200 km), hence numerical errors in description of tidal reflection are negligible. The model simulations post barrage shows no significant reflection. The seaward boundary of the model is located far downstream so it can be assumed that the tidal variation at the boundary is unaffected by the barrage. Consequently the hydrodynamics of both the existing and post-barrage conditions are adequately described in the model.

J. NICHOLSON, University of Liverpool

Two questions concerning the mud transport model of the River Usk:

(i) The flow of fluid mud transverse to the river axis is modelled. Does the same hold true for the flow of fluid mud along the axis of the river?

(ii) Is the model run for a complete Spring-Neap-Spring tidal cycle and, if so, was it necessary to repeat this process for several cycles in order to achieve dynamic equilibrium?

A. J. PARFITT, Rendel, Palmer & Tritton

The points raised by Dr Nicholson are both of interest and importance. The model does not explicitly consider the movement of 'fluid mud' along the axis of the river, although what constitutes 'fluid mud' may be a separate issue. We have monitored the erosion of 'fluid mud' and have no unequivocal evidence for its flow along the channel axis. Erosion is abrupt and general, although it may lag by one tide between different reaches of the lower estuary. The model can be run over several Spring-Neap-Spring cycles and remains stable, the only limitation being storage capacity for result files. Runs from high Springs through Neaps, low Springs, then Neaps to high Springs form a complete cycle, but the estuary re-equilibrates during one Spring-Neap-Spring cycle and so it was not felt necessary to run for longer than this.

D. J. TRUMPER, Acer Sir Bruce White

Our conference opening speaker suggested that tidal power might well be moving from a twinkle in the eye to reality. However, we have many examples of submerged tunnels constructed both in Europe and beyond. Is

Mr Billington able to give any examples of where composite caisson structures have either been placed or at least tenders submitted for such schemes.

C. J. BILLINGTON, *Billington Osborne-Moss Engineering Ltd*

I am not able to give examples of where composite caissons have been installed, although I know that a scheme was prepared for the Conwy crossing in North Wales on a composite basis and the final design did incorporate a steel skin, primarily to eliminate leakage. The objective of the project is to bring the technology to a level where it can be confidently proposed on a competitive basis with either steel or concrete.

Dr G. UNDERWOOD, *University of Bristol*

Mr Parker and Mr Parfitt showed a good slide of diatom films stabilising surface sediments in the Usk. Was it possible to include a biological component in the models of sediment transport, and if not, how important do you feel the biological effects of sediment stability are, and how valid are models that do not take biological factors into account?

A. J. PARFITT, *Rendel, Palmer & Tritton*

Dr Underwood has raised a very important point, as none of the generally used transport equations include a 'biological' term, yet it has long been apparent that, particularly within the photic zone, the role of microfaunas is crucial to seasonal variations in sediment mobility. We believe that the intertidal microbiology of the Usk does influence the seasonal modulation of the fine sediment population and, as such, is very important. This has been accounted for in an elementary manner in a seasonally varying coefficient in the suspended solids input algorithm. Erosion or deposition functions for fine sediment are difficult to determine and we are somewhat sceptical of small scale apparatus used for in-situ determination of erosion functions, since they do not appear to take into account the effect of the test procedure on the elastico-viscous properties of the substrate.

The validity of models that ignore biological effects will vary according to the relative importance of these factors. They may be incorporated by other means, but there is a very serious difficulty in incorporating effects which have been qualitatively observed but remain mechanistically unquantified. In the Usk, we do not believe that biological effects are dominant with respect to fluid mud. The model represents very well the modulation and phase relationships between tidal energy, suspended solids and fluid mud (as defined in the paper). It does not include a specific biological term, but this may be included via the in-situ investigations of 'fluid mud' stability.

19. The Rance Tidal Power Station: a quarter of century

M. RODIER, Electricité de France

A quarter of a century old, La Rance tidal power station still brings a valuable contribution to the generation of energy in France. Such a long time in service allows us to form an opinion on the operation conditions of a tidal power plant. The Paper comments on the main features and operational data connected to 25 years operation.

Introduction
Numerous papers dealing with the Rance Tidal Power Station have been published by Electricité de France (EDF) and commented upon. It does not seem useful to recall all data connected with the plant which can be found in an abundance of literature. The tidal power plant, the operation of which started in 1966 is a quarter of a century old and EDF has a good experience in service. The purpose of the Paper is to summarize the main features and comment on the more recent data regarding the operation of the plant.

Civil engineering works
Characteristics of the works
From the left bank to the right bank

- a lock 13 m wide
- the power plant: 390 m long, 53 m wide and 33 m high
- the rock fill dike: 160 m long
- the 115 m gate structure dam equipped with 6x 15x 10 m sluice gates

The volume of concrete is 350,000 m^3 for 16,000 t of steel and the surface of the facings exposed to the sea water is 90,000 m^2.

Reinforced concrete structures
The preliminary experiments, the precautions taken during construction, rock treatment, sand washing, the quality of aggregates and cements, carefully performed pervibration etc., have been such that after twenty five years exposure to the sea atmosphere the behaviour of the civil engineering works is practically perfect, the appearance of the facings is impeccable.

(a) No new cracking has appeared since construction. The small superficial deterioration in the tidal fluctuation zone range from 8 to 12 m is of no immediate consequence.

(b) There is no apparent deformation of the structure.

(c) The periodic under water investigations show the scouring to be slight and hardly evolutive.

(d) The watertightness of the structure is excellent. Since the grouting treatments performed between 1968 and 1972, and between 1981 and 1982, using acrylic polymerzable resin grout, leakage is no more than 5 litres per minutes.

(e) Prolonged work (over a week) within a united shaft, necessitates the removal of the shellfish deposits on the walls. This precaution prevents the workmen from being poisoned (30 man days per 1,000 kg of deposit).

Behaviour of electro-mechanical equipment
Characteristics

The power station is equipped with 24 turbo-alternator bulb sets consisting of

- a kaplan turbine: for a flow of 275 m^3/s
- a 10 mw generator: rotation speed 93.75 rpm, voltage 3.5 kv, operating in air under 2 bars absolute pressure.

Each assembly of 4 units discharges into the tertiary of a 3.5/220 kV transformer unit of 80 MVa connected directly to the 220 kV network by oil filled cable under a pressure of 3.5 bars. There are therefore 6 assemblies and 3 transformer units, perhaps somewhat complicated terminology but so familiar to the Rance operators.

If one is to have a good understanding of the electrical and mechanical stresses to which the equipment is subjected, it is most important to remember that the Rance generating sets were built for a quite unusual job.

As the operating cycles use pumping, these sets must be able to function as both generators and pumps. If one adds to that their use as open sluices, to speed up the operations of emptying and filling, and the use for the double acting cycle, we have the following types of functions:

(a) direct turbine functioning, reverse pumping, coupled, emptying estuary

(b) direct open sluice, uncoupled, emptying estuary

(c) reverse turbine functioning, direct pumping, coupled, filling estuary

With reversions in the direction of rotation and start-ups under full load. And since 2 high tides are experienced every 24 hours, the number of shutdowns and start-ups recorded is considerable.

Turbines

Numerous precautions have been taken to prevent corrosive attack

(a) the use of 17% Cr, 4% Ni steel for 12 sets of runner blades, the runner hubs, the distributors
(b) the use of 18% Cr, 8% Ni, for runner rings
(c) the use of a copper alloy, cupro-aluminium, for 12 other sets of runner blades.

The dismantling of machinery, which was carried out from 1975 to 1982, allowed us to state that after approximately 200,000 hours immersion - of which 140,000 was operative - no appreciable oxidization had occurred.

The result is implied by the association of three components: cathodic protection, paint and high quality metals.

During disassembly, all watertight seals have been changed. As the units age, i.e. greater than 25 years, a complete overhaul of the machines is planned to be carried out in the future.

Generators

Stators: Following verification of reduction in the air gaps in 1975, a programme to rebuild the 24 stators was undertaken between 1975 and 1982 making all possible provision to limit the energy losses and to reduce the duration of the work.

This work was required following the rupture of the pins locking the magnetic circuit to the frame. This must certainly be caused by the stresses and strains linked to asynchronous start-ups.

Rotors: In 1968 all 1350 poles were dismantled for the bruising of the shock absorber bars in order to prevent electroerosion.

Gates

The six sluice gates, for the accelerated emptying and filling of the estuary, work perfectly. These are in good order and this result is due to the use of a cathodic protection and appropriate painting. In service for about 30 years, the gates will require a complete overhaul in the future.

Control Equipment

The control equipment did not present any operating problems, maintenance and the finding of spare parts became progressively more difficult.

In July 1988 a new structure for the automatic control of the plant was commissioned (using computere and programmable controllers). This choice now best responds to the new operating and performance needs; it also relieves EDF of the need for permanent supervisory staff in the control room.

Cathodic protection

All the positive results shown above depend on cathodic protection. This technique has played an important part in the functional success of the Rance tidal power plant. For each set, the cathodic protection ot the machinery is composed of 3 rings of 12 anodes of 50 micron platinum plated tantalum:

- (*a*) a ring on the runner casing, sea-side
- (*b*) a ring on the distributor
- (*c*) a ring on the anchor ring (8) and the access shaft (4).

That of the gates is ensured by 45 micron platinum plated titanium tubes per gate. That of the lock by 16 tubes also platinum plated titanium.

All these anodes are fed through 3 rectifiers under a 10V voltage, a 20 a current. There annual consumption is 150,000 KWh.

Investigation of all these protected parts demonstrates the efficiency of this protection, and the right choice made at the time of construction.

Furthermore, everywhere the presence of a calcomagnesium deposit of a few microns is observed on all the painted surfaces which, if scratched, show a sound and bright metal, still with non trace oxidization. It too, therefore becomes an additional means of protection.

The voltage and current measurements are taken over by the new control equipment.

Operating results

The Rance operation cycles

At the time of its construction, and even well before at the design stage, the operating cycles of the Rance were the subject of precise studies. The technology of the time - now 40 years old - already permitted much to be achieved, to such an extent that, without the knowledge of the results or necessities, the most complete cycles were foreseen. Some have been simplified, according to the experience in service and well determined economic and technical approaches.

Single acting cycle (with or without pumping): It uses the tide in one direction only, the estuary towards the sea, in no pumping cycle: it corresponds to the system used by the Rance's ancestors: the tidal mills.

> *Filling the estuary*: The sets are functioning as open sluices (OS)(machinery uncoupled), mobile dam gates open. Pumping is possible, according to the energy requirements and energy values of pumping and of restitution. It is also worth finding the best adjusted sea-estuary levels to turbine under a head of around 6 m. Every effort is made to store the greatest possible quantity of water in the estuary, especially during Neap tides.

The overfilling of the estuary level by pumping can reach the following values

Amplitude of the tide	Topping up of the estuary level	Maximum pumping head at the end of cycle
7 to 10 m	0.5 to 1.25 m	1 to 2.5 m
6 to 7 m	1.25 m	2.5 m
5 to 6 m	1.75 m	3.5 m

The pumping head at the end of the cycle, when the pump is stopped, is the difference between the levels of the estuary and the sea.

Standing: Its length depends on the network active or reactive power needs. The start-up time of the sets can, as a result, be moved toward or delayed, with a correlatively lesser or greater active power availability (and the reverse for reactive power).

Emptying the Estuary: It is made by direct turbining (DT) and the outage occurs when H< 1.20m.

Double acting cycle (with or without pumping): It uses the tide in both directions: estuary-sea, and sea-estuary.

Filling the estuary: It is made by reverse turbining (RT) and the outage occurs for < H = 1.70 m. It is worth noting here that, owing to the reverse turbining the maximum estuary level is lower than the maximum sea side level. It therefore follows that if this operation is not succeeded by direct pumping (DP) a reduction in the estuary level, and thus in energy production results. The estuary is filled by opening the gates and operating the sets in open sluice (OS) mod, or the accelerating pump started up for H = 30 cm.

Standing: With the same conditions as above plus even more advantagious potential uses.

Emptying the estuary: Direct turbining (DT) as in the previous case, operating in open sluice (OS) mode and opening the gates at the end of the cycle. Functioning in accelerated pump (AP) mode with reverse pumping (RP) for over-emptying the estuary, can be used. This provides a better head for working in reverse turbine (RT) for the following cycle.

Actual operations of the Rance Power Station

The constructive turbine (efficiency) data favours direct turbining (DT) and direct pumping (DP).

Up to the beginning of the 1980s the important maintenance and repair work, undertaken between 1975 and 1982, allowed limited operations. During this period, direct pumping (DP) has been used less (3%). This is simply because EDF wished to prevent machinery not yet repaired being subjected to too many electro-mechanical stresses. Direct pumping now represents 18% to 20% of plant operating time; i.e for the 24 machines some 28,000 hours of annual operation - no negligible figure.

Direct pumping (DP) is considerably more significant, but what could be more natural? It represents approximatively 60% of the total operating time, or around 93,000 annual operating hours.

As for the two reverse functions, the reverse turbining (RT) and the reverse pumping (RP) although not abandoned, they are very little used: only representing 3% and 0% respectively of the total operation - a negiigible machinery for the advantages derived to be substantial. Indeed, the operating staff judge certain functions to be too detrimental to the material and that it is wise to take this into account. (It is worth noting that the 240 MW of the Rance represents but a small percentage of the installed capacity in the French network and that for a plant of different dimensions, reverse turbining can, with well chosen timing, provide a valuable power contribution. Even at the Rance, reverse turbining is used in certain network configuration, approximately 8,000 hours per year for for 24 sets.

The open sluice function is a natural operation. It is simply the complement of the opening of the mobile dam gates, performing an identical role, that of allowing a maximum of water into the reservoir. In this case, the distributor and runner blades are locked fully open for a maximum flow, the speed of the set is less than its normal speed (30 to 60 rpm instead of 94 rpm for 0.30 m$<<$ H < 1.20 m).

It is at this point that is used the variable speed operation to throw the sets into direct pumping (DP) thus preventing the stresses induced by asynchronous start ups. Anticipation of the pumping also enables a substantial gain in stored volume.

To complete the picture, let it be added that the various operating modes proposed are also a function of the amplitude of the tides and that for high amlitude tides, in the order of 7-10 m there is little pumping. Both the stored volume and the head are, in effect, sufficient for the plant equipment.

Energy Evaluation

The Rance tidal power plant gross production is roughly 610 GWh. Among other things, it is a function of the mean of the tides. It is worth noting, for example, that from now - probably until the year 2,000 - the natural production will diminish by about 5%. But this decrease will not be noticeable except at constant cost functioning, which will no doubt not be the case.

If one considers the years prior to 1982 (the plants total capacity has only been available since June 1982), the energy consumed by pumping is roughly

100 gwh, or the same 18% to 20% mentioned in the last section. But as one well knows - pumping GWh do not have the same value as general GWh.

As far as the availability of the plant is concerned, it might be interesting to note: it has been nearly 97% since 1982. As an indicator during the maintenance and large repairs of the generators, it varied from 71% to 94% between 1976 and 1982.

Valuation of the production

The output capacity at the Rance is low in relation to the French network installed capacity, it is thus managed as marginal to the national generetion system.

Its operation is determined in such a way as to maximise the sum of the instantaneous values of generation. These instantaneous values are calculated as a product of generated or consumed energy, at a particular moment, but the marginal cost of energy at that same moment. (In reality, it is the cost of the replacement combustible which is entered into the programme).

Of interest both to operators and designers of new tidal power plants is

(*a*) the choice of the most valuable type of functioning (single or double acting, with or without pumping) for energy generation
(*b*) the adjustment of marginal costs which could be the ratio of energy cost in peak hours to energy cost in slack hours.

A series of simulations carried out using adjustment ratios and allowing to respond to situations other than those found in France, lead us to specify the following

(*a*) With constant energy costs, the difference between the single action without pumping and the single action with pumping is approximately 10%. The double action only improves it by about 1%.
(*b*) In value, the difference between the single action without pumping and that with pumping varies from 10% to 20% depending on the values chosen. The double action with pumping provides an additional gain of 2% to 10% when compared with the single action with pumping.
(*c*) The energy efficiency of pumping can vary between 157% and 100% depending on the adjustment ratios, whereas the value efficiency of pumping can take on a great significance depending on the times at which pumping is performed (at low cost) and energy is generated (at high cost).

Environmental aspects
Ecological aspects

The total closing of the estuary between 1963 and 1966 during the construction of the plant caused the almost complete disappearance of the original species.

The commissioning since 1967 has enabled the stocking of normal and diverse species. A new ecological equilibrium as rich as that which preceded it has been installed. This stability is closely tied to the regularity of the operating conditions at the Rance. Numerous studies have been peformed and result published by the Laboratoire Maritime de Dinrd.

The dams existence has not accelerated the silting up of the estuary, but may have led to a sedimentary redistribution. Relevant studies are in progress.

Socioeconomic aspects

The presence of a lock in the project has enable pleasure boating to be developed in the estuary. Depending on the year, from 16,000 to 18,000 boats cross the lock between the sea and estuary.

The originality from the very outset, of transforming this structure into a bridge has greatly contributed to the economic development of the western side of the estuary. The already existing tourism has been still further reinforced. Numerous people visit the plant, either via the open circuit (300,000 to 400,000 per year), or within guided visits (5,000 per year).

Conclusion

A quarter of a century old, La Rance Tidal Power Station continues to supply competitive energy to the French system.

The different problems which occured at the beginning of the operation have been successfully resolved and the plant availability and reliability are acceptable.

The various operating possibilities are used in order to provide the greatest benefit according to an appropriate utilisation of the generating and pumping periods. The operetion is controlled through a new automatic system which relieves EDF of the need for permanent supervisory staff.

As far as environmental aspects are concerned the presence of the barrage has involved a profound but not negative change in the ecosystem, the new biological equilibrium depends on the regularity of the operation conditions. It allows a large development of tourism and pleasure activities, which become more and more attractive along the estuary. La Rance Tidal Power Station appears to be as well suited to its industrial purposes as to its many surrounding activities. It has still many years to go.

The Rance Power Station operating results (Years 1980-1990)

Year	Production (GWh)			Period (total for the 24 sets)						Hours
	Gross	Net	True	DT	RT	DP	RP	0	Total	
1980	503	495	485	64	5	3	0.1	28	100	116800
1981	570	562	500	61	2	16	0.0	21	100	134700
1982	607	599	511	59	2	18	0.0	21	100	151000
1983	610	601	503	57	6	17	0.1	20	100	155900
1984	609	601	494	58	3	18	0.0	21	100	157500

20. Tidal Power in Russia

Dr L. B. BERNSHTEIN, Hydroproject Association, Moscow

Tidal power plants in Russia were first proposed in 1935. In 1968 an experimental 400 kW tidal power plant was commissioned at Kislaya Guba. This plant was installed in a concrete caisson which was constructed off site and then floated into position. The turbine is capable of bi-directional pumping, generating and sluicing and has in practice met the design specifications.

Recent feasibility studies have re-examined the potential for large tidal power barrages on the north west and far east Russian shores. The most promising of these are at Tugur and Mezen bays.

Introduction

The first conceptual designs for tidal power plants in Russia appeared in 1935 - 1940 in connection with the development of electrification in the USSR. These included a proposal by I. A. Potenkhin for Mezenskaya Guba.

From 1938 to the present, all tidal power survey and design/investigation work has been carried out under the leadership of Dr L.B. Bernshtein of the Hydroproject Association. During this period the design methodology has been developed, the tidal coastline has been reconnoitred, alignments have been selected, diagrams and assessments of possible power plants have been drawn up and an experimental tidal power plant has been constructed at Kislaya Guba.

The experimental tidal power plant at Kislaya Guba

While searching for a technical solution providing cheaper construction of tidal power plants in the USSR, Dr L.B. Bernshtein in 1959 suggested a thin-walled design of power house with construction using floating caissons. With this method no expensive cofferdam is needed and the project becomes cheaper because the main construction work is transferred to a more accessible site with favourable conditions. Furthermore, the power house construction and foundation work may proceed concurrently.

The location of the experimental tidal power plant in Kislaya Guba (a bay) proposed in 1938 was decided by its nearness to an industrial centre (the city of Murmansk) and the existing power lines, while the configuration of the narrows connecting the basin with the sea allowed the experiment to be carried out at relatively low cost.

TIDAL POWER

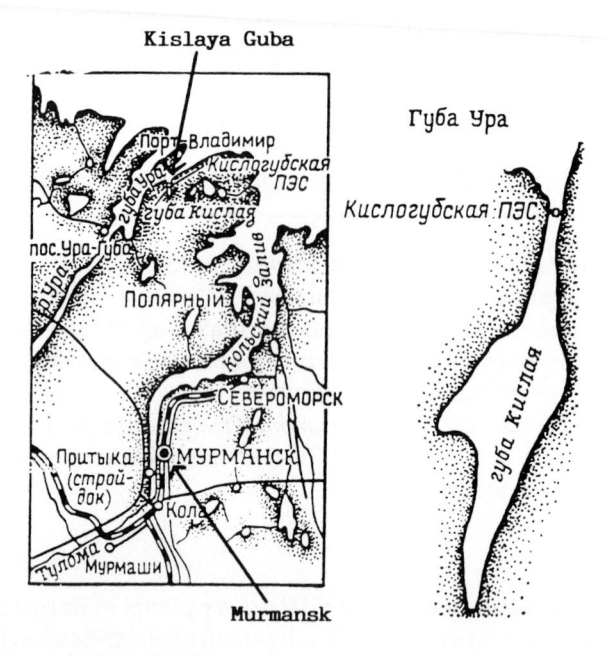

Fig. 1 . Location of Kislaya Guba

A narrow bottleneck connects Kislaya Guba, situated on the coast of the Kolsk peninsula, 60 km westwards from Murmansk, to the gulf of Ura (Fig. 1). This allowed a tidal power basin to be formed by constructing a short barrage, cutting the basin off from the sea. The low tidal range (1.1 m-3.9 m) made it possible to test the power unit for operation at minimum heads.

At the bottleneck a wide scoop (150 m) narrows abruptly to 35 m at low water. The total length of the bottleneck is 450 m. To the south it widens, connecting with the large basin (surface area 1.1 km^2 and depth up to 35 m) of Kislaya Guba.

The climate in the region of Kislaya Guba is comparatively mild in winter and cool, cloudy and humid in summer. The average annual temperatures is -0.4°C, with maximum +3°C and minimum-35°C. The annual average precipitation is 515 mm.

Since the Kislaya Guba is oriented in a north-south direction and is protected at its exit from the north by high rocks, the design wave height is taken to be 2 m (4% provision) with a maximum observed wave 2.6 m high. Before the construction of the tidal power plant, the bottleneck restricted the tidal current and a cross-head was formed in the narrowest gap. The maximum flood and ebb tide velocities reached 3 m/s and 3.68 m/s 1.5 hours before high and low water respectively and the maximum flow rate exceeded 300 m^3/s.

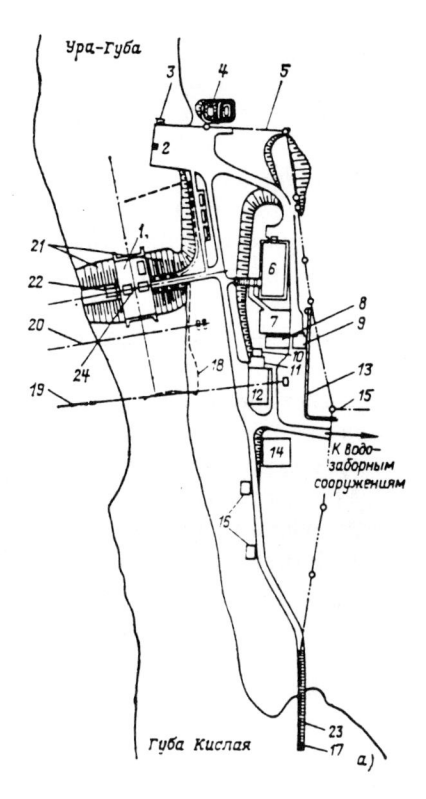

Fig. 2. Layout of Kislaya Guba tidal power scheme

The tide is of a regular semi-diurnal nature, with amplitudes $A_{\text{mean spring}}$ = 3.23 m, and $A_{\text{mean neap}}$ = 1.61 m. The water surface area of Kislaya Guba varies from 0.97 km² to 1.5 km², reaching 1.7 km² after pumping in and falling to 0.91 km² after pumping out.

Due to the warming effect of the Gulf Stream, no icing is observed at the approaches to the Kislaya Guba. Shore icing 60 cm - 70 cm thick is found in the gulf itself. A sheet of floating ice, 50 cm - 100 cm thick, occupies most of the basin surface with an opening formed near the approaches to the station.

Sediments at the bottleneck of Kislaya Guba comprise large cobblestones with sandy gravel and partial shell-rock filling up to 7 m deep.

The favourable natural lie and configuration of the shores forming a narrow neck with a steep cliff on the western shore and bluff developing into a small plateau on the eastern shore allowed a fairly convenient layout of the tidal power scheme (Fig. 2). A natural scoop in front of the inlet to the Kislaya Guba forms a convenient approach area with a berth built in it.

The shore site contains open-type switchgear, an accommodation block providing comfortable apartments for attending personnel, storehouses, a

TIDAL POWER

garage, a water main supplying water from a mountain lake and tide-gauge installations. The throat is dammed by a single caisson and connecting embankments up to 15 m high and 35 m long. The caisson block of the Kislaya Guba tidal power plant which was floated to site measures 36 m x 18.3 m in plan and is 15.35 m in height (Fig. 3).

Originally, two bulb turbine generators were to be mounted in the caisson, one of which was delivered by the firm Neyrpic-Alsthom, while the other was to be a Soviet-made unit (with variable speed of rotation). However, during the study, a decision was made to realize the variable speed of rotation at the former generating unit by temporarily replacing the synchronous generator with an asynchronous machine. The penstock of generating unit No 2 was instead used as a sluice.

The space above generation unit No 2 is used to site the electrical devices and the control board, since one power generating unit aisle was enough for testing the combined configuration. The shape of the water passageway was specified by the turbine supplier and develops into a rectangular cross-section at the inlet and outlet. Situated at the ends of the passageway are the slots for the gates. The length of the water passageway was dictated by the bi-directional operating conditions and was taken equal to 11 x turbine runner diameter (D1).

The stay ring, main support and the upstream inlet shaft transmitting loads from the turbine to the building structure are re-embedded in the monolithic mass.

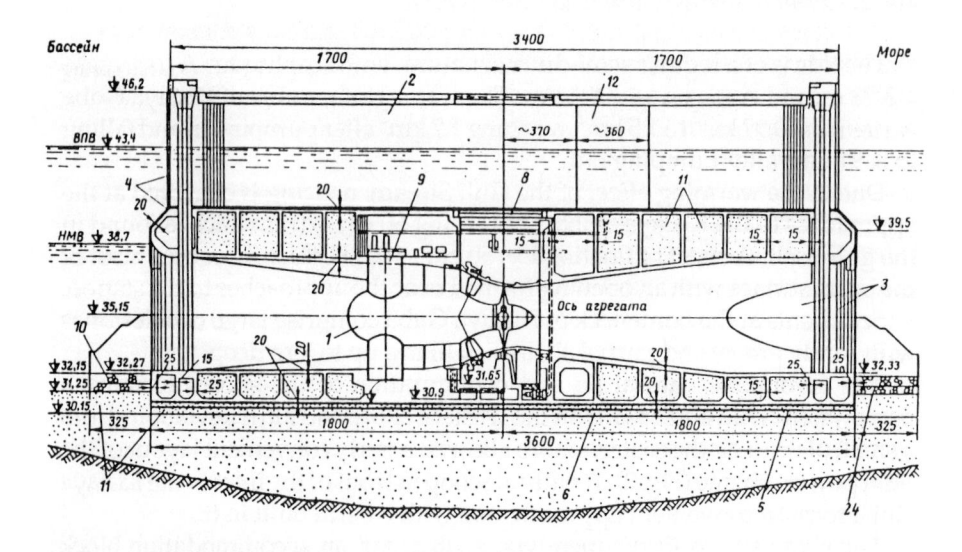

Fig. 3. Kislaya Guba caisson

The power unit building is a thin-walled reinforced concrete box of the caisson type. The structure comprises a bottom slab 20 cm thick and laid at an elevation of 30.65 m. The slab carries bulkheads, 15 cm thick, in parallel with its end edge. The bulkheads are spaced at a distance of 1.5 - 2 m. Two side vertical slabs, 15 cm thick, whose height extends to an elevation of 46 m, run along the side edges of the bottom slab.

A spillway slab is laid at a height of 7.5 - 8 m, above the first power generating unit. A ceiling slab is laid above the other generating unit. At the upstream and downstream ends of the caisson, within the zone of generating unit No 2, the slab at an elevation of 39. 4 m together with the middle and outer piers forms a room accommodating the control panel, monitoring and measuring facilities and cathodic protection station. The slab of the upper spillway has a watertight opening for access to dismantle the generating unit. Under this unit, the ceiling slab at elevation 38.15 m carries the cooler of the oil pump installation and the regulation system equipment. Embedded in the slab is a shaft to permit access to the generating unit bulb.

The above mentioned layout and structural design made it possible to create an open structure of high strength, providing a compact layout of the power generating equipmemt and allowing its installation, maintenance and easy operation.

A structural analysis of the caisson was carried out, considering it as a complex cross-section girder on an absolutely rigid base.

In order to eliminate thermal stresses in the upper part of the caisson, the wall was protected by a 5 cm coating of thermal insulation of foamed epoxy resin reinforced with glass cloth.

The caisson was constructed from 1965 to 1968 in a construction dock on the shore of Kolsk Bay, near Murmask. After testing the caisson for watertightness, the dock pit was filled with water and the caisson floated having a design draught of 8.32 m with displacement of 5,200 t. Pontoons were attached to enable the structure to clear the dock sill. On 28 August 1968 the caisson structure with a draught of 6.5 m, was towed some 60 miles over a period of 19 hours, by two 2,000 hp tugs and one special tug boat to the waterway of Kislaya Guba (Fig. 4). Then it was positioned on site with the aid of winch mounted on the caisson unit and cables fastened to eyebolts secured in shoreline rocks. By taking on water ballast, later replaced by sand ballast, the caisson unit was then sunk into position onto a prepared underwater foundation protected by an embankment. Preparation of the foundation was accomplished using a floating crane to excavate blast loosened rock to formation level, after which the pit was filled with a 0.5 m bed of specially selected aggregate levelled underwater by means of specially devised scheme.

Dynamic tests with the turbine generator running have shown the low frequencies of vibration undergone by the caisson lie within the range 30 Hz

TIDAL POWER

Fig. 4. Transport of caisson

to 90 Hz, under a concentrated load of 50 kN displacements being approximately 0.4 mm.

Hydrodynamic loads applied to the spillway slab and elements of the turbine penstocks under service conditions were studied with water flow rates of 10 m^3/s per metre length, at heads ranging from 0.5 m to 2.6 m in both penstocks and the surface spillway simultaneously or in different combinations. These studies have shown that the pressure fluctuation, being a stationary random process with distribution close to a normal one, produces a maximum fixed value of pressure variance equal to 3.2 kPa; the fluctuations represent a low frequency process, the maximum energy of which lies within the angular frequency range of 0s^{-1}–0s^{-1} and the spectrum of pressure fluctuations at 0 the pier have their maximum also within the frequency range of 50 s^{-1} – 70s^{-1}. The dynamic coefficient computed for the middle of the spillway slab is equal to 1.004, which confirms the very high rigidity of the structure and negligible vibration of it within the range of disturbing frequencies excited by passing water.

Site investigations, diving examination and observations with the aid of piezometers carried out during operation have shown the reliability of the foundations as to filtration and piping stability.

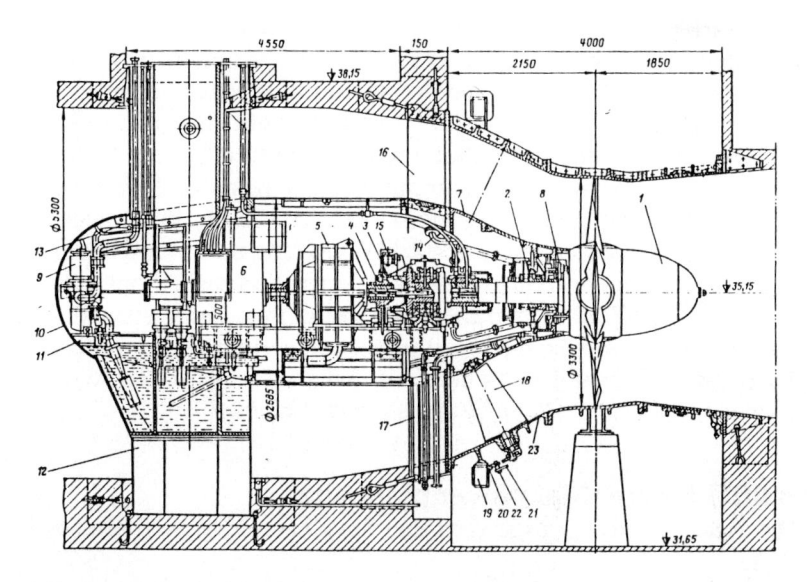

Fig. 5. Bulb turbine generator unit

Changes in the sea and basin water levels are quickly indicated by the piezometers which indicate no "standstill" and "settling" zones in the foundation. The head gradients varied on an average within 0.08 to 0.15, while the filtration flow rate under the caisson unit was of 0.12 m^3/s - 0.15 m^3/s. Three years after installation the settlement of the caisson unit measured at the corners was less than 20 with a maximum difference between the corners of less than 3 mm. The settlement rate today is 1 mm per year.

The Kislaya Guba power unit is equipped with a bulb turbine generator unit (Fig. 5) designed by the French firm Neyrpic-Alsthom(D_1 = 3.3 m, N = 400 kW) with four adjustable S-like blades providing six operational modes (bi-directional turbining, pumping and sluicing). In contrast to the Rance generating units, the turbine is coupled to the generator by a step-up gearbox increasing the speed from 72 rpm to 600 rpm. The step-up gearbox is designed by the firm Krupp. It is a coaxial single stage type with three planetary pinions and a built-in clutch mechanism. This design was dictated by the very low head at the unit (H = 0.5 m - 2.5 m, H_o = 1.28 m). Flexible operation of the power unit has been proved by testing the generator and studying its power output. When operating to obtain maximum energy capture, the energy recovered per year is 1.2 GWh which is 10% higher than predicted by computer calculations. Operation of the water passageway openings has an important influence on the energy recovery; it is increased by 15.3% for two-way generation and by 11.4m for one-way (ebb) generation.

TIDAL POWER

The machine efficiency generally corresponds to the firms guarantee, though there is a slight increase in the efficiencies obtained. Thus, during flood generation, which is the predominant mode for 2/3rd of the time, a higher efficiency is obtained (91%) compared with the guaranteed value (86%). Operation in ebb generation mode yields an efficiency lower by 1% and in ebb pumping and flood pumping modes the natural efficiency is 1% higher. In the flood generation mode, at H = 1.28 m, the power output exceeded the guaranteed value of 400 kW, yielding up to 410 kWh except at the initial head. The head ensuring stable operation turned out to be equal to 0.8 m instead of 0.18 m guaranteed by the manufacturer. Also, due to the operation of the turbine over-speed protection system, direct no-load sluicing turns out to be possible at a head of 0.8 m in place of 1.28 m and reverse sluicing is feasible at a head of 1 m instead of 1.62 m.

Potential future tidal power plant sites

The most promising large sites that have been identified are shown in Table 1.

Schemes in the Lumbovsk Bay

The Lumbovsk Bay is situated on the Murmansk coast of the Barents Sea, 60 km eastward from the peninsula of Svyatoy Nos. The mean regular semi-diurnal tide range is as follows: $A_{mean} = 4.2$ m, $A_{max} = 7.4$ m and

Table 1

Site	Mean Spring Tidal Range m	Area Of Basin km²	Installed Capacity GW	Annual Energy Yield TWh
Lumbovsk	4.20	70	0.67	2.0
Mezen upstream	5.66	2,640	21.00	15.0
Mezen downstream	7.53	6,451		92.0
Penzhin North	6.20	6,788	21.40	71.4
Penzhin South	6.20	20,530	87.40	191.3
Tugur	7.25	1,120	6.80	16.2

$A_{min} = 2.9$ m. The bay has an oval shape extended along the coastline and is convenient for use. The bay is 17 km- 25 km long and 4 km - 6 km wide. Its water surface area varies from 56 km^2= to 70 km^2 at high water.

During the period 1977 - 1983 studies were made of a potential scheme of total capacity 670 MW. This allowed 64 heavy-duty generator units to be installed in position without underwater excavation ($D_1 = 10$ m, $N = 670$ MW, $E = 2$ TWh). However, the project could not be justified economically due to the modest tidal range ($A_{mean} = 4.2$ m).

Mezen Bay

Mezen Bay is located near the mouth of the White Sea and provides the main resource of exploitable tidal power on the European coastline of Russia. Upstream the bay has tidal range of $A_{mean\ spring}= 5.37$ m, and downstream $A_{mean\ spring} = 7.53$ m ($A_{max} = 10$ m). The water surface area of the bay is 6.45 km^2. Its potential energy capacity has been estimated at 92 TWh.

One possible scheme at site IX provides for cutting off a bay area of 2,640 km^2 by a dam with a power house 85.6 km long, having an asymmetric configuration to provide a power house 19 ham long without underwater excavation. The technical potential of barrage IX (with $A_{av} = 5.66$ m) is 15 TWh/year and with installed capacity of 21 GW. To make use of such a potential, 1,110 machines each rated at 19 MW would be required.

The Sea of Okhotsk

As a result of studying alternative possible locations and site investigation of those preferred, it was proposed to concentrate further work on two bays of Tugur and Penzhin (Fig. 6). Other barrage alignments were rejected because of unfavourable conditions and insufficient tidal ranges.

Penzhin Bay

The energy potential at the mean spring tidal range of 6.2 m, with a basin area of 6,788 km^2 for a northern (Manechin) barrage or alternatively 20,530 km^2 for a southern (Bozhedomov) barrage, is estimated at 35 GW rated capacity and 105 TWh for the former and 100 GW and 300 TWh respectively for the latter.

However, complete utilization of this potential is limited by the difficulty of placing the turbine caissons without expensive underwater rock dredging. In the southern barrage, between the capes of Povorotnyi and Dal'nii, 72 km long, a section of 51 km has considerable depths (up to 67 m). With 4,416 generator units, each of 19.9 MW capacity, the total output capacity of the station would be 87.4 GW with an annual energy yield of 191.3 TWh. The total amount of concrete required would be 65,000,000 m^3 and of aggregates 160,000,000 m^3.

TIDAL POWER

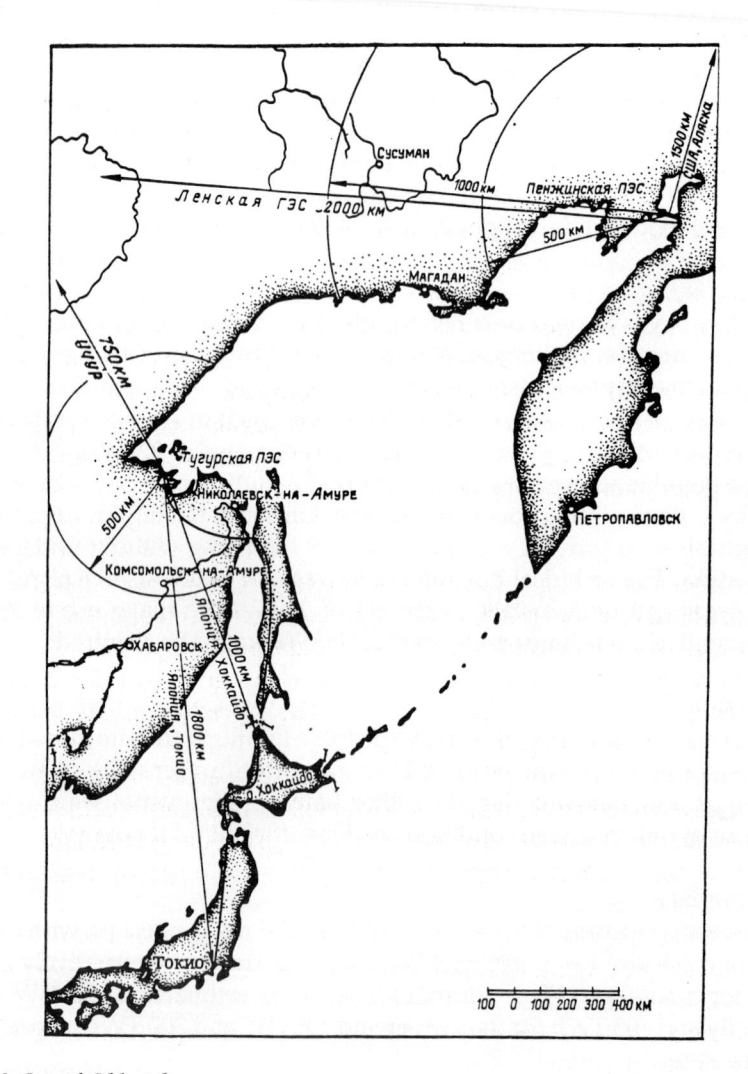

Fig. 6. Sea of Okhotsk

The northern barrage (between the Capes Srednii and Vodopadnyi) could accommodate 568 turbine generators with $D_1 = 10$ m, each having capacity of 19.8 MW, over a section 32. 2 km long, 26 m deep and 920 units with $D_1 = 7.5$ m, 11 MW capacity each, laid out in shallower water (21 m). The total capacity of the station would be 21.4 GW with an annual output of 71.4 TWh.

However, Penzhinskaya Guba has severe climatic conditions and unusual tides. The average annual temperature is -6.5°C and the minimum ranges from -40°C to -50°C. Winter has a duration of 220 days. This results in extremely heavy ice formation, which makes the Guba the "kitchen" of icing in the north-east part of the Sea of Okhotsk.

The Bay of Penzhin is the only one of those studied on the coasts of the world ocean where indirect diurnal tides of mixed nature occur. Although the maximum diurnal tidal range here reaches the record height for the coastline of Russia (13.4 m), its frequency and variability are of a fairly intricate nature with inequalities, the period of which is equal to 18.66 years.

Thus, at the barrage site of Cape Srednii, the equinoctial tidal ranges within a 19 year period vary from 5 m to 13 m and those of the solstitial tides from 3.2 m to 6.1 m per year with maximum tides and from 2.6 m to 5.8 m per year with minimum tides. Therefore, operation of a tidal power scheme under such conditions would need power compenstion in a 19 year period, rather than just an intersyzygial lunar period. Regulation to meet this might be achieved by joint operation of a tidal power plant with a powerful river hydroelectric station having large water storage.

Tugur Bay

The selection of Tugur Bay for a tidal energy barrage was based on site investigations and hydrological observations carried out between 1972 and 1984 which made it possible to state that natural conditions in the Tugur Bay are relatively favourable for construction of a barrage. Tugur Bay is situated in the southern part of the Sea of Okhotsk. The bay is 7.4 km long and 37 km wide at its entrance (Fig. 7).

Because of these natural conditions and a severe power shortage in the area, a government assignment was adopted in 1987 on the initiative of Dr L.B. Bernshtein. It was decided to undertake during the period 1988 - 1991 a Technical and Economic Appraisal of the proposed Tugur Barrage, at a cost of 8,500,000 roubles.

Three alignments were proposed for further studies; the northern barrage, the southern barrage and a barrage between the Bays of Konstantin and Tugur as the first stage of one of the others.

During the initial stage of designing the Tugur Barrage, the latter version involving construction of a power house on the neck between the Bays of Konstantin and Tugur was considered. It was supposed that use would be made of the head at the turbines installed in the power house on the neck, formed by the difference between the tidal ranges 3.7 m - 2.88 m and the shift of their phases. After a series of observations, however, this scheme was rejected because of the insufficient head resulting from the small phase difference between the tide in both bays.

The southern barrage site then presented itself preferable, because its greater tidal range than at the northern barrage made it possible to obtain

TIDAL POWER

Fig. 7. Tugur Bay

the same energy yield with almost half the dam size and a smaller number of generator units.

A decrease in the basin area at this barrage site to 1,120 km^2, compared to the northern barrage site (1,800 km^2) is compensated for by a material increase in the tidal range A $_{mean\ spring}$ = 7.25 m, A $_{mean}$ = 5.63m, A$_{qu\ tide}$= 3.71 m) which makes it possible to install a capacity of 6.8 GW to give a yield of 16.2 TWh per year. An optianum layout has been selected on the basis of economic appraisal, fuel saving and partial replacement of ths capacity of a thermal power station.

Consideration was given to one-way and two-way generation schemes, with combined and separate sluices and a power plant with bulb generator units having runner diameters of 7.5 m, 8.5 m and 10 m. The analysis of 90 alternatives with the aid of the Silakov's algorithm made it possible to identify an advantage common to all versions. This advantage is the use of a bulb turbine generator of the Rance type, having a runner with a maximum possible diameter (equal to 10 m) designed and built by the manufacturing company Leningrad Metalische Zhavod. The energy yield versus installer capacity for a turbine with a runner of this diameter shown in Fig. 8 was brought to light.

On the basis of this dependence, versions with output of 12 TWh per year were subjected to further consideration. Taking into account the fact that two-way operation of the turbines is better for the environment and also the necessity of obtaining maximum amount of power from the plant with improved flexibility, the choice was made of a version with two-way generation and combined turbine/sluice caissons. In spite of the advantage provided by the deep water layout of sluices which reduces the possibility of their packing with ice, the version with separate sluices was rejected, on the grounds of increased civil works and higher capital cost, due to the necessity of building additional caissons. If sluices are omitted, the annual

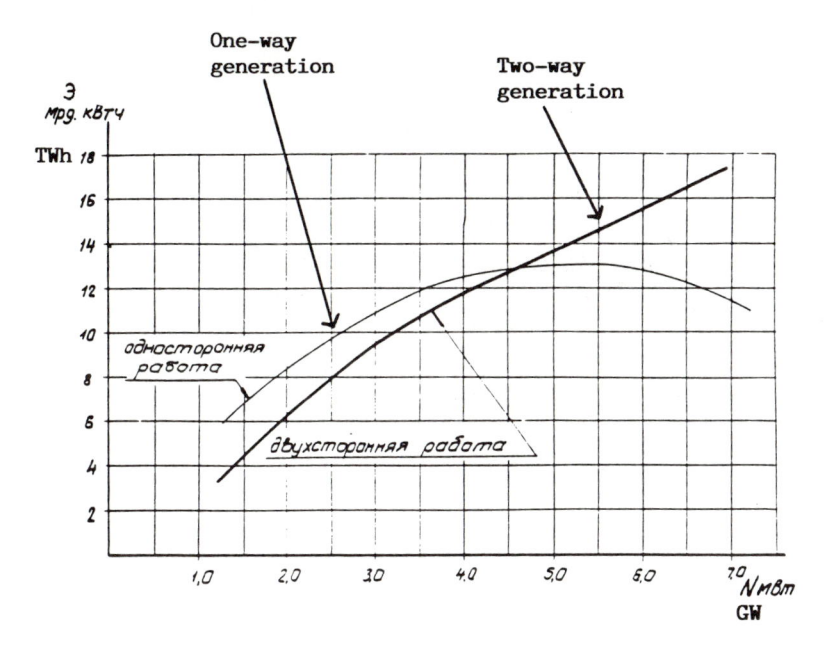

Fig. 8 . Tugur energy yield against installed capacity (sluices replaced by turbines with increasing capacity)

output would be reduced by 2.0 TWh. To protect the sluices and turbines against packing with ice, an ice-holding boom, consisting of metallic pontoons secured by anchors, would be positioned 600 m northwards from the barrage

TIDAL POWER

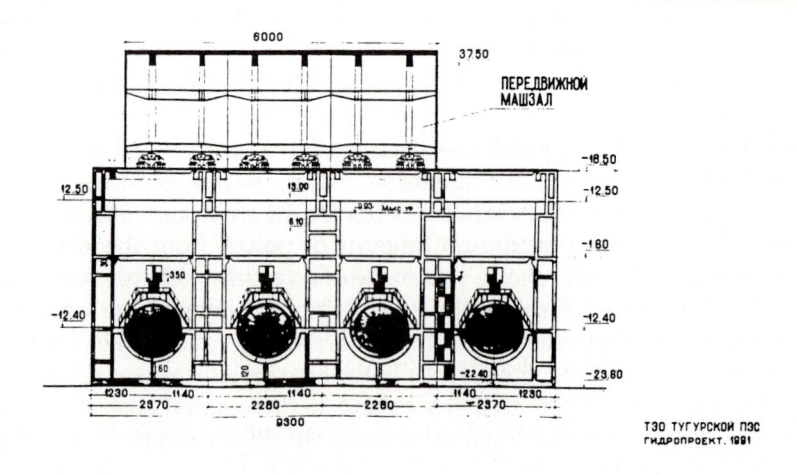

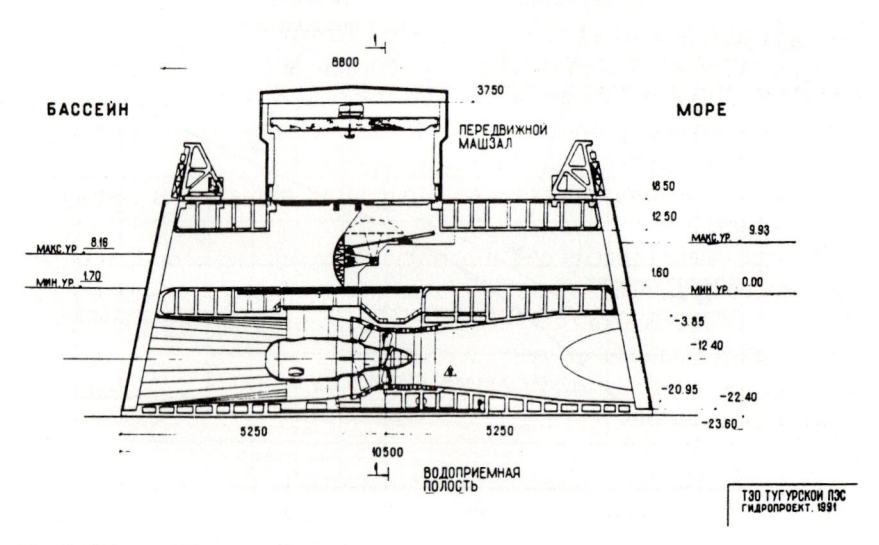

Fig. 9. Proposed Tugur caisson

Following a process of optimisation the choice was made of a two-way generation scheme with a layout consisting of 105 caissons with four turbines each, i.e. 420 generator units, each of 16.2 MW capacity. Each caisson would measure 105 m x 93 m 62 m and have a concrete volume of 74,000 m^3.

The caisson (Fig. 9) has a modernized version of that used at the Kislaya Guba station, with a movable gantry crane above the penstock, as shown in the Authorship Certificate (registered design) of 1959. This travelling gantry crane is used for servicing and maintenance of the generator unit. The legs of the crane are sheathed with thermal insulation material. The caissons would be constructed in a construction dock laid out near the eastern landfall of the dare. The dock measures 1.565 m x 290 m. In the dock the caissons would be concreted only to a height of 42 m and unfinished weight of 147,000 tons, draught of 14 m. At the end of the 8th year of construction work, after the caisson have been installed and although there would still be a 2 km channel left. A head would be created which would be sufficient to drive, at partial power output the 16 turbines that had been installed. By the end of the 9th year, in the same way but at a higher head, 32 more turbine generator units would be in operation. The remaining turbine generators would be brought into operation as they were installed. Thus, the investment in finished turbine generators would not be kept idle.

The power station, 10,500 m long, would be connected with the shores by embankments at the eastern and western ends. The principal quantities needed for construction of the Tugur Barrage are given in Table 2.

Use of energy and efficiency of the Tugur Barrage

The possibility of increasing the proportion of barrage energy which would be utilised and the growth (with time) of its economic viability, were shown by studying a promising model of the development of power systems for the far East region of the country.

The results of computing the economic return of the Tugur barrage are given in Table 3.

The equivalent capacity of the barrage in the full compensation regime (P_g) is given by $E/8760$, where E is its annual energy output. For the Tugur Barrage, this equals about 2.0 GW. In practice, the displaced capacity of the thermal power stations will depend on the part losses during pumping and in the power transmission lines, and on the extra power installed at the hydroelectric stations, and also on the weekly regulation and reserve.

For a typical 24-hour schedule, annual curves of load duration and power generation for each zone of the operating power $E = f(N_{op})$ were plotted. Then, by applying the tidal power station schedules to the loads supplied, with the aid of the hydroelectric station compensator, the fraction of the barrage power output accumulated by the hydroelectric station was finally defined.

As a result, with an additional expansion of the power capacity of the hydroelectric station by 2.400 GW, the mean value of the coefficient of weekly regulation and reserve of the order of 1.2 and the fraction of power losses caused by the pumping and accumulating transmission lines (of the order

Table 2. Quantities for principal construction of the Tugur Barrage

	Description	Volume
1	Excavation of soft ground, $m^3 \times 10^6$	33.4
2	Excavation of rocky ground, $m^3 \times 10^6$	1.9
3	Embankment of rocky mass, $m^3 \times 10^6$	49.0
4	Embankment of sand/gravel mixture, $m^3 \times 10^6$	29.5
5	Concrete (power house, protection of basin banks and slopes), $m^3 \times 10^6$	9.9
6	Laying of cloth material, $m^2 \times 10^3$	5,973
7	Assembly of metal structures, $t \times 10^3$	561

of 15%), the probable decrease in capacity supplied by the thermal power stations to the power system was calculated to be 1.5 GW - 1.8 GW, or about 20% - 25% of the barrage capacity.

The fuel saved through the use of tidal energy was calculated both for the commissioning years and for the 30-35 years of the subsequent normal service. The computations took into account the yearly real increase (3%) in cost of the saved fuel, which is the case for most countries of the world during the last few decades. At the directive discount rate of 8%, the benefit/cost ratio was 0.71 and the simple pay-back time come to 21 years, compared to thermal power stations and fossil fuel bases.

However, for State projects of national economic importance having a long period of construction, decreasing the discount rate to 5% is justified. The resultant benefit/cost ratio computed at the specified discount rate at a period of time exceeding 40 years, which increases to 1,22. This ensures an annual return for the tidal power scheme of the order of 6.6%, i,e. the pay-back period will decrease to 15.1 years.

The surveys carried out, nature investigations and mathematical models it possible to predict the absence of any cardinal factors that might affect its environment to a degree which might prevent the construction and operation of the Tugur Barrage.

To preserve the natural rhythm of tidal movement is the principal environmental objective. Elimination of extra levels will reduce bank erosion, silt content and turbulence. Reduction of the tidal range will admittedly lead to a slight reduction in water exchange, (between the basin and the sea) to 80%

No	Description	Unit of Measurement	By Project			Alternative Version		
			Tidal Power Scheme	Coupled	Total Plants	Thermal Station	Fuel	Total
1	Installed capacity	10^6 kW	6.8	2.4		2.74		
2	Annual energy output	10^9 kWh	16.02	0.7	16.72	2.3		1.7
3	Capital cost	10^9 roubles	13.77	1.33	15.1			2.3
4	Capital cost per installed kW	rouble/kW	2,020	550	2,570	840		
5	Capital cost, reduced to the design basis year (15th)*	10^9 roubles	20.06	1.42	21.48	2.33		2.33
6	Annual expenses	10^9 roubles	248	35	283	195	1,095	1,290
7	Annual expenses, reduced to design basis year (15th)*	10^9 roubles	295	39	334	198	1,405	1,603
8	Unit cost of energy	kopecks/kWh	1.55	5	1.69	1.17	8.25	9.42
9	Pay-back time	years			15.1			

* At 5% discount rate

Table 3. Economic efficiency of the Tugur Tidal Power Plant

of the present natural exchange, but this will not lead to significant changes in salinity.

Naturally, the construction and operation of the Tugur and other barrages cannot be accomplished without leaving a trace on the environment and the resultant influences will be both of a positive and negative nature. However, there can be no comparison between these effects and the great harm that might be done to the population of the adjoining regions by environmental pollutions caused by the generation of the equivalent amount of power by thermal power stations operating on fossil fuel.

Therefore, construction of the Tugur Barrage by the end of this century is dictated by the above studied circumstances and is warranted.

Pilot commercial Kolskaya tidal power scheme

To jump from the small Kislaya Guba plant, of 0.4 MW capacity to the gigantic Tugur and Mezen barrages requires an intermediate stage within the scope of existing commercial power plants. A unit with runner diameter of 10 m, and its different modifications (connection through step-up gear, reduced or variable frequency of rotation, propeller-type runner) should be tested in the prototype. Also, the effectiveness of double-effect (two-way barrage operation and other construction, technological and ecological solutions should be checked. For this purpose construction of a pilot-commercial barrage at Kolskaya on the Murmansk coast of the Barents Sea is suggested.

The area of the bay (5 km^2 and values of half-day tide ($A_{av\ spring}$ = 3.3 m, A_{av}= 2.36 m $A_{av\ neap}$= 1.8 m) are sufficient to conduct the desired trials.

A generating unit, of 16.2 MW capacity is proposed, with direct drive bulb turbine of the Rance type with six-cycle operation with variable frequency rotation of 30 + 41.7 rev/min (as an alternative for the second unit connection through planetary step-up gear. The runner would have a diameter 10 m with S-shaped blades. The generator would be asynchronous with converter of variable frequency of rotation into the conventional current frequency (as an alternative - asynchronous, synchronous without frequency converter).

The cost of constructing the Kolskaya Barrage turns out to be high (155 million roubles at 1984 price levels), in relation to the energy produced. This is a result of the modest tidal range. It does not provide a cost effective barrage (4,940 roubles per installed kW, which is twice as expensive as for the Tugur barrage).

However, its economic viability may be achieved if we take into account the significant reduction of Tugur and Mezen costs, which may be expected after approval of modified and new types of units. The saved billions provide the justification for spending 156 million roubles on construction of the Kolskaya Barrage.

Thus, long-term studies and investigations, also design elaborations conducted lately, show that to provide energy for the economic development of two most important regions of the country (north-western and far eastern areas), construction of two tidal power plants of high capacity (first Tugur and then Mezen) is technically possible and quite justified from an ecological point of view, as an alternative to the construction of thermal power plants, operating on fossil fuel and to nuclear power plants whose construction in the near future seems not to be possible.

Discussion on Papers 19 and 20

E. HAWS, Consultant

Two questions concerning La Rance:

(i) From a brief visit, it appeared that heavy equipment from the power station is passed under the lock to the unloading bay. Is this in fact the case, and has experience shown it to be satisfactory?

(ii) How does actual energy capture from any particular tide compare with predictions made by modelling during preparation of the project?

This question is independent of availability of plant, which can be assumed as 100% for the comparison.

E. A. WILSON, Mersey Barrage Company

It is apparent that flood pumping is regularly in use at La Rance. Are there any environmental or legal constraints upon the extent of flood pumping or is maximizing energy yield the principal influence? If the latter, to what extent does the relative cost of importing power further influence the amount of flood pumping undertaken?

A. V. HOOKER, Consultant

M. Rodier confirms that two-way power generation is now seldom used at La Rance. In estuaries such as the Severn and the Mersey, flood generation would not be acceptable as it would reduce the level of high water in important navigation channels. However, direct pumping, which requires variable pitch turbines, has been used to a significant extent at La Rance and my question relates to the economics of the flood pumping mode.

In La Rance, as in the Severn estuary, many Spring tides have their time of high water within a period of peak demand. When this occurs it would seem neither economic nor desirable to take power from the grid at a time when its value is high. Flood pumping also reduces the duration of the high water stand when most shipping movements take place and increases siltation and the cost of dredging.

A factor which has not been mentioned is the reduction in the capital cost of the mechanical and electrical plant made possible by installing turbines with fixed blades. Any reduction in the plant cost would also reduce the annual interest charge and therefore the running cost.

It will be of great interest to have M. Rodier's opinion as to the significance of the factors to which I have referred and whether, with hindsight, simple ebb generation might have been an option for La Rance Barrage?

S. E. GODDARD, *Nuclear Electric*

What is the economic performance of La Rance and how does the generation cost in centimes/kWh compare with French nuclear generation?